JOURNAL OF CHROMATOGRAPHY LIBRARY – volume 63

the proteome revisited

theory and practice of all relevant electrophoretic steps

On the cover:
Parameters governing packing of dodecylsulfate anions into micelles. Geometry of the SDS micelle is effected by head group area ($a_\circ$) and volume (v). Maximum and mean hydrocarbon tail length (l_{max} and l_c) are indicated. The interfacial region (ifr) can move along the Z axis as the head group bobs in and out of the micelle surface (through the XY plane). The hydrophilic mantle region (m) is relatively constant in size due to inflexibility at the first CH_2 group. Contributed by Gary B. Smejkal, Cleveland State University and David Segrist, WebMolecules.com. Star added by Eaton Publishing.

Reprinted from BioTechniques Vol. 29, No. 6, December 2000, with permission of Eaton Publishing.

JOURNAL OF CHROMATOGRAPHY LIBRARY – volume 63

the proteome revisited

theory and practice of all relevant electrophoretic steps

Pier Giorgio Righetti
Department of Agricultural and Industrial Biotechnologies, University of Verona, Verona, Italy

Alexander V. Stoyanov
Department of Biomedical Sciences and Technologies, University of Milan, Milan, Italy

Mikhail Y. Zhukov
Department of Mechanics and Mathematics, Rostov State University, Rostov-na-Don, Russia

2001

ELSEVIER
Amsterdam – London – New York – Oxford – Paris – Shannon – Tokyo

ELSEVIER SCIENCE B.V.
Sara Burgerhartstraat 25
P.O. Box 211, 1000 AE Amsterdam, The Netherlands

First edition 2001

Library of Congress Cataloging in Publication Data
A catalog record from the Library of Congress has been applied for.

ISBN: 0 444 50526 1
ISSN: 0301-4770

∞ The paper used in this publication meets the requirements of ANSI/NISO Z39.48-1992 (Permanence of Paper).
Printed in The Netherlands.

Introduction

It might be argued that electrophoresis was born as soon as Volta described the pile, *i.e.* the first power supply able to deliver continuous electricity. In fact, there are rumours that Reuss, an officer in the Czar army, was caught stealing sand on the banks of the Moskva river, instead of fighting against the Napoleon army, for filling a U-tube with which he was performing electrophoresis, utilizing as a power supply a voltaic pile composed of 92 silver roubles and an equal number of zinc plates. His was a rich man pile, indeed. In his memoirs, by the title "Notice sur un nouvel effet de l'electricité galvanique", dated 15 April 1808, he reported a curious phenomenon of water transport at the negative pole, upon passage of the current; he had discovered electroendoosmosis!

By all means, though, 1808 could hardly be labelled as the birth of modern electrophoresis. It was the twentieth century that made fundamental contributions to the field and, perhaps, it was the elegant work of Arne Tiselius, a 1948 Nobel laureate, that laid the foundations of present-day electrokinetic methodologies. Arne himself, though, realized that his instrumentation would not have carried us that far in the field. All previous work, including his own, was performed in free solution, which was anathema to any separation of macromolecules. The latter, in fact, had the nasty habit of sedimenting in the electric field, as soon as they were physically separated from each other and surrounded by zones of pure electrolyte, for the simple reason that such zones were denser than the surrounding liquid. Therefore, Tiselius' electrophoretic cell was constructed in such a way (here too a U-cell with ascending and descending limbs) that only boundary separation among the various protein zones would occur. The sample zone was thus rather large and would fill up all the bottom, as well as parts of the limbs, of the U-cell. Thus, as the current was applied, ascending and descending boundaries would be detected in a schlieren observation chamber, but no complete physical separation of pure zones could ever be achieved. Thus, Tiselius himself searched for other means for achieving a "zone separation", in which each component would be allowed to form a zone separated by others by empty regions.

A host of stabilizing media for zone electrophoresis were soon described, since the electrophoretic chamber had to be filled with a micro capillary system able to suppress convective flows, as well as to prevent protein sedimentation. Paper, fabrics, such as pure cotton, silk, potato starch, cellulose powder, glass powder and plastics, such as polyvinyl chloride, pevikon C-870 and even minerals, such as asbestos, were tested. Definitely, though, modern zone electrophoresis was born with the advent of hydrophilic gels, such as Sephadex, agarose and polyacrylamide. For proteins, polyacrylamide turned out to be a unique zonal support, offering full transparency in the visible and near UV region, elasticity, flexibility and a full range of porosities which could be engineered at whim by altering either the total monomer content (%T) or the cross-linker value (%C), or both. In polyacrylamide support media, three major events concurred in shaping modern technologies:

• The description of disc electrophoresis by Ornstein [1] and Davies [2], a method which dramatically increased resolution by introducing in the matrix and buffers a series of discontinuities able to sharpen up the bands and form incredibly thin starting zones;

• The discovery of sodium dodecyl sulphate (SDS) electrophoresis [3], by which detergent-saturated proteins would be separated in a polyacrylamide gel mostly via their mass, rather than by combined mass and charge effects, as customary in conventional electrophoresis;

• The description of isoelectric focusing in polyacrylamide gels [4], by which proteins would be separated essentially on the basis of their net charge (at a given value of their titration curve, the isoelectric point).

Now the elements for the next major quantum leap were laid out on the table. Three groups took advantage of that and reported, simultaneously and independently, in 1975, the creation of two-dimensional maps, by combining orthogonally a first dimension, based on pure charge fractionation, with a second dimension, based on size discrimination [5–7]. With that, modern proteome analysis was born, although it took, of course, many years of developments and refinements, for bringing the technique to the present-day very sophisticated level. Instrumental to that were major contributions from the field of informatics, who had to develop new algorithms for mapping the field, acquiring the images, cleaning the background, matching different maps among themselves. Informatics has also tremendously contributed by laying out a format for protein databases, many of which today are available and which represent a formidable tool in protein characterization [8]. We biochemists owe a big tribute to them; without their contribution, present-day two-dimensional map analysis would be meaningless. Another major contribution, of course, was the introduction, in 1982, of immobilized pH gradients [9], which offered a new, most powerful view of the field, guaranteeing much increased resolution and much higher reproducibility in spot position. Just as fundamental was the introduction, in the early nineties, of mass spectrometry as a tool for sequencing small peptides and for identifying proteins. Especially its version of MALDI-TOF (matrix-assisted, laser desorption ionization, time of flight) coupled to delayed extraction, has been instrumental in protein recognition and expanding databases [10]. With this trident (2-D maps, protein databases and mass spectrometry) we can venture in the ocean, like the Greek God Triton, equipped to catch the Marlin of our life, like the old character in the famous novel of Hemingway.

At the start of the third millennium, we are now faced with a dramatic growth of proteome analysis, due also to the fact that we have reached near completion in the major undertaking of the molecular biology of the last decade, namely the sequencing of the human genome as well as of a number of other genomes. Now that the code is decoded, we are faced with the fact that the vast majority of proteins are still a *terra incognita*, a huge field to be explored and mapped. This is perhaps the starting of a new stampede, a rush towards the new gold field, the forty-niners running after the gold of the third millennium, mining the proteome. Pharmaceutical companies, universities, venture capitals, geneticists, physicians, they are all engaged into this race, which promises a good harvest. Therefore, it is felt that a book in the field was sorely needed, to accompany us in this search. It is true, plenty of books have appeared recently in the field of proteome analysis, but they also become rapidly obsolete, as major advances appear almost every day. The present book also has a distinct flavour: it combines not only the practice, amply described in three major chapters (Chs. 12–14), but also the theory, dealt with *in extenso* in the first eleven chapters. Present-day books, most unfortunately, simply ignore all the theoretical developments, and often are reduced to mere cook books. But science is not, and cannot be, relegated to pure recipes: in order to progress, science has to be nourished by theory, by predictions, by description of basic phenomena. So, we hope that this unique combination will make the book more palatable to present-day audience. Perhaps, it might become the *vademecum*, the map of the gold fields for the new miners, proteome scientists, ready to go underground and dig. Dig you may, of course, and we hope you will also find the mother lode.

REFERENCES

1. L. Ornstein, Ann. N.Y. Acad. Sci., 121 (1964) 321–349.
2. B.J. Davis, Ann. N.Y. Acad. Sci., 121 (1964) 404–427.
3. A.L. Shapiro, E. Vinuela and J.V. Maizel, Biochem. Biophys. Res. Commun., 28 (1967) 815–820.
4. P.G. Righetti and J.W. Drysdale, Biochim. Biophys. Acta, 236 (1971) 17–24.
5. P.H. O'Farrell, J. Biol. Chem., 250 (1975) 4007–4021.
6. J. Klose, Humangenetik, 26 (1975) 231–243.
7. G.A. Scheele, J. Biol. Chem., 250 (1975) 5375–5385.
8. A. Bairoch, in M.R. Wilkins, K.L. Williams, R.D. Appel and D.F. Hochstrasser (Eds.), Proteome Research: New Frontiers in Functional Genomics, Springer, Berlin, 1997, pp. 93–148.
9. B. Bjellqvist, K. Ek, P.G. Righetti, E. Gianazza, A. Görg, W. Postel and R. Westermeier, J. Biochem. Biophys. Methods, 6 (1982) 317–339.
10. J. Godovac-Zimmermann and L.R. Brown, Mass Spectrometry Reviews, 20 (2001) 1–57.

PIER GIORGIO RIGHETTI (Verona)
ALEXANDER V. STOYANOV (Milano)
MIKHAIL Y. ZHUKOV (Rostov-na-Don)

Contents

PART I

Isoelectric Focussing: Fundamentals. Perspectives and Limits. Optimization of the Separation Process

Introduction

CONTENTS

This part of the book does not pretend to be a complete description of all aspects of isoelectric focussing. It is devoted first of all to those aspects that have still not yet found the appropriate reflection in the scientific literature. Firstly this concerns the question of the theory of polyvalent electrolyte dissociation. This is a matter of great importance since for modelling the polyelectrolyte behaviour at any process of electrophoretic separation one needs an appropriate pH-mobility relation, and it is extremely important to use a correct dissociation scheme.

The first section is devoted to the fundamentals of IEF. In the first two chapters some different aspects of the dissociation theory are considered, in the third one the kinetic aspects of acid–base equilibria are briefly discussed, but there the disadvantages of the traditional description are only emphasised. The fourth and fifth chapters concern the basic principles of the pH gradients. The steady-state IEF is discussed in Chapter 6, some aspects of the dynamics are dealt with in the following chapter (Chapter 7).

The second section deals with various topics under the title 'optimisation of electrophoretic separation'. Here, the two dimensional methods are also analysed. Chapter 11, entitled 'The limitation of the method of IEF', deals with important aspects like the resolving possibility limits for the isoelectric focussing technique, the perspectives of IEF whilst also the questions of microheterogeneity of biopolymers are discussed there.

ISOELECTRIC FOCUSSING: PRINCIPLES AND HISTORICAL ASPECTS

It is hardly possible to overemphasise the importance of isoelectric focussing for analytical biochemistry. Isoelectric focussing (IEF) is the method of electrophoretic separation of amphoteric substances based on the difference in their isoelectric points (pI). In the electric field any charged particle is subjected to the force $F = QE$ and in the isoelectric point, where its charge by definition equals zero, it should be immovable.

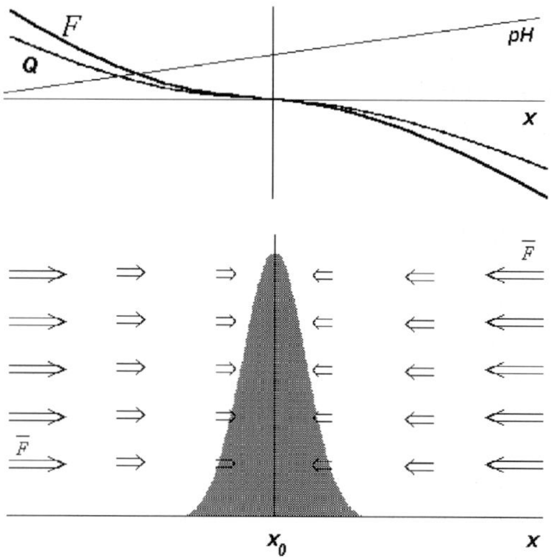

Fig. 1. The main principle of IEF.

Schematically the main principle of IEF is illustrated in Fig. 1. When a force acting on particles changes its sign at some point these particles should be concentrated in the vicinity of that point. For IEF, some special media are necessary which can provide the so-called pH gradient pH $=$ pH(x). Since the electric charge of an amphoteric molecule depends monotonously on the pH, there exists an opportunity of selecting the appropriate pH range where $F(x) = Q(\mathrm{pH}(x))E$ passes though the zero value (the corresponding value of the pH at which the charge equals zero is called isoelectric point (pI)). Due to diffusional forces which prevent a high concentration growth the resulting equilibrium distribution is established. (Low panel at Fig. 1 shows the vector field of external forcc F).

In reality such equilibria in systems with multi-dissociating amphotetic electrolytes have a little bit more complex character, but what is important is that the isoelectric focussing is an equilibrium method. This makes IEF very suitable in handling, provides a very high resolving possibility and gives an opportunity of clear and evident result interpretation.

No doubt, the discovery of isoelectric focussing, indeed, has opened a 'new dimension' for biochemists in characterising proteins and other biopolymers. Nature is created so that the latter have their isoelectric points rather 'uniformly' along the important pH range. More precisely, each pH interval, even a very small one, contains an enormous number of different substances and what is important, for biopolymers any small variation in composition (change in monomer unit — amino acid or nucleotide; or even single chemical modification) can result in a considerable pI shift.

The earliest work in the field of electrophoresis that is often referred to is the one of Reuss [1] so the history of the isoelectrofocussing started no later than in 1912 [2] or even before that time [3]. Ikeda and Suzuki had isolated glutamic acid from the digest of vegetable proteins. They described the construction of the electrolyser they used which was chambered into a set of compartments with ion penetrable membranes, and this construction with further modifications was used by other investigators [4–6], while the work of Hittorf can be considered as one of studying the problem of electrolysis of electrolyte solutions. It should be mentioned here that the works of Tiselius [7] were performed almost one hundred years after.

It is possible to say that the principles of isoelectric focussing were formulated by Williams and Waterman [8]. Nevertheless, perhaps, it has become possible to speak about the method of IEF by itself, after it was understood that a special medium is needed for maintaining the necessary pH gradient in which the focussing of amphotetic substances becomes possible, and most of the early examples can be treated as steady electrolysis of complex mixtures. Although someone definitely can consider even the electrolysis of one component solution as 'focussing' of this substance in the pH gradient which is created by its non-uniform distribution.

It was Kolin who proposed to separate proteins within a continuous pH gradient [9–12], and also introduced the term of 'Isoelectric line spectra'. Kolin used pH gradients created due to a diffusion of two different pH solutions. Soon after, two theoretical works appeared devoted to the mechanism of IEF [13,14].

The gradients developed by Kolin were quite unstable during prolonged electrolysis although their lifetime was sufficient enough for the component separation. For these gradients, Svensson [15] proposed to use the term 'artificial' in contrast to 'natural' ones, the latter being formed as a steady-state distribution during the electrolysis of an amphoteric compound mixture solution, he also defined the basic properties of the ampholytes to be used for gradient creation and for separation of proteins [16]. With the Vesterberg method of polyamino-polycarboylic acids synthesis by random polymerisation of pentaetylethylene hexamine with acrylic acid [17] it has become possible to produce a multicomponent mixture of amphoteric compounds covering any desired pH range with their p*I*s being spaced rather uniformly. So the IEF became a standard analytical procedure [18].

The next important development was the invention of the immobilised pH gradients (IPG) containing buffer substances covalently bound to the gel matrix [19–21]. This technique has proved to be capable of increasing resolution up to 0.001 pH unit (for two samples to be separated). Nevertheless, the authors of this book suppose that the potential of isoelectric focussing has still not been completely realised and a new breakthrough can be achieved in the future.

Only the main events have been briefly summarised above; there are many other important contributions without which the IEF would never have been able to reach its present-day high-developed state. One can obtain his own complete impression about the history of IEF by reading the appropriate sections in the monographs [22–25] and reviews devoted to this matter, e.g. [14,26–30].

REFERENCES

1. F.F. Reuss, Mémoires de la Société Impériale des Naturalistes de Moscou 2 (1809) 237.
2. K. Ikeda and S. Suzuki, US Patent 1015891.

3. G.W. Hittorf, Poggendorff's Ann., 89 (1853) 177–180.
4. A.A. Albanese, J. Biol. Chem., 134 (1940) 467–471.
5. J.l. Cox, H. King and C.P. Berg, J. Biol. Chem., 81 (1929) 755–762.
6. G.l. Foster and C.L. Shmidt, J. Am. Chem. Soc., 48 (1926) 1709–1713.
7. A. Tiselius, Svensk. Chem. Tidsskr., 58 (1941) 305–310.
8. R.R. Williams and R.E. Waterman, Proc. Exp. Biol. Med., 26 (1929) 56–61.
9. A. Kolin, J. Chem. Phys., 22 (1954) 1628–1629.
10. A. Kolin, J. Chem. Phys., 23 (1955) 407–410.
11. A. Kolin, Proc. Natl. Acad. Sci., 41 (1955) 101–110.
12. A. Kolin, Naturwissenschaften, 42 (1955) 367.
13. W.G. Kauman, Classe des Sciences de l'Académie Royale de Belgique, 43 (1957) 854–886.
14. E. Shumacher, Helv. Chim. Acta, 40 (1957) 2322–2340.
15. H. Svensson, Acta. Chem. Scand., 15 (1961) 325–341.
16. H. Svensson, Arch. Biochem. Biophys., Suppl., 1 (1962) 132–140.
17. O. Vesterberg, US Patent 3485736, 1969.
18. O. Vesterberg and H. Svensson, Acta Chem. Scand., 20 (1966) 220–824.
19. V. Gasparic, A. Rosengren and B. Bjellkqvist, Swedish Patent 7514049-1, 1975.
20. A. Rosengren, B. Bjellqvist, V. Gasparic, US Patent 4130470, 1978.
21. A. Rosengren, B. Bjellqvist and V. Gasparic, German Patent 2656162, 1981.
22. V.G. Babskii, M.Y. Zhukov and V.I. Yudovich, Mathematical Theory of Electrophoresis, Consultants Bureau, New York, 1989.
23. O. Gaal, G.A. Medgesi and L. Vereczkey, Electrophoresis in the Separation of Biological Macromolecules, Wiley, Chichester, 1980.
24. P.G. Righetti, Isoelectric Focusing: Theory, Methodology and Applications, Elsevier, Amsterdam, 1983.
25. P.G. Righetti, Immobilized pH Gradients: Theory and Methodology, Elsevier, Amsterdam, 1990.
26. A. Kolin, in N. Catsimpoolas (Ed.), Isoelectric Focusing, Academic Press, New York, 1976, pp. 1–11.
27. A. Kolin, in B. Radola (Ed.), Isoelectrofocusing and Isotachophoresis, Academic Press, New York, 1976, pp. 3–33.
28. H. Rilbe, Electrophoresis, 16 (1995) 1354–1359.
29. H. Rilbe, Ann. N.Y. Acad. Sci., 209 (1973) 11–21.
30. H. Rilbe, in N. Catsimpoolas (Ed.), Isoelectric Focusing, Academic Press, New York, 1976, pp. 14–52.

PART I.I

Isoelectric Focussing: Fundamentals

CHAPTER 1

Electrolyte Dissociation in Water Solution. Simple Electrolytes

CONTENTS

1.1. INTRODUCTION

We shall start our consideration from a simple bivalent electrolyte. In this chapter the properties of parallel and stepwise dissociation schemes will be considered. We shall obtain the appropriate expressions for hydrogen ions concentration and buffer capacity and, also, try to discuss briefly such basic definitions as the isoelectric point of an amphoteric molecule, electrophoretic mobility and conductivity. More rigorous treatment will be performed in further chapters devoted to the multi-dissociating systems and the kinetic aspects of the acid–base equilibria.

1.2. STEPWISE AND PARALLEL DISSOCIATION SCHEMES FOR A BIVALENT PROTOLYTE

Let us start our consideration with a bivalent protolyte since some important results can be obtained with this simple example, and further extension to multi-dissociating protolytes will be made in the next chapter.

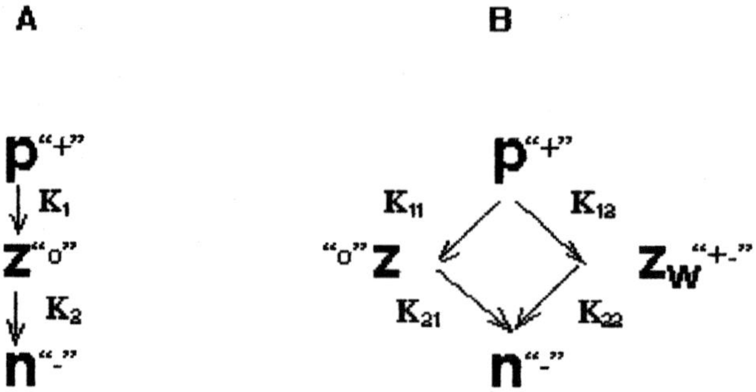

Fig. 1.1. Dissociation scheme for ampholyte with two ionogenic groups when the process is described by the stepwise 'A' and parallel 'B' mechanisms. p, z and n denote the positively charged, uncharged and negatively charged ampholyte states, respectively.

There are only two possible schemes for the dissociation of any bivalent protolyte: stepwise and parallel (Fig. 1.1). We shall consider the ionisation reactions both for basic and acidic groups in terms of proton loss (Eqs. (1.1) and (1.2)),

$$K_a = [(A^-)(H^+)]/(A_0) \tag{1.1}$$

$$K_b = [(B_0)(H^+)]/(B^+) \tag{1.2}$$

so formally, these two schemes may be applied to mono–mono-valent and bivalent protolytes, but at first let us suppose that our protolyte has one acidic and one basic group, i.e. we are dealing with an ampholyte. It is not so difficult to see that our two systems differ from each other by the number of species: system A contains particles of three types (without water ions) and for system B the total number of kinds is four (i.e. there is no zwitterion state in scheme A).

The dissociation constants K_{11} and K_{22}, K_{12} and K_{21} are not equal to each other in a common case. An assumption that they are equal and $K_{11} = K_{22} = K_a$ and $K_{12} = K_{21} = K_b$ allows us to treat the process in frames of 'independent dissociation', and that, for multi-dissociating ampholytes (see Chapter 2), essentially simplifies the theoretical treatment.

1.3. RELATIVE CONCENTRATION OF DIFFERENT PROTOLYTE FORMS FOR STEPWISE AND PARALLEL SCHEMES

For the system corresponding to scheme A we have the set of equations, which represents the chemical equilibrium Eqs. (1.3) and (1.4)

$$z(H^+) = K_a p \tag{1.3}$$

$$n(H^+) = K_b z \tag{1.4}$$

and the expression for the analytical concentration C, which is

$$p + z + n = C \tag{1.5}$$

By solving this linear system (relative to the concentration of each protolyte form), we shall have expressions for their concentrations though equilibrium constants values and hydrogen ion concentration Eqs. (1.6)–(1.9)

$$p = \frac{C(\mathrm{H}^+)^2}{(\mathrm{H}^+)^2 + (\mathrm{H}^+)K_\mathrm{a} + K_\mathrm{a}K_\mathrm{b}} \tag{1.6}$$

$$z = \frac{C(\mathrm{H}^+)K_\mathrm{b}}{(\mathrm{H}^+)^2 + (\mathrm{H}^+)K_\mathrm{a} + K_\mathrm{a}K_\mathrm{b}} \tag{1.7}$$

$$n = \frac{CK_\mathrm{a}K_\mathrm{b}}{(\mathrm{H}^+)^2 + (\mathrm{H}^+)K_\mathrm{a} + K_\mathrm{a}K_\mathrm{b}} \tag{1.8}$$

For system B we can use any three relations of chemical equilibrium, e.g. Eqs. (1.9)–(1.11)

$$z(\mathrm{H}^+) = K_\mathrm{b}p \tag{1.9}$$

$$\mathrm{zw}(\mathrm{H}^+) = K_\mathrm{a}p \tag{1.10}$$

$$n(\mathrm{H}^+) = K_\mathrm{b}\mathrm{zw} \tag{1.11}$$

$$n(\mathrm{H}^+) = K_\mathrm{a}z \tag{1.12}$$

and write the ampholyte concentration as Eq. (1.13)

$$p + z + \mathrm{zw} + n = C \tag{1.13}$$

That gives us Eqs. (1.14)–(1.17)

$$p = \frac{C(\mathrm{H}^+)^2}{(\mathrm{H}^+)^2 + (\mathrm{H}^+)(K_\mathrm{a} + K_\mathrm{b}) + K_\mathrm{a}K_\mathrm{b}} \tag{1.14}$$

$$z - \frac{C(\mathrm{H}^+)K_\mathrm{b}}{(\mathrm{H}^+)^2 + (\mathrm{H}^+)(K_\mathrm{a} + K_\mathrm{b}) + K_\mathrm{a}K_\mathrm{b}} \tag{1.15}$$

$$\mathrm{zw} = \frac{C(\mathrm{H}+)K_\mathrm{a}}{(\mathrm{H}^+)^2 + (\mathrm{H}^+)(K_\mathrm{a} + K_\mathrm{b}) + K_\mathrm{a}K_\mathrm{b}} \tag{1.16}$$

$$n = \frac{CK_\mathrm{a}K_\mathrm{b}}{(\mathrm{H}^+)^2 + (\mathrm{H}^+)(K_\mathrm{a} + K_\mathrm{b}) + K_\mathrm{a}K_\mathrm{b}} \tag{1.17}$$

For the system B the concentration ratio of the zwitterion to the unionised state is defined by the dissociation constants ratio — Eqs. (1.9) and (1.10) — and is expressed as

$$\mathrm{zw}/z = K_\mathrm{a}/K_\mathrm{b} \tag{1.18}$$

References p. 21

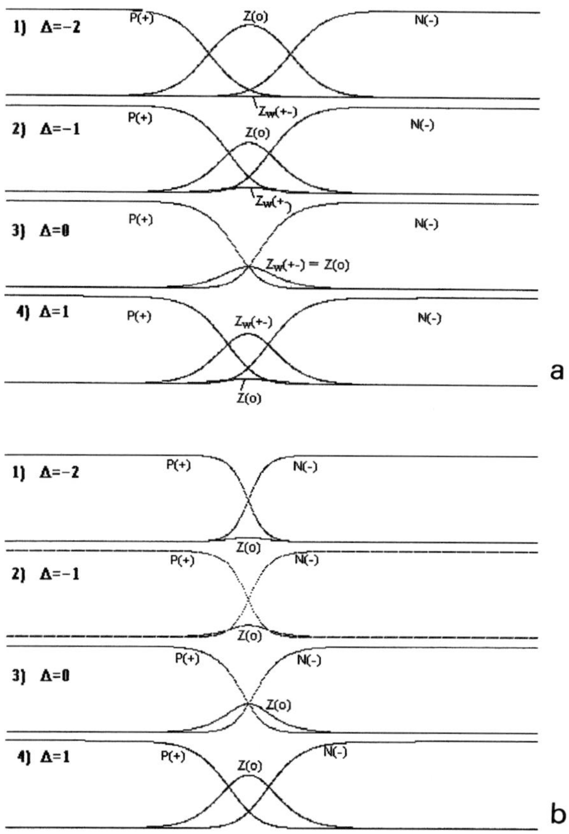

Fig. 1.2. Relative concentration of each ampholyte state, as a function of pH for stepwise (A) and parallel (B) dissociation, at different values of the ΔpK parameter.

Often the situation when $\Delta pK \gg 1$ is analyzed gives us $zw/z \gg 1$ but there are no formal restrictions to consider the case of $\Delta pK = 0$ or even $\Delta pK < 0$.

Fig. 1.2 shows the relative concentration of each ampholyte state, as a function of pH for stepwise (A) and parallel (B) dissociation, at different values of the ΔpK parameter.

1.4. HYDROGEN IONS CONCENTRATION AND BUFFER CAPACITY

As we can see, the electroneutrality equation, which allows us to determine the pH, is an equation of four degrees relative to (H^+).

$$p + (H^+) = n + \frac{Kw}{(H^+)} \tag{1.19}$$

Since we have already seen that we are dealing with different values of concentrations for p and n (Eqs. (1.11) and (1.12)) or (Eqs. (1.14) and (1.15)), depending which scheme (A or B) is used, we must obtain the different values of pH. Although the numerical

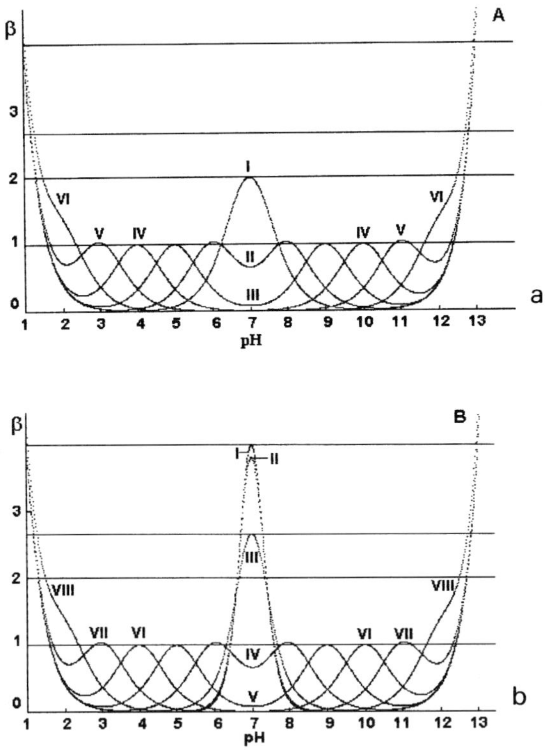

Fig. 1.3. Relative buffering capacity of the two systems with 10 mM ampholyte concentration in a wide pH interval, calculated with the assumption of a parallel (A) or stepwise dissociation mechanism (B). The ampholyte's p*I*s are 7.0 for all the families of compounds with different Δp*K*s. Curves I, II, III, IV, V, VI, VII and VIII in (B) correspond to Δp*K* values of: −4, −2, 0, 2, 4, 6, 8 and 10, respectively. Curves I, II, III, IV, V and V in (A) correspond to absolute Δp*K* values of: 0, 2, 4, 6, 8 and 10 (in case of the parallel dissociation scheme, the buffer capacity is a function of the absolute Δp*K* difference). Note that, whereas in the case of stepwise dissociation (A) the relative β value is 2.7 for $\Delta pK = 0$, in the case of a parallel dissociation scheme the relative β is 2 for $\Delta pK = 0$.

calculations demonstrate a rather small difference in pH, the difference in the buffer capacity β may become considerable.

The expression for β is obtained by direct application of buffer power definition to the electroneutrality equation, rewritten in presence of a strong titrant. Performing the differentiation (this procedure is essentially more easy than solving Eq. (1.19) relative to (H)), we obtain Eq. (1.20) and Eq. (1.21) for the systems A and B respectively:

$$\beta^{A} = \frac{2.303C[4(H^+)^2K_aK_b + (H^+)K_b((H^+)^2 + K_aK_b)]}{[(H^+)^2 + (H^+)K_b + K_aK_b]^2} \qquad (1.20)$$

$$\beta^{B} = \frac{2.303C[4(H^+)^2K_aK_b + (H^+)(K_a + K_b)((H^+)^2 + K_aK_b)]}{[(H^+)^2 + (H^+)(K_a + K_b) + K_aK_b]^2} \qquad (1.21)$$

Fig. 1.3 represents the buffer power of our two systems, calculated for the 10 mM solution in a wide pH interval (we ought to take into account that this is an abstraction

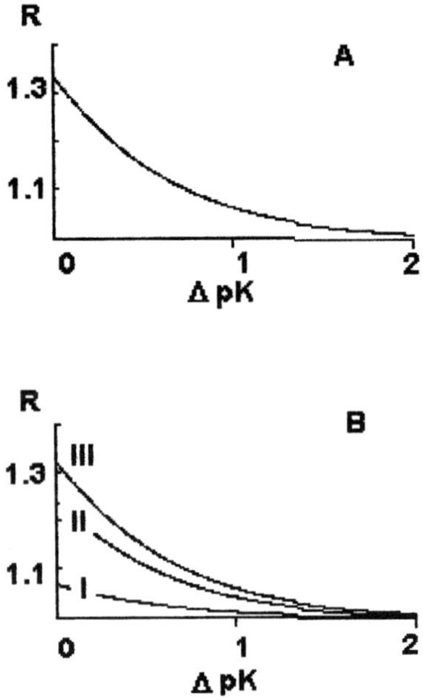

Fig. 1.4. Buffering capacity ratio (β_a/β_b) for the two systems (with stepwise and parallel dissociation mechanism) as a function of ΔpK. (A) Ratio of the ampholytes' buffer capacity only (neglecting the water contribution). (B) Buffering capacity contribution of water is plotted when the ampholyte concentration is progressively increased: $C = 0.001$, $C = 0.01$ and $C = 0.1$, curves I, II and III, respectively. The calculations are performed at $pI = 3$.

and the real physical sense is the value of β at the point of real pH). But since the buffer power is an additive value [1,2], it is not useless to consider the contribution of each component separately as a function of pH. We can see a considerable discrepancy when the ΔpK difference is small or negative. Fig. 1.4 demonstrates the β_a/β_b ratio as a function of ΔpK.

1.5. IONISATION COEFFICIENT

The convenient definition of the ionisation coefficient [3] as

$$k = (p + n)/C \tag{1.22}$$

also has a meaning of 'net degree of dissociation'.

We have for our two systems, respectively:

$$k^A = \frac{(H^+)^2 + K_a K_b}{(H^+)^2 + (H^+)K_b + K_a K_b} \tag{1.23}$$

and

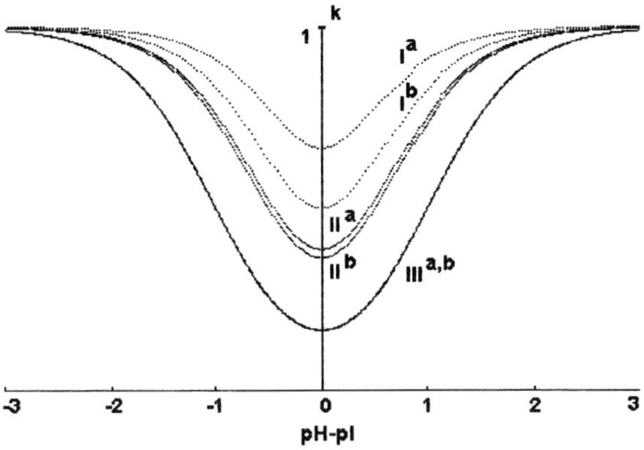

Fig. 1.5. Ionisation coefficient as a function of pH–pI. The two upper curves represent the ionisation coefficient for stepwise and parallel dissociation (I^a and I^b, respectively) at $\Delta pK = 0$. The curves II^a and II^b correspond to $\Delta pK = 1$. With $\Delta pK = 2$ these curves practically coincide – $III^{a,b}$.

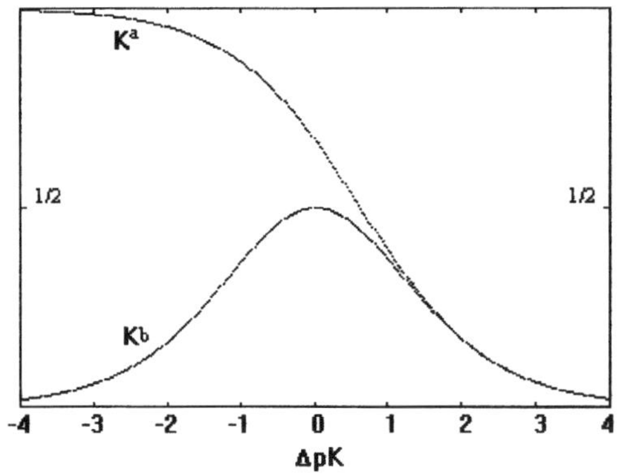

Fig. 1.6. Ionisation coefficient value at the point pH $= 1/2[pK_a + pK_b]$ as a function of ΔpK. The superscript indices 'a' and 'b' denote stepwise and parallel dissociation mechanisms, respectively.

$$k^B = \frac{(H^+)^2 + K_a K_b}{(H^+)^2 + (H^+)(K_a + K_b) + K_a K_b} \tag{1.24}$$

where the upper indices 'A' and 'B' denote the 'stepwise' and 'parallel' dissociation mechanisms, respectively.

Fig. 1.5 illustrates the ionisation coefficients behaviour in the vicinity of the isoelectric point (as earlier we should again emphasise that the real physical sense will have only a point of the real pH value). E.g. for 10 mM ampholyte solution with p$I = 3$ and $\Delta pK = 0.5$, the pH value is around 3.07 and the ionisation coefficients are $k = 0.53$, and $k = 0.46$ for A and B schemes, respectively. The value of the ionisation coefficient at the point of pH $= 1/2[pK_a + pK_b]$ as a function of ΔpK is represented in Fig. 1.6. We can see that the upper limit of the ionisation coefficient is $2/3$ for the stepwise dissociation scheme, while for the parallel one it is $1/2$.

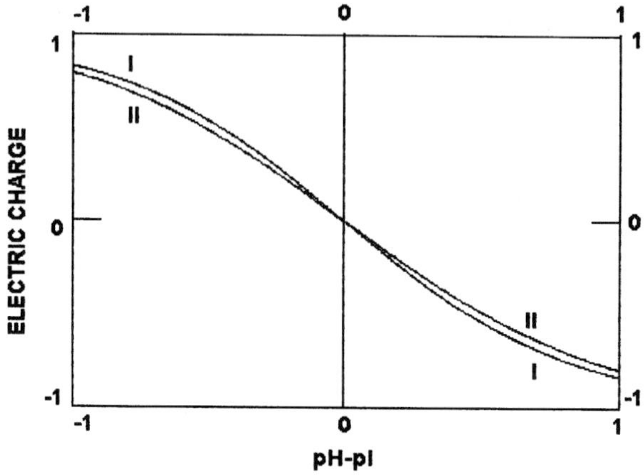

Fig. 1.7. Electric charge versus pH for a bivalent ampholyte molecule, calculated assuming stepwise and parallel dissociation model – curves I and II, respectively ($\Delta pK = 0.5$).

1.6. ISOELECTRIC POINT

With a known concentration of each microstate one can easily determine the average charge of the molecule, and when the above value equals zero (for the case of a bivalent ampholyte there are the same concentrations of the positive and negative forms) this state is called 'isoprotic'. Together with the 'isoprotic point' the term 'isoionic point' is used when one wants to take into account other ions capable of binding to the ampholyte molecule (other than hydrogen ones).

The definition of the isoelectric point is the pH of the zero electrophoretic mobility, but the term 'mobility' in its turn needs some explanation. These are not strictly accepted definitions [4–8]. When it is unnecessary to emphasise the difference between the isoionic (isoprotic) state and the isoelectric state any term is used.

For the two different dissociation schemes an example of the electric charge versus the pH behaviour is given in Fig. 1.7.

1.7. MOBILITY OF PROTOLYTE MOLECULE

While the appropriate set of thermodynamic constants is sufficient for a complete description of the acid–base properties, for modelling the electrophoretic mobility (and conductivity, as well) we need to have at least some information about the kinetics of the interconversion between the microstates present in the solution. In particular, under the conditions of the same ionisation coefficient, a faster conversion rate will result in a lower net mobility.

There is no possibility of obtaining experimentally all the precise data, so only some simplified models can be analyzed.

1.8. NON-ADDITIVE SUM FOR BUFFER CAPACITY IN CASE OF STEPWISE DISSOCIATION

The stepwise dissociation mechanism possesses one very important characteristic feature, namely, its buffering power is the result of 'non-additive sum' [9]. That means that over all of the pH region the buffer capacity of a dissolved protolyte is not equal to the sum of the two monovalent protolytes with the same pKs. The above effect is clearly visible starting from a ΔpK difference of approximately 1.5 units. Moreover, getting into the negative range of ΔpK, we approach to maximal limit, practically we achieve the limiting curve at Δp$K \approx 3$, with a four-fold excess of the maximal value of the monovalent protolyte at the point of pH $= (\text{p}K_1\text{p}K_2)^{1/2}$ (see Fig. 1.3).

For an arbitrary two-step reaction, generally speaking, the demand of p$K_1 \gg$ pK_2 (or least p$K_1 >$ pK_2) is not obligatory. For any system with an 'inverse' dissociation constant order (i.e. the second step is more preferable in comparison with the first one) we shall have one very important property, namely, a very small concentration of the intermediate product (see Fig. 1.2). With an n increase of ΔpK, the concentration of the intermediate state (in our case, having zero charge) increases correspondingly, thus we have a decrease in the charged forms, which contribute to the buffering power (see Figs. 1.2A and 1.3A).

Now let us look at the system with two independently dissociating groups (Fig. 1.1B). Here we have two uncharged states (true zero and zwitterion). For each protolyte state we have the following set of equations, expressing its relative concentration (Eqs. (1.9)–(1.12)).

As we can see from these equations, at any pK difference there is a considerable amount of uncharged forms in the vicinity of the isoelectric point (pH $= (\text{p}K_1\text{p}K_2)^{1/2}$, see Fig. 1.3B. With Δp$K = 0$ the concentrations of 'zero' and 'zwitterionic' states are identical. By reordering the dissociation constants, we only mutually exchange the concentrations of the two uncharged states, while the concentrations of p and n remain the same (Fig. 1.2B).

The buffering capacity for such a system is given by Eq. (1.21) and it depends not on ΔpK, but on abs$\{\Delta\text{p}K\}$ only. Thus for this system there is no difference whether we are dealing with a positive or a negative ΔpK see Fig. 1.3B.

We can analyze, also, these two systems in terms of the ionisation coefficient (it is defined as the relative concentration of charged forms). Its value coincides with the one of the buffer capacity (with accuracy of $C \ln 10$ factor) at the isoelectric point. In Fig. 1.4 the behaviour of the ionisation coefficient for the cases of stepwise and parallel dissociation at the point of pH $= (\text{p}K_1\text{p}K_2)^{1/2}$, as a function of ΔpK, is visualised. The system with a parallel dissociation mechanism has a symmetrical profile of the ionisation coefficient (with a maximum of $1/2$), while for the stepwise dissociation it grows with a negative ΔpK difference, asymptotically approaching the unity limit.

In the previous sections we analyzed carefully the intrinsic properties of the stepwise dissociation scheme, but the question of the correctness of applying any scheme to real substances was not considered. The structure formula of any substance enables us to make our choice between the parallel and stepwise dissociation mechanisms (or to use the necessary combination of them). In a general case, for multi-potential ampholyte, the parallel scheme is very complicated, since it operates with 2^N different constants (N is the number of dissociating groups). An independent dissociation represents a 'degenerative' case of parallel dissociation, it is essentially easy in operation, but in many cases its application is justified and gives a good correlation with experimental results.

According to Svensson's (Rilbe) interpretation, there is a limitation placed on the dissociation constant values, namely:

$$\Delta pK = pK_b - pK_a > 0.6 \tag{1.25}$$

This condition was formulated by Svensson rather late (in 1971, [10]), after his concept of carrying ampholytes had been reported (in 1961–1962, [11,12,3]). The reason to perform this correction was not to overexceed the value of 1/2 for the ionisation coefficient (calculated at the isoelectric point), since it makes the buffering capacity of a bivalent protolyte greater than two (in relative units). With this limitation the buffer capacity will never exceed two, but this restriction does not eliminate the phenomenon of non-additive summing, it only makes it not so evident.

The pH–charge relationship of any system dissociating according to the parallel scheme may be described with the help of the so-called 'macroscopic' constants (see next chapter). For the case of two ionisation centres scheme B (Fig. 1.1) is to be replaced with a less complicated scheme A, and this procedure needs the appropriate formulas connecting the new set of dissociation constants with the old one. The latter are derived in a simple manner and the necessary restriction on the macroscopic constants is imposed automatically.

1.9. NON-AMPHOTERIC COMPOUNDS AND BUFFER CAPACITY IN 'ISOPROTIC STATE'

The circumstance that, while using the scheme of the stepwise dissociation we obtain a non-additive summing of buffering powers is not a contradiction by itself. The contradiction arises only with an attempt of applying this scheme, but also using the dissociation constant values determined by the assumption of a quite different mechanism. For any true stepwise process we have really the situation of a non-additive sum, but this effect is visible only in the case of a 'not so big difference in ΔpK', see above section. Let us consider Fig. 1.8, which represents the buffering capacity of some hypothetical acid, which is dissociated in two steps and possesses rather small $\Delta pK = 1$, in a wide pH region.

In the vicinity of $pH = (K_1K_2)^{1/2}$ (isoprotic point according to Rilbe's definition, [13]) we have, indeed, an appreciated excess of the buffer capacity in comparison with the arithmetical sum of the contributions, that each ionogenic group would give separately (see dotted line), although the so-called 'critical' value of twice-excess is not quite reached. But, in the case of non-ampholytes, the excellent buffering power in the so-called 'isoprotic point' will not have any relation to the one of real solution since the pH of the solution would be situated far from $(pK_1 + pK_2)/2$ (see Fig. 1.9).

An important question is arisen: is there any limit on ΔpK in true stepwise processes?

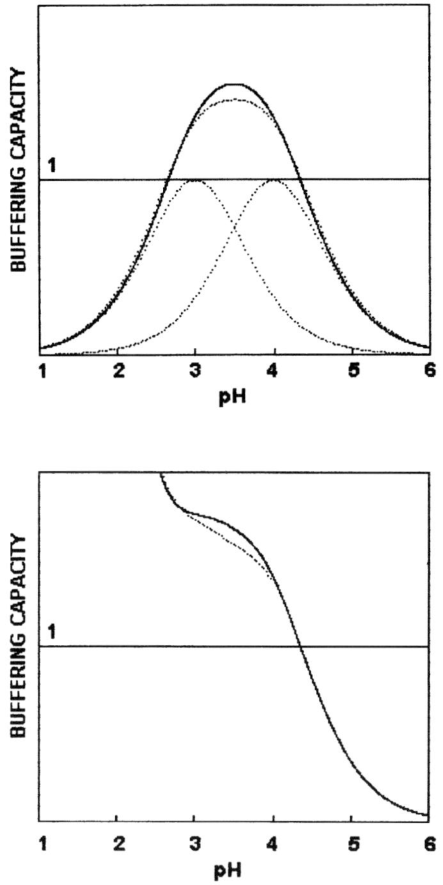

Fig. 1.8. Buffer power contribution of a hypothetical two-step dissociating acid (pK_1 = 2, pK_2 = 3 (ΔpK = 1)). The dotted line corresponds to the 'additive sum', in the case of parallel independent dissociation. (B) The calculations are performed taking into account the contribution of water.

(Note, that now the case of non-ampholytes is discussed.) The answer is: yes, certainly. Whenever the system loses (or obtains) a proton, it receives an appropriate additional charge, thus any second step is considerably more unfavourable energetically, since it will be connected with acquiring the charge of the same name. What about the value of this limit? It can be evaluated theoretically and must be dependent on a concrete substance; but if we recollect the tabulated values for substances, for which we have no doubts in a stepwise dissociation mechanism (e.g. non-organic acids), we come to the conclusion that this limit considerably exceeds the value given by Eq. (1.25) — the appropriate constants differing from each other by several times in order [14].

Different dissociation schemes result in different properties of a system we describe, in particular, with respect to a number of microstates. In case of the stepwise scheme there is a direct way to the final state which goes through only one intermediate state,

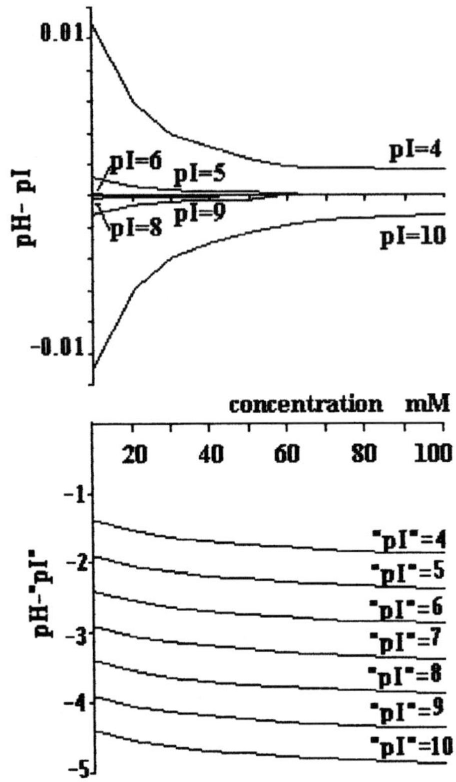

Fig. 1.9. pH deviation from the 'isoprotic point' value pH $= 1/2[\mathrm{p}K_1+\mathrm{p}K_2]$ as a function of $1/2[\mathrm{p}K_1+\mathrm{p}K_2]$ and protolyte concentration for a biprotic ampholyte and non-ampholyte (acid), respectively. In the first case the real pH of the solution approaches the 'isoprotic point' as the concentration increases (A), whereas in the case of non-amphotere (biprotic acid) this difference grows with the concentration of the substance in the solution (B).

while in case of the parallel dissociation there are two routes to go from the initial to the final state. It is possible to describe a parallel process in terms of a stepwise scheme. There are many different dissociation schemes of other types and their descriptions can also be done in terms of stepwise processes, but some part of the information will be lost. These aspects will be considered in the following chapter.

1.10. NOTATIONS

K_a, K_b	dissociation constant of the acidic and the basic ionogenic groups
p (positive)	relative concentration of the positively charged ampholyte form
n (negative)	relative concentration of the negatively charged ampholyte form
z (zero)	relative concentration of ampholyte form with both ionogenic groups uncharged

zw (zwitterion)	relative concentration of ampholyte in the zwitterionic form
(H^+)	hydrogen ion concentration
β	buffering capacity
k	ionisation coefficient

1.11. REFERENCES

1. J.N. Butler, Ionic Equilibrium, Addison-Wesley, Reading, MA, 1964.
2. J.E. Ricci, Hydrogen Ion Concentration. New Concept in a Systematic Treatment, Princeton University Press, Princeton, NJ, 1952.
3. H. Svensson, Acta Chem. Scand., 16 (1962) 456–466.
4. R.A. Alberty, in A.H. Neurath and K. Bailey (Eds.), The Protein Chemistry, Biological Activity, and Methods, Academic Press, New York, 1953, Vol. 1, pp. 461–594.
5. R.K. Cannan, Chem. Rev., 30 (1942) 295.
6. D.J. Davidson, Chem. Educ., 32 (1955) 550.
7. H. Rilbe, Ann. N.Y. Acad. Sci., 209 (1973) 11–22.
8. S.P. Sorensen L, K. Linderstrom-Lang and E. Lund, Compt. Rend. Trav. Lab. Karlsberg, 16 (1926) 5.
9. A.V. Stoyanov and P.G. Righetti, Electrophoresis, 19 (1998) 187–191.
10. H. Rilbe, Acta Chem. Scand., 25 (1971) 2768–2769.
11. H. Svensson, Arch. Biochim. Biophys. Suppl., 1 (1962) 132–140.
12. H. Svensson, Acta Chem. Scand., 15 (1961) 325–361.
13. H. Rilbe, pH and Buffer Theory — A New Approach, Wiley, Chichester, 1996.
14. R.C. Weast (Ed.), CNC Handbook of Chemistry and Physics, CRC Press, Boca Raton, FL, 1987.
15. H. Rilbe, in N. Catsimpoolas (Ed.), Isoelectric Focusing, Academic Press, New York, 1976, pp. 51–63.

CHAPTER 2

Dissociation of Polyvalent Electrolytes

CONTENTS

2.1. INTRODUCTION

Different theoretical approaches could result in different values for the microscopic parameters of the system under analysis, as well as for its integral characteristics (such as conductivity and buffering power). We analysed previously the properties of parallel and sequential dissociation schemes and now some examples of a hybrid type scheme will be considered and the expressions relating the macroscopic and microscopic constants will be obtained.

Further, the problem of the proton binding curves (so-called titration curves) modelling will be discussed. The information about the equilibrium constants only gives an opportunity of obtaining the values of each microconcentration at steady state and nothing more, that is we have no information about the life time of microstates. On the other hand, for the purpose of describing some properties of such systems, we do not really need even the precise scheme of the general reaction. For example, for any polyvalent substance, the complete information about proton binding reactions taking place in the solution and their constants (equilibrium) gives us an opportunity of obtaining a comprehensive description of the entire system (average microstate concentrations). But, for obtaining the pH–electric charge relationship (and the buffering capacity as well) there is no necessity of knowing the complete set of microconstant values. Also, when the other task of interpreting the experimental protein binding isotherm or 'titration curves' is considered, only limited information about the so-called intrinsic dissociation constants may be derived from the appropriate data. This fact was pointed out as early as in 1925 by Adair [1], who analysed the system with a set of parallel reactions and concluded that only 'apparent' or 'macroscopic' constants could be experimentally measured (the term of Adair constant is also used). Except for the case of very simple compounds, there is no possibility of obtaining a precise information about the intrinsic dissociation constants, so for titration curves modelling, which is a matter of great importance in some separation technologies (electrophoresis, ion exchange chromatography), usually we need to adopt a simplified approach. This may be either an assumption of independent dissociation or some other way of postulating all the necessary microscopic constant values, but the result will depend on the reaction scheme which is selected for description. In this chapter some general properties of possible dissociation schemes and some problems of interpretation of experimental titration curves will be analysed. We will also discuss which scheme should be used in order to have a correct description of the entire system and of some of its particular properties as well.

2.2. ACID–BASE EQUILIBRIA, MACROSCOPIC AND MICROSCOPIC CONSTANTS

A system containing a set of groups (N), capable of parallel dissociation of protons, results in a rather high number of microscopic states, the latter growing rapidly with the number of dissociating groups. For such a system, a number of possible reactions, in turn, increases more rapidly. Consider the most simple case of only two ($N = 2$) ionogenic groups (Fig. 2.1A): here we have four possible microstates and the same number of chemical reactions. Here we are not specifying the nature of dissociating group (acid or base); any reaction corresponds to a proton loss. The symbols 'P' and 'U' are used to designate the fully protonated and fully unprotonated states, respectively; the other intermediate states are numbered with small letters and the same horizontal level implies the same charge value. If we analyse a process with a pure type of parallel dissociation, for $N = 3$ the number of microstates becomes equal to eight with twelve different reactions among them (see Fig. 2.1B). Obviously, for calculating all the relative microconcentrations, we do not need all of the microconstant values, e.g. for the system with only two ionogenic groups (Fig. 2.1A) there is one additional relation connecting microconstant values:

$$k_{\mathrm{Pa}}k_{\mathrm{aU}} = k_{\mathrm{Pb}}k_{\mathrm{bU}} \tag{2.1}$$

and thus any three microconstants provide all the information about the relative concentration of each microstate.

For the system in Fig. 2.1B one can easily obtain five independent relations, connecting the dissociation constant values; for example we may take them in the

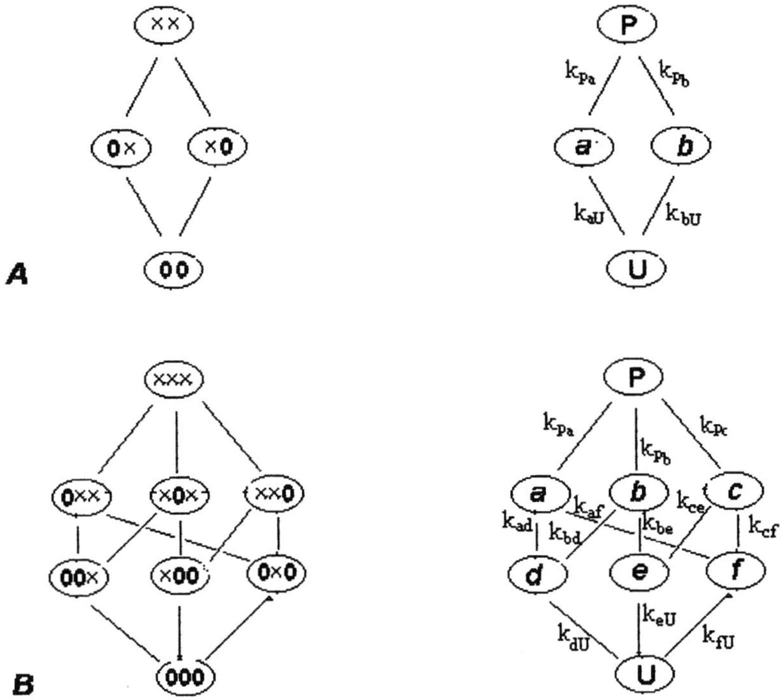

Fig. 2.1. Reaction schemes of parallel dissociation for the cases of two (A) and three (B) ionogenic groups. The nature of the dissociating group (acid or base) is not specified, since any reaction corresponds to a proton loss. X's and zeroes are used to designate the presence or the absence of a proton (left panel). The two extremes 'P' and 'U' correspond to the fully protonated and fully unprotonated states, respectively; the other intermediate states are designated by small letters. The double subindexes of the microscopic constants reflect the two microstates involved in the microreaction (right panel). The appropriate relations connecting macroscopic and microscopic constants are given in the text.

following form:

$$k_{\mathrm{ad}}k_{\mathrm{dU}} = k_{\mathrm{af}}k_{\mathrm{fU}} \tag{2.2}$$

$$k_{\mathrm{pa}}k_{\mathrm{ad}}k_{\mathrm{dU}} = k_{\mathrm{pb}}k_{\mathrm{bd}}k_{\mathrm{dU}} \tag{2.3}$$

$$k_{\mathrm{pa}}k_{\mathrm{ad}}k_{\mathrm{dU}} = k_{\mathrm{pb}}k_{\mathrm{bc}}k_{\mathrm{cU}} \tag{2.4}$$

$$k_{\mathrm{pa}}k_{\mathrm{ad}}k_{\mathrm{dU}} = k_{\mathrm{pc}}k_{\mathrm{ce}}k_{\mathrm{eU}} \tag{2.5}$$

$$k_{\mathrm{pa}}k_{\mathrm{ad}}k_{\mathrm{dU}} = k_{\mathrm{pc}}k_{\mathrm{cf}}k_{\mathrm{fU}} \tag{2.6}$$

and, starting from only seven ($7 = 12 - 5$) independent equations of dissociation we arrive at a complete description of our system. (These equations may be selected, for example, by following the simple rule: "All of the possible reaction paths, obtained by specification of microreactions, do not form closed trajectories.") As we see, in order to know the concentrations of each microstate, we need only seven microconstants. Note that the general form of a titration curve does not depend on whether acidic or basic groups are involved (provided the set of microconstants remains the same), their balance influences only the isoelectric point value. Further, if we want to obtain the proton binding isotherm only, we do not need the precise distribution between all the possible microstates in our system, it is sufficient to know only the 'total' concentrations of microstates with the same charge values. The pH–charge relationship for the system shown in Fig. 2.1A may be described with the help of two macroscopic constants only. (In order to write the expression for the electric charge we need to specify the nature of ionogenic groups. In this and in the following examples all of them are taken as acid ones.)

$$Q[(\mathrm{H}^+)] = \frac{1 + K_1/(\mathrm{H}^+)}{1 + K_1/(\mathrm{H}^+)1 + K_1K_2/(\mathrm{H}^+)^2} \tag{2.7}$$

$$K_1 = k_{\mathrm{pa}} + k_{\mathrm{pb}} \tag{2.8}$$

$$K_2 = \frac{k_{\mathrm{aU}}k_{\mathrm{bU}}}{k_{\mathrm{Pa}} + k_{\mathrm{Pb}}} \tag{2.9}$$

A three dissociating group system (Fig. 2.1B) is characterised by three macroconstants:

$$K_1 = K_{\mathrm{pa}} = K_{\mathrm{pb}} + K_{\mathrm{pc}} \tag{2.10}$$

$$K_2 = \frac{k_{\mathrm{Pa}} + k_{\mathrm{ad}} + k_{\mathrm{pb}}k_{\mathrm{be}} + k_{\mathrm{Pc}}k_{\mathrm{cd}}}{k_{\mathrm{Pa}} + k_{\mathrm{Pb}} + k_{\mathrm{Pc}}} \tag{2.11}$$

$$K_3 = \frac{k_{\mathrm{Pa}} + k_{\mathrm{ad}} + k_{\mathrm{dU}}}{k_{\mathrm{Pa}}k_{\mathrm{ad}} + k_{\mathrm{pb}}k_{\mathrm{be}} + k_{\mathrm{Pc}}k_{\mathrm{cd}}} \tag{2.12}$$

And the expression for the electric charge is given by

$$Q[(\mathrm{H}^+)] = \frac{1 + K_1/(\mathrm{H}^+) + K_1K_2/(\mathrm{H}^+)^2}{1 + K_1/(\mathrm{H}^+)1 + K_1K_2/(\mathrm{H}^+)^2 + K_1K_2K_3/(\mathrm{H}^+)^3} \tag{2.13}$$

In the case of a pure parallel dissociation process of N groups we are dealing with 2^N microstates (this result obtained by Wyman, see [2]). The number of possible microreactions (M) increases very rapidly with the number of ionisable groups:

$$M = N!\sum_{k=0}^{N-1} 1/(k!(N-k-1)!) \tag{2.14}$$

Indeed, for any true parallel dissociation process with N groups capable of dissociating a proton consider the general scheme as a 'N-stage' process. Each 'stage' includes all of microreactions of one proton loss, so if we denote by k the number of deprotonated groups the total number of microstates (n_k) for given k is obtained by binomial coefficient

$$n_k = C_k^N = N!/(k!(N-k)!)$$

TABLE 2.1

NUMBER OF POSSIBLE MICROREACTIONS AND THEIR INCREASE WITH THE NUMBER OF IONISABLE GROUPS (N)

N	Number of microstates (2^N)	Number of reactions ($N!\sum 1/[k!(N-k-1)!]$)	Number of independent microconstants ($2^N - 1$)
$n = 1$	2	1	1
$n = 2$	4	4	3
$n = 3$	8	12	7
$n = 4$	16	32	15
$n = 5$	32	80	31
$n = 6$	64	182	63
$n = 7$	128	348	127
$n = 8$	256	744	255

Any of these microstates can loose a proton in a number of ways which can be expressed by another binomial coefficient:

$$n_{k-1}^{k} = C_1^{N-k} = (N-k)!/[1!(N-k-1)!]$$

Thus each stage (k) includes N_k reactions, where:

$$N_k = n_k \times n_{k-1}^{k} = C_k^N \times C_1^{N-k}$$

And the number of all possible reactions (M) is given by the sum in Eq. (2.14).

These data on the numbers of possible microstates and microreactions are summarised in Table 2.1 (from one to eight steps of dissociation). Fortunately, in order to find all the equilibrium concentrations we do not need to know all of the microconstant values, it is sufficient to have only $2^N - 1$ independent constants.

Moreover when someone is interested in proton binding curves only, a further reduction of necessary parameters is possible. For an arbitrary number N of ionogenic groups, acting parallely, it is sufficient to operate with N macroscopic constants. Indeed, since we are interested in only the total concentration of microstates with the same charge value, in order to know the concentration distribution between the $N+1$ possible 'macro-states' (which are connected with the normalisation equation), we need only N macroscopic constants.

2.3. DISSOCIATION SCHEMES OF A HYBRID TYPE

This conclusion can be easily generalised to an arbitrary closed process, no matter what general scheme of dissociation is taking place [3], but obviously, the expressions for macroconstants must differ from the case of pure parallel dissociation. As an example let us take a scheme of 'hybrid type', which is shown in Fig. 2.2A. In this system two groups dissociate consecutively and the third one may dissociate irrespective of which states the two first groups are in. There are six microstates and seven chemical reactions in such a system, with the relative concentrations of each microstate being

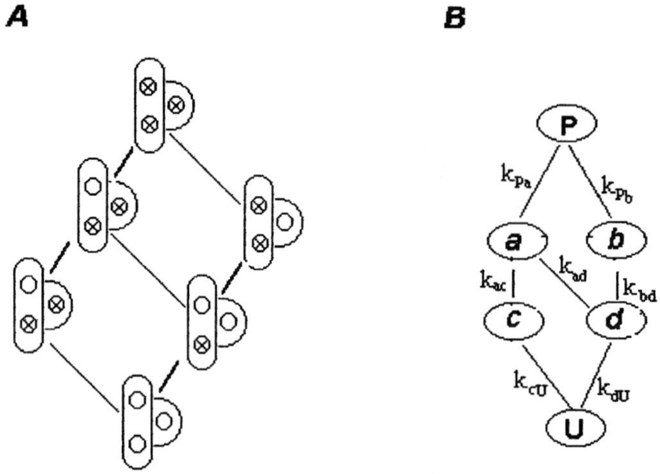

Fig. 2.2. Dissociation scheme of a 'hybrid type', where two groups dissociate consecutively and the third one may dissociate either in a parallel or a consecutive fashion, regardless of which states the two first groups are in. Prolonged ovals symbolise a subsystem of two groups dissociating stepwise, while a semicircle corresponds to one independently dissociating group. The expressions for macroconstants are given in the text.

shown in Fig. 2.2B. The appropriate expressions for the three macroconstants will be:

$$K_1 = K_{\rm pa} + K_{\rm pb} \tag{2.15}$$

$$K_2 = \frac{k_{\rm Pa}k_{\rm ad} + k_{\rm pa}k_{\rm ac}}{k_{\rm Pa} + K_{\rm Pb}} \tag{2.16}$$

$$K_3 = \frac{k_{\rm Pa}k_{\rm ad}k_{\rm dU}}{k_{\rm Pa}k_{\rm ad} + k_{\rm pa}k_{\rm ac}} \tag{2.17}$$

A system composed of two sets (each one representing a true two-phase stepwise process) is a particular case of reaction with four 'macrostages' (see Fig. 2.3). This system includes nine microstates and the number of chemical reactions is twelve. The appropriate expressions for the four macroconstants are:

$$K_1 = k_{\rm Pa} + k_{\rm Pb} \tag{2.18}$$

$$K_2 = \frac{k_{\rm Pa}k_{\rm ac} + k_{\rm pa}k_{\rm ad} + k_{\rm pb}k_{\rm bc}}{k_{\rm Pa} + K_{\rm Pb}} \tag{2.19}$$

$$K_3 = \frac{k_{\rm Pa}k_{\rm ac}k_{\rm cf} + k_{\rm Pb}k_{\rm be}k_{\rm eg}}{k_{\rm Pa}k_{\rm ac} + k_{\rm pa}k_{\rm ad} + k_{\rm pb}k_{\rm be}} \tag{2.20}$$

Should we be interested in the concentration of each microstate for this scheme, obviously, we need to specify eight independent microconstants.

Other examples of four macrostages of the hybrid type are given in Fig. 2.4A, where two groups dissociate sequentially and the other two may act in their own way. Here we

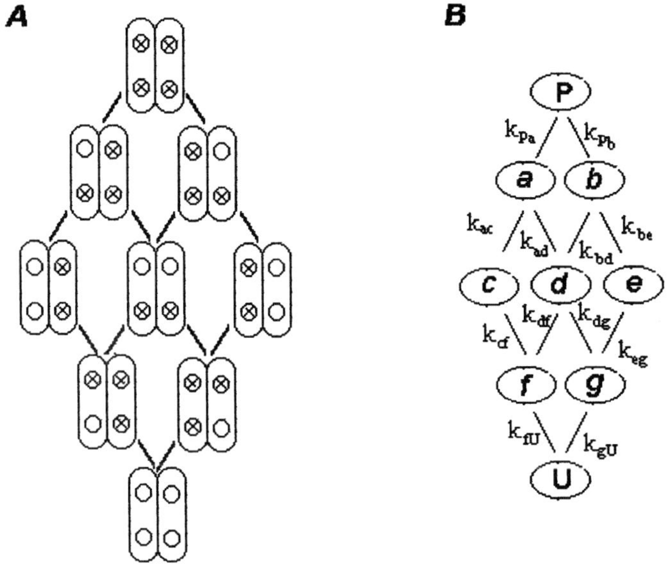

Fig. 2.3. Dissociation scheme of a 'hybrid type', describing the case of two subsystems, in which two groups dissociate consecutively, but these subsystems, in turn, may act parallely (two stepwise plus two stepwise). The symbols used are the same as in Figs. 2.1 and 2.2. The expressions for macroconstants are given in the text.

have twelve microstates and twenty reactions. In contrast with the pure parallel process (Fig. 2.4B) this scheme not only contains a different number of microstates, but it is also a non-symmetrical one.

2.4. PROTON TRANSFER TAUTOMERISM

In the above section we have considered some examples of 'hybrid schemes' with this term being used to designate any combination of a true parallel dissociation process and a true stepwise one.

One can imagine also some systems where there is a combination of dissociation reactions with other chemical reactions. As an example let us consider a system including a compound where a proton transfer tautomerism takes place. Suppose, as a result some neutral substance (Neu) converts in an amphoteric (Amph) one (the tautamerisation constant may be in particular a function of temperature), i.e.

$$\text{Neu} \overset{K\tau}{\longleftrightarrow} \text{Amph}$$

For this system all the formulas previously derived for bivalent ampholyte can be applied, provided the expression $C/(1 + K\tau)$ is used, instead of analytical concentration (C) in appropriate formulas (Eqs. 1.13 to 1.17 in Chapter 1).

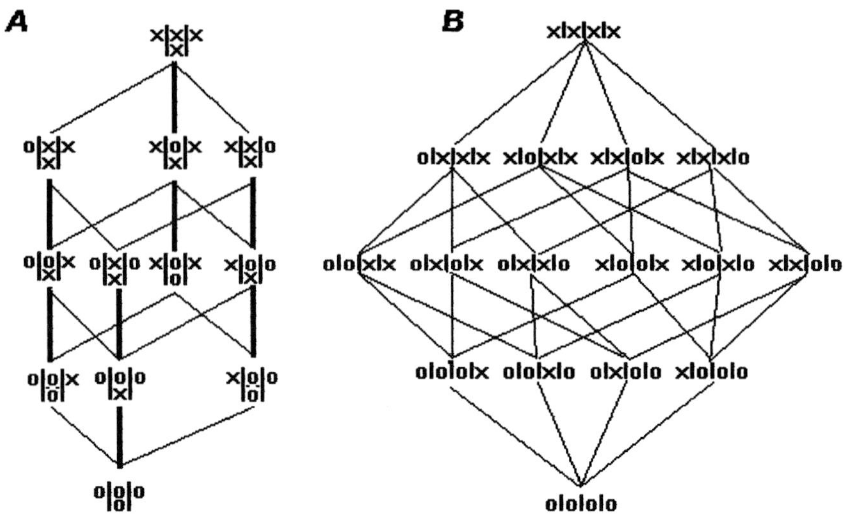

Fig. 2.4. Dissociation scheme of a 'hybrid type', where two groups dissociate consecutively and the other two act in parallel (two stepwise plus two parallel) (A). Two symbols, one above the other, between vertical lines correspond to 'stepwise' subsystems. The four-step scheme with pure parallel dissociation (all the four ionogenic groups act in parallel) is shown in (B). The expressions for macroconstants are not given.

2.5. SCHEMES WITH INDEPENDENT DISSOCIATION

The assumption of independent dissociation essentially simplifies the treatment; in this case the set of apparent constants uniquely defines all of the intrinsic constants. For a process of parallel dissociation with an arbitrary number M of ionogenic groups, acting independently, as shown by Simms [4], the $M - 1$ macroscopic constants (K_i) are connected with the microscopic ones (k_i) as follows:

$$K_1 = k_1 + k_2 + \cdots + K_M = \sum_i k_i$$

$$K_1 K_2 = k_1 k_2 + k_1 k_3 + \cdots + k_{N-1} k_N = \sum_{ij} k_i k_j \tag{2.21}$$

$$\ldots$$

$$K_1 K_2 K_3 \ldots K_N = k_1 k_2 k_3 \ldots k_N$$

$$K_1 K_2 K_3 = k_1 k_2 k_3 + k_1 k_2 k_4 + \cdots = \sum_{ijl} k_i k_j k_l \tag{2.22}$$

For any 'hybrid' scheme the appropriate relations are different from Eq. (2.22), see for example a transformation of the set Eqs. (2.10)–(2.12) to Eqs. (2.15)–(2.17). Regardless of a concrete dissociation scheme, the assumption about independent dissociation for each ionogenic group leads to a 'one-to-one' correspondence between the sets of intrinsic (microscopic) and apparent (macroscopic) constants.

2.6. TITRATION CURVE MODELLING

Although the titration curve of any polyelectrolyte may be described in terms of a limited number of macroconstant values, the use of macroconstants is not a very suitable instrument for titration curves modelling, based on the information about chemical composition. For this we may operate only with intrinsic constants, since the macroscopic (apparent) constants are not invariable. For example, a single change (or an addition) of one intrinsic constant may completely change the initial set of apparent constants. Certainly the necessary intrinsic dissociation constant values are found as tabulated data or evaluated by some other way; the values of the macro- or 'apparent' constants are easily calculated also. An experimental binding isotherm offers an opportunity of obtaining the macroscopic constant (by means of the best interpolation using the theoretical proton binding equation), so even in the case of simple substances we need other independent measurements in order to have microconstant values.

Complete prediction of the protein binding curve may be achieved only if we know all of the intrinsic constant values. The assumption of independent dissociation enormously simplifies the theoretical treatment, but unfortunately this approach is not always legitimate, and usually an evaluation of structural formulas is sufficient to understand if it is possible to use this approach or not. For example, for the description of polycarboxylic acids ionisation the model of independent dissociation is incorrect, since ionogenic groups are located rather close to each other. The same for amino acids, where we may not neglect the interaction between α-amino and α-carboxyl groups. Just opposite is the situation in the case of polypeptide chains, where the amino acid side chains, capable of ionisation, may be treated as acting independently.

Unfortunately, there is a serious mistake, when a truly parallel dissociation process is treated as a stepwise one, but the dissociation constant values are taken as intrinsic. This may lead to essential errors in calculating the isoelectric point and also brings about unusual slopes of the final titration curves obtained, which might exhibit some parts with very high slopes. This is clearly visible in the case of several pKs with the same or rather close values.

One should also pay attention when interpreting some maxima on the titration curve slope (buffering capacity maximum). The latter does not necessary correspond to the presence of ionogenic groups with appropriate pKs [3]. That result is valid only for pKs which are rather widely separated in the case of pure parallel dissociation (with the assumption of an independent mechanism). It is easy to evaluate the minimal ΔpK distance, by which it is possible to detect two separated maxima on the buffering capacity curve. With an assumption of independent dissociation, one can describe the buffering capacity as a function of (H^+), by the equation:

$$\beta = 2.303(\mathrm{H}^+)\left(\frac{Ck_1}{[(\mathrm{H}^+)+k_1]^2}\frac{Ck_2}{[(\mathrm{H}^+)+k_2]^2}\right) \tag{2.23}$$

where, for simplicity, we used $k_1 = k_{\mathrm{pa}}$ and $k_2 = k_{\mathrm{pb}}$.

In order to find the conditions for the existence of two maxima, we need to take a second derivative:

$$\frac{d}{d(pH)}\beta = 2.303^2(H^+)C\frac{d}{d(H^+)}\left(\frac{Ck_1}{[(H^+)+k_1]^2} + \frac{Ck_2}{[(H^+)+k_2]^2}\right) \tag{2.24}$$

and by supposing its value equal to zero, we my easily derive the appropriate limitation on the pK difference. Bringing to a common denominator we obtain an equation of the fourth degree relative to (H^+), which easily reduces to a quadratic equation. By analysing its discriminant we come to:

$$k_1/k_2 = 7 + 4\sqrt{3} \approx 13.93$$

The latter gives

$$\Delta pK = 1.144$$

Often we do not specify any difference between the true 'titration curve', i.e. the quantity of strong titrant versus pH and the 'dissociation curve' (Q versus pH). That is correct only if we operate "not very far from neutrality" [5].

2.7. LINDERSTRØM-LANG EQUATION

The idea of connecting such different matters, at a first glance, as standard charge deviation of multi-dissociating systems with the form of the proton binding curve belongs to Linderstrøm-Lang (see [2], p. 462).

$$\frac{\partial\bar{q}}{\partial pH} = 2.303[\overline{q^2} - (\bar{q})^2] \tag{2.25}$$

The treatment given there is connected with the consideration of a parallel process, but may be generalised to any other, which is formally reducible to a stepwise scheme, by an appropriate selection of the macroscopic constant values. As any formal stepwise process is defined by a limited number of parameters, it is clear that an essential part of information about such a complex system is lost.

2.8. CALCULATION OF THE COMPLETE SET OF MICROCONSTANTS

Although a reduced number of equilibrium constants (N constants for N dissociation steps) is sufficient for modelling the pH–charge relationship, in this way all the information about concentration distribution between microstates with the same charge value is lost. As a simple example consider the glycine molecule. The tabulated values of dissociation constants may be easily found in any handbook of chemistry, e.g. [6], and, obviously, they should be considered as macroscopic (or apparent) ones. The appropriate pK values are pK_1 = 2.35 and pK_2 = 9.78. Due to a high ΔpK difference, when describing the proton binding curve in terms of 'reduced' intrinsic constants, one obtains practically the same values. The assumption of independent dissociation allowed us to determine all the intrinsic constants, since it made it possible to reduce the number of independent constants

Let us suppose: $k_{\mathrm{Pa}} = k_{\mathrm{bU}} = k_1$, $k_{\mathrm{Pb}} = k_{\mathrm{aU}} = k_2$ and $k_1 = 10^{-2.35}$, $k_2 = 10^{-9.78}$. Such a system is characterised by a predominant concentration of a zwitterionic microstate and by a negligible concentration of a 'true zero' state (see Fig. 2.1). At the same time, another approach, in which the dissociation constant value of glycine ethyl ester (2×10^{-8}) is used to approximate the appropriate (k_{Pb}) microconstant value, leads to the following set of microconstants: $k_{\mathrm{Pa}} = 10^{-2.35}$ ($k_{\mathrm{bU}} = 10^{-4.43}$), $k_{\mathrm{Pb}} = 10^{-7.8}$, $k_{\mathrm{aU}} = 10^{-9.78}$ [7,8]. In the latter case, despite the same general pattern, the ratio 'zero/zwitterion' which is pH independent, although being very low, nevertheless increases by a factor of more than two hundred times.

For two dissociation step system the two macroscopic constants K_1 and K_2 (see Eqs. (2.7)–(2.9)) are quite sufficient to derive a pH–charge relationship. But we can say nothing about micro-concentration distribution within an intermediate 'macrostate'. This leaves us an opportunity of modelling, for example, such important characteristics as dipole moments. But one can easily find that any other set of new three microconstants designed according to the rule $k_{\mathrm{Pa}} = k^1 \times \theta$, $k_{\mathrm{Pb}} = k^2 - k^1 \times (\theta - 1)$ and $k_{\mathrm{aU}} = k^3/\theta$, where θ is an arbitrary parameter, will result in the same macroscopic constants.

There is one simple case when the set of macroscopic constants is sufficient to determine all the intrinsic constants. This is the case of 'symmetric' biprotic protolyte, that is the two ionogenic groups are identical (the groups of the same name in the same environment). Despite of interaction between the two groups, due to the symmetry of the system $k_{\mathrm{Pa}} = k_{\mathrm{Pb}}$ and $k_{\mathrm{aU}} = k_{\mathrm{bU}}$. Such compounds can be found among bicarboxylic acids.

Obviously, any ampholyte molecule never possesses such symmetry. Note that for ampholytes, on the contact, we assumed $k_{\mathrm{Pa}} = k_{\mathrm{bU}}$ and $k_{\mathrm{Pb}} = k_{\mathrm{aU}}$ in order to neglect the interaction between the dissociating groups.

As an example of a three dissociation step system let us consider glutamic acid. The

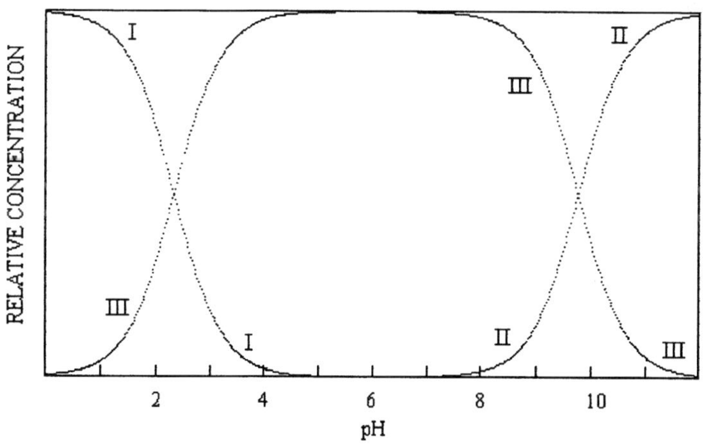

Fig. 2.5. Relative concentrations of different microstates for the glycine molecule as a function of pH. Curves I, II, and III correspond to positive, negative, and zwitterionic states, respectively. Microconstant values used are: $k_{\mathrm{Pa}} = k_{\mathrm{bU}} = k_1$, $k_{\mathrm{Pb}} = k_{\mathrm{aU}} = k_2$ and $k_1 = 10^{-2.35}$, $k_2 = 10^{-9.78}$. By these conditions the 'zwitterion/zero' concentration ratio is maintained at 2.7×10^7.

someone could ask for a theory that needs some minimal number of universal parameters. Since precise and exhaustive information is often not available and it is moreover impossible or unreasonable to obtain such information, we need to take advantage of some simple general model, which can provide rather good qualitative agreement, but for analyzing some complicated phenomena it may not be useful.

In particular, the question exists as to what extent it is justified that the kinetic aspects of the acid–base equilibria are essentially ignored? The appropriate evaluations are easily obtained for concrete situations, but to create a model taking into account the effect of appropriate mobility changing (acceleration or retardation to a new value, after conversion from one microstate to another) is not easy. For this purpose we need, also, a precise distribution of substances between all possible microscopic states; this task, probably, can be solved correctly in some limited cases only.

3.5. REFERENCES

1. A.V. Stoyanov and P.G. Righetti, J. Chromatogr. A, 853 (1999) 35–44.
2. F. Daniels and R.A. Alberty, Physical Chemistry, Wiley, New York, 4th ed., 1974.
3. M. Eigen and L. DeMeyer, in A. Weissberger (Ed.), Technique of Organic Chemistry, Vol. VIII, Part II, Chapter XVIII, Wiley–Interscience, New York, 1963.
4. M. Bier (Ed.), Electrophoresis, Theory, Methods, and Applications, Academic Press, New York, 1959.
5. D.J. Shaw, Electrophoresis, Academic Press, New York, 1969.
6. A.V. Stoyanov and P.G. Righetti, J. Biochem. Biophys. Methods, 46 (2000) 21–30.
7. M. Almgren, Chem. Scripta, 1 (1971) 69–75.
8. O.A. Palusinski, T.T. Alligyer, R. Mosher, M. Bier and D.A. Saville, Biophys. Chem., 13 (1981) 193–202.
9. H. Svensson, Acta Chem. Scand., 16 (1962) 456–466.
10. J.R. Brown, Fed. Proc., 35 (1976) 2141–2144.

CHAPTER 4

Natural pH Gradients

CONTENTS

4.1. INTRODUCTION

In the previous chapters we considered the acid–base equilibria in solutions containing polyelectrolytes. In frames of a simplified approach, ignoring to a certain extent the kinetic aspects (Chapter 3), the information about thermodynamics of the dissociation process is quite sufficient for describing the electrophoretic behaviour of the analysed molecule. We proceed now with electrophoresis and we shall start our consideration from the steady electrolysis. In this chapter some simple cases of natural pH gradients will be considered in order to understand better the general principles of pH gradient formation in multi-compound systems.

4.2. SIMPLEST EXAMPLES OF NATURAL pH GRADIENTS

Since a detailed treatment of the electrophoretic process is impossible (the most general approach for describing the electrophoretic phenomena, which takes into account all that is known of the present-day physical phenomena, can be found in [1,2]), one can take the advantages of some simplified approach. For practical purposes we need a system of equations that can be solved, and here the art consists in how to perform the simplifications but not to lose the general pattern, or at least not to create a contradictive model or to consider a definitely unreal situation.

References p. 53

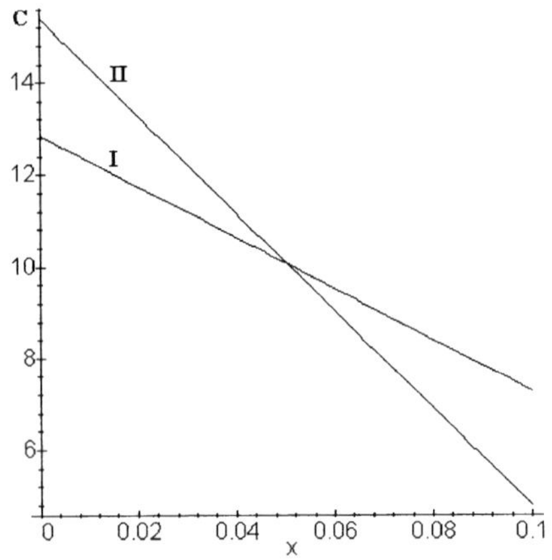

Fig. 4.1. Concentration course for steady-state electrolysis of strong acid. Result of simulations by taking into account the diffusional current (curve I) and by neglecting it (curve II). Current density is 0.1 A/m. In each case the initial concentration corresponds to 10 mM of uniform concentration.

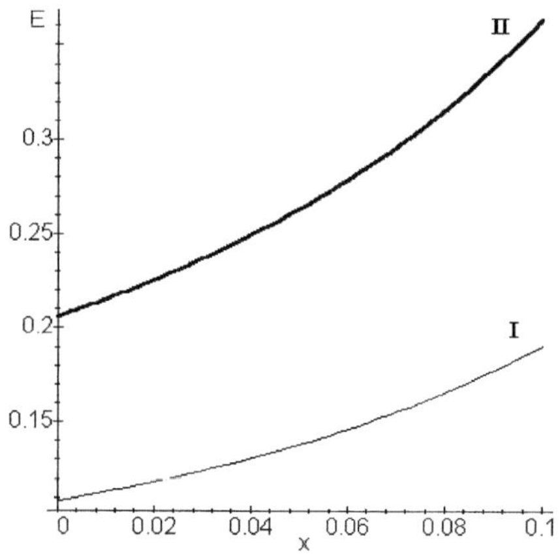

Fig. 4.2. Calculated electric field strength for the corresponding systems depicted in Fig. 4.1. When the electrolysis is set by voltage stabilisation the identical concentration profile should be achieved with the J_1/J_2 ratio being stabilised at $(1 + \mu/\mu_H)/2$.

Thus the concentration profile is linear and surprisingly the slope does not depend on the diffusion coefficient of the acid subjected to the electrolysis. It should be noted, that neglecting the diffusional current, one comes to

$$\frac{dC}{dx} = -\frac{J}{(1 + \mu/\mu_H)RT\mu_H} \tag{4.12}$$

that gives, approximately, a two times higher concentration curve slope (see Figs. 4.1 and 4.2). Thus, in this example, by neglecting the diffusional current we obtain a wrong result.

When a complete dissociation is not assumed, a little bit more complex system is obtained

$$\frac{\mathrm{d(H)}}{\mathrm{d}x} - \frac{1}{r}\frac{\mathrm{d}c^-}{\mathrm{d}x} + \left(1 + \frac{1}{r}\right)\frac{\mathrm{d}C}{\mathrm{d}x} = -\frac{J}{FD_{\mathrm{H}}} \tag{4.13}$$

$$E = \frac{D}{\mu c^-}\frac{\mathrm{d}C}{\mathrm{d}x} \tag{4.14}$$

where r is used to designate the mobilities ratio $r = \mu_{\mathrm{H}}/\mu$, and c^- is expressed as:

$$c^- = C\frac{K}{(\mathrm{H}^+) + K} \tag{4.15}$$

The hydrogen concentration value is easily calculated from

$$(\mathrm{H}^+) = \frac{-K + \sqrt{K^2 + 4KC}}{2} \tag{4.16}$$

Here we also neglect the hydroxyl ion concentration that is valid if $\mathrm{H} \gg 10^{-7}$.

Performing the necessary simplifications we come to the final equation for analytical concentration C

$$\frac{K(1 - 1/r)}{\sqrt{K^2 + 4KC}} + \left(1 + \frac{1}{r}\right) = -\frac{J}{FD_{\mathrm{H}}}\frac{\mathrm{d}x}{\mathrm{d}C} \tag{4.17}$$

By separation of the variables one obtains the solution

$$x = \frac{1}{\alpha}\left(-aC + \frac{b}{2}\sqrt{K^2 + 4KC}\right) + Const \tag{4.18}$$

where $\alpha = J/FD_{\mathrm{H}}$; $a = 1 - 1/r$; $b = 1 + 1/r$; and the *Const* is defined from the initial conditions. This equation allows us to express the concentration of the coordinate relationship in an evident form.

Let us now consider the electrolysis of an amphoteric electrolyte solution. We assume that its isoelectric point is in the mild acidic range. Repeating the same procedure one obtains

$$-D\frac{\mathrm{d}C}{\mathrm{d}x} + (c^+ - c^-)\mu E = 0 \tag{4.19}$$

$$-D_{\mathrm{H}}\frac{\mathrm{d(H)}}{\mathrm{d}x} - D\frac{\mathrm{d}}{\mathrm{d}x}(c^+ - c^-) + (\mu_{\mathrm{H}}(\mathrm{H}) + \mu c^+ + \mu c^-)E = \frac{J}{F} \tag{4.20}$$

For the sake of simplicity we assume the same dissociation constant value for the acid and base groups. We restrict our consideration to the case of a high ampholyte concentration, this allows the hydrogen ions conductivity contribution to be neglected. Under these conditions the equation for (H^+) should be (see Fig. 4.3):

$$(H^+) = K - 2K^2/C \tag{4.21}$$

and the system reduces to:

$$-D\frac{\mathrm{d}C}{\mathrm{d}x} - \left(1 - \frac{2K}{C}\right)K\mu E = 0 \tag{4.22}$$

References p. 53

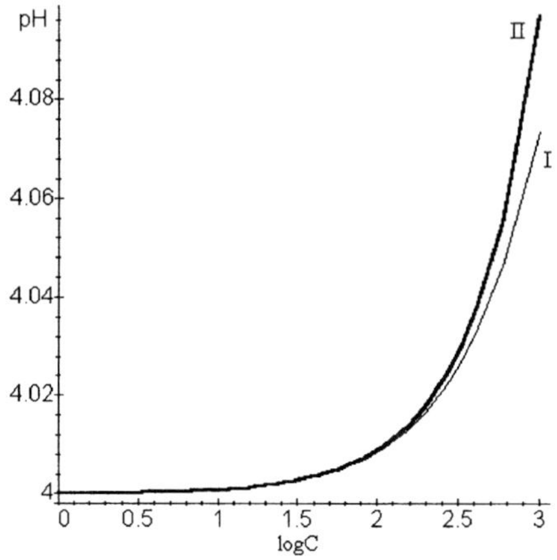

Fig. 4.3. pH versus ampholyte concentration; curve I corresponds to the approximate solution (Eq. (4.20)), curve II to the exact one.

$$-\frac{2K^2D}{C}(r-1)\frac{dC}{dx}-\left(\frac{C}{2}+\frac{2K^2}{C}\right)\mu E=\frac{J}{F} \tag{4.23}$$

As we already made the assumptions of $C \gg K$, and $C \gg H^+$, it is clear that the first term on the right side of Eq. (4.23) corresponding to the diffusional currents can be neglected.

Thus

$$\frac{dC}{dx}=\left(\frac{2K}{C}-1\right)\frac{K\mu E}{D} \tag{4.24}$$

$$E=\frac{-J}{F\mu\left(\frac{C}{2}+\frac{2K^2}{C}\right)} \tag{4.25}$$

and the concentration course is defined by the following equation

$$\frac{dC}{dx}=-\alpha\frac{1-\frac{2K}{C}}{\frac{C}{2K}+\frac{2K}{C}} \tag{4.26}$$

where $\alpha = J/FD$.

The solution of Eq. (4.23) can be easily obtained, although in a non-evident form

$$-\alpha x=\frac{C^2}{4K}+C+4K\ln(C-2K)+Const \tag{4.27}$$

At high concentrations we can use the following approximate expression

$$C=\sqrt{C_0^2-4K\alpha x} \tag{4.28}$$

As compared to a non-amphotere the electrolysis under the same conditions results in a considerably more flatter concentration profile. More precise consideration taking into account the diffusional currents can also be easily performed; in this case, instead of Eq. (4.27), one obtains:

$$-\alpha x = \frac{C^2}{4K} + C(1+r) + 4K\ln(C-2K) - 2(r-1)\frac{K^2}{C} + Const \tag{4.29}$$

To derive this equation the following assumptions were made, $H \gg 10^{-7}$ and $C \gg K$, thus the hydrogen ion concentration can be expressed as Eq. (4.21).

4.3. pH GRADIENTS CREATED WITH A MULTI-COMPONENT MIXTURE OF AMPHOTERIC COMPOUNDS

The most important effect defining the conductivity mechanism of natural pH gradients, at least in the case of not so high molecular weight compounds, is connected with the reactions of chemical equilibria between the microstates obtained due to any polyelectrolyte dissociation. This effect strongly depends on the so-called 'ionisation coefficient' in the vicinity of the ampholyte isoelectric point.

Besides that, there is a contribution of hydrogen and hydroxyl ions, and any electric current passing through a water solution should be accompanied by water decomposition. The existence of diffusion currents makes the situation still more complex.

At the initial stages of the ampholyte mixture electrolysis a sharp conductivity decrease is observed, since the ampholytes lose their charge while approaching to their pIs (at pH = pI, an average electric charge of the ampholyte molecule equals zero, although some fraction is composed of charged microstates; Chapters 2 and 3). Then the current tends to become stabile, although the concentration distribution of each component may be far enough from the final one. The peaks, corresponding to the neighbouring ampholytes should overlap to a certain extent. The final degree of overlapping will be defined both by the concrete composition of the ampholyte mixture and the current density value. Although, when the number of ampholytes per pH unit is high and they are distributed rather uniformly, the degree of ionisation should be close to that in the isoelectric state. Supposing the conductivity to be proportional to the ionisation degree (see, also, the discussion in Chapter 3), one can expect the conductivity course to be close to uniform provided the water ions contribution may be neglected (an assumption of close ΔpK differences should also be made). Any gaps that can be observed in commercial carrier ampholytes (CA) products [5–8] should be explained by rather high ΔpK values ('bad' ampholytes) or a nonuniform pI distribution. In the latter case the ampholytes, which are more 'non-isoelectric', that is removed from their isoelectric points, can provide a more essential contribution to the conductivity in the areas with low component density. Unfortunately, the detailed composition of commercially available CA is unknown, perhaps, even for the manufacturers.

For evaluating the properties of multicomponent natural pH gradients some simplified models can be used. For instance, they may be approximated by some superposition of some concentration functions, e.g. Gaussian [9], or one can take the set of thin layers of uniform ampholyte concentrations. Despite such approaches do not take into account

References p. 53

the fact that the system is to obey the corresponding differential equations characterising the steady state, they are able to give quite reasonable results. Also the properties of natural pH gradients as well as the questions of gradient formation and stability in some idealised and real systems were studied by numerical simulations [10–13,3,14,15], and the numerical simulation is the only way to consider any real complicated system.

How important the water ions (hydrogen and hydroxyl) contribution is depends on the ampholyte concentrations. Since for a characteristic ampholyte concentration of 1% the appropriate molar concentration should be around 10 mM, water ions contribution should be taken into account starting from pH 4, or even before (pH 11 for hydroxyl); here one should remember about at least one order excess in mobility for hydrogen (hydroxyl) ions with respect to even small amphoteric compounds. (Also, few amphoteres possess an ionisation coefficient approaching 50%.) Thus at pH extremes the bulk water conductivity contribution becomes most important. This case is rather interesting to be analysed in more detail.

Neglecting the diffusion currents we can restrict our consideration to a system of N differential equations (for N component system). Supposing the main contribution to conductivity is due to the presence of hydrogen ions, we can write:

$$\frac{\mathrm{d}c_1}{\mathrm{d}x} = Ac_1\frac{Q_1(c_1, c_2 \ldots c_N)}{\mathrm{H}(c_1, c_2 \ldots c_N)}$$

$$\frac{\mathrm{d}c_2}{\mathrm{d}x} = Ac_2\frac{Q_2(c_1, c_2 \ldots c_N)}{\mathrm{H}(c_1, c_2 \ldots c_N)}$$

$$\ldots$$

$$\frac{\mathrm{d}c_N}{\mathrm{d}x} = Ac_N\frac{Q_N(c_1, c_2 \ldots c_N)}{\mathrm{H}(c_1, c_2 \ldots c_N)} \tag{4.30}$$

We also assume identical mobilities for each component, which gives us an opportunity to put the same factor A at the right side of the equations ($A = J/\mu RT$), so we make an assumption about the same size for CA.

For this system it is easy to find that it should have an integrable combination. Indeed, summing Eq. (4.30) one obtains

$$\frac{\mathrm{d}(c_1 + c_2 + \cdots + c_N)}{\mathrm{d}x} = A\frac{c_1Q_1 + c_2Q_2 + \cdots + c_NQ_N}{\mathrm{H}(c_1, c_2 \ldots c_N)} \tag{4.31}$$

Taking into account the electroneutrality equation, which for our specific conditions has the form:

$$(\mathrm{H}^+) + \sum_{i=1}^{N} c_iQ_i = 0 \tag{4.32}$$

We can rewrite Eq. (4.31) as

$$\frac{\mathrm{d}(c_1 + c_2 + \cdots + c_N)}{\mathrm{d}x} = -A \tag{4.33}$$

That is,

$$\sum_{i=1}^{N} c_i = Const - Ax \tag{4.34}$$

So, the sum of ampholyte concentrations obeys Eq. (4.34), and this result is valid for an arbitrary composition of the CA mixture, since no other additional assumptions except the same molecular size were made. In particular, the equivalence of ΔpK (the same 'quality') for all ampholytes has not been assumed. In the case of a low A parameter the concentration sum keeps its value close to the constant while being slightly linearly modulated.

4.4. REFERENCES

1. V.G. Babskii, M.Yu. Zhukov and V.I. Yudovich, Mathematical Theory of Electrophoresis, Consultants Bureau, New York, 1989.
2. V.G. Babskii, M.Yu. Zhukov and V.I. Yudovich, Mathematical Theory of Electrophoresis, Naukova Dumka, Kiev, 1983.
3. O.A. Palusinski, T.T. Alligyer, R.A. Mosher and M. Bier, Biophys. Chem., 13 (1981) 193–202.
4. W.G. Kauman, Cl. Sci. Acad. R. Belg., 43 (1957) 854–868.
5. J.R. Cann, Biophys. Chem., 11 (1981) 249–264.
6. J.S. Fawcett, in Progress in Isoelectric Focussing and Isotachophoresis, P.G. Righetti (Ed.), Elsevier, Amsterdam, 1975, pp. 25–37.
7. S.J. Fredriksson, J. Chromatogr., 135 (1977) 441–446.
8. J.W. Gelsema, C. Ligny and N.G. Van der Veen, J. Chromatogr., 154 (1978) 161–174.
9. M. Almgren, Chem. Scripta, 1 (1971) 69–75.
10. M. Bier, R.A. Mosher and O.A. Palusinski, J. Chromatogr., 211 (1981) 313–335.
11. R.A. Mosher and W. Thormann, Electrophoresis, 11 (1990) 711–723.
12. R.A. Mosher, W. Thormann, R. Kuhn and H. Wagner, J. Chromatogr., 178 (1989) 39–49.
13. R.A. Mosher, D. Saville and W. Thormann, The Dynamics of Electrophoresis, VCH, Weinheim, 1991.
14. K. Shimao, Phys.-Chem. Biol., 24 (1980) 191–199.
15. K. Shimao, Phys.-Chem. Biol., 25 (1981) 59–64.

CHAPTER 5

Immobilised pH Gradients

CONTENTS

5.1. CLASSICAL IMMOBILISED pH GRADIENTS CREATED WITH LINEAR DENSITY GRADIENT

The term 'immobilised pH gradients' is usually assumed to mean a pH gradient 'immobilised' on a polyacrylamide matrix. The first success was reported by the 'Aminkemi' group (a small company associated with LKB) in 1975 [1–3], although the idea to create e such gradients was conceived even before, perhaps just when it became evident that it was extremely difficult to overcome the disadvantages of conventional natural pH gradients, connected with their instability. This event was an essential improvement of the isoelectric focussing technique, the resolving power was increased at least one time in order [4], and in a few years, immobilised pH gradients (IPG) gained their enormous popularity.

One should expect that, at least with respect to the buffer properties, there is no principal difference between any true water solution containing charge species and some porous media with 'immobilised' charges filled with water, provided the concentration of charge centres is sufficiently high and they are distributed uniformly.

Thus the main idea of IPG is to graft some preliminary created pH profile onto a space stabile matrix. In the case we describe, this task was solved by immobilising non-amphoteric substances (with different carboxyl groups or tertiary amino groups) to a gel pore surface.

For creating the pH gradient to be immobilised, the gradient mixture of two solutions was used (one light and the other heavy), and so the pH gradient was created with the help of a linear density gradient. Later, a multi-chamber device was described [5], nevertheless the main idea remains the same. Namely, to try to create any desired pH gradient (usually close to linear) by the appropriate choice of the composition of one pair (or more) of vessels with solutions of a different density. The gradient obtained can be effectively incorporated in the gel during polymerisation, the only thing we need is a higher temperature as compared with the simple polyacrylamide gel preparation procedure.

There are appropriate Tables for the chemicals commercially available, and one can find a recipe for preparing a lot of linear (also non-linear) gradients, narrow and wide, covering all the important pH regions. Computer programs were developed to help construct any desired pH interval and to optimise its buffer properties [6–8].

5.2. LINEAR pH GRADIENTS WITH NON-LINEAR GRADIENTS OF CONCENTRATION

The IPG technology brings about a new opportunity of modulating buffer properties. When working with a buffer containing one or more electrolytes dissolved in water, we are able, to a certain extent, to vary the pH of the solution by changing the concentrations, the latter defining also the conductivity value. In the case of immobilised electrolytes (immobilines), the conductivity is determined by the pH value only. Thus the buffer capacity to conductivity ratio can be maintained, theoretically, as high as we need.

As described above, the classical way to prepare a pH gradient is associated with a linear concentration gradient, so for such a pH gradient there are some relations connecting the concentration of all immobilines (c_i) at any point along the x-axis (collinear to the gradient):

$$c_i(x) = a_i x + b_i \tag{5.1}$$

where a_i and b_i are some constants, specific to each immobiline component. Thus the problem of any pH gradient (linear or non-linear) creation is the one of best interpolation by the set of Eq. (5.1). The way above is simple and suitable for use in practice, but leaving this restriction, one gains an opportunity to solve the problem of an arbitrary pH gradient profile construction precisely.

A linear pH gradient

$$(\mathrm{H}^+) = \mathrm{A}\exp(-x \ln 10/d) \tag{5.2}$$

(here d reflects the gradient slope; for $d = L$ we have a unit increment in the pH when encompassing the interval of L; in the midpoint of such an interval $(-L/2, +L/2)$ (H^+) equals A), may be created with the help of one immobiline only (Fig. 5.1).

Another simple example is a two-component gradient (Figs. 5.2 and 5.3). For such a

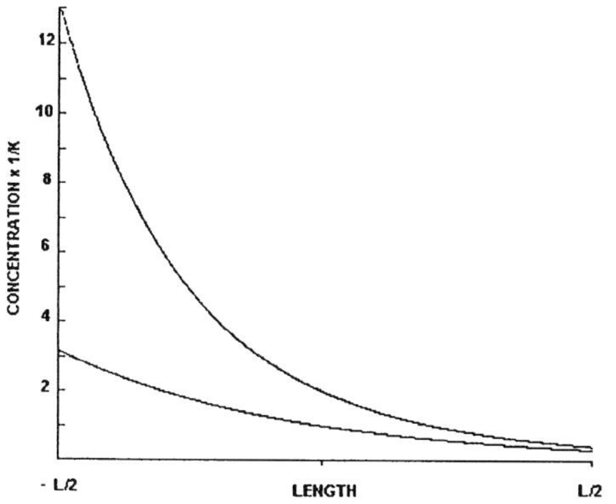

Fig. 5.1. Immobiline concentration along the x-axis for creating a linear pH gradient. The lower curve corresponds to an extremely acidic immobiline $K_i \gg A$ (in this case, the immobiline concentration coincides with the hydrogen ion concentration). The upper curve represents the case of A weak acid ($K_i = A$). Supposing A to be 10^{-3}, one obtains the pH gradient 2.5–3.5.

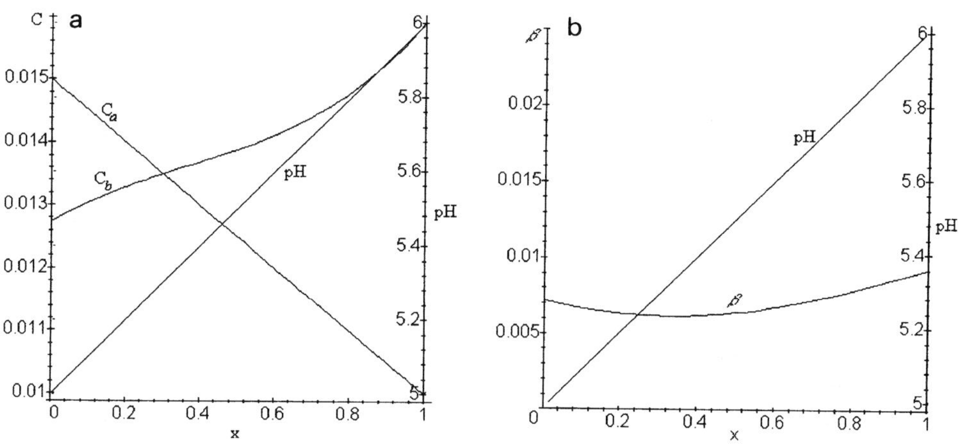

Fig. 5.2. A linear pH gradient (5–6) obtained with the help of two immobilines, pK 4.4 (acid) and pK 6.2 (base). The concentration course of the first one (C_a) is defined as a linear (15 to 10 mM), and $C_b(x)$ is to be calculated to obtain, finally, the necessary pH profile; (b) shows the buffer capacity profile.

two-component system there exists a possibility of obtaining any necessary gradient in a wide concentration range by selecting the appropriate concentration distribution for the second component, provided the concentration profile for the first one is known, as it is evident from the electroneutrality equation.

5.3. BUFFERING AND CONDUCTIVITY PROPERTIES OF IMMOBILISED pH GRADIENTS

The electroneutrality equation for any system with immobilised charges does not differ from the one containing the same ionogenic groups as mobile ions, and the

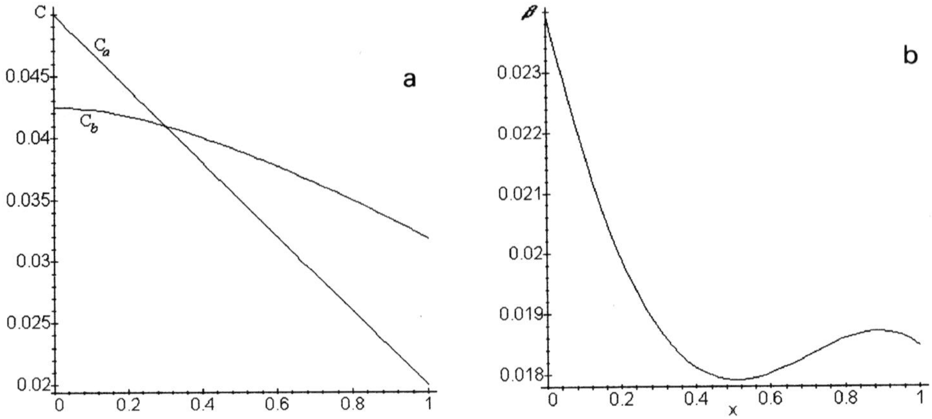

Fig. 5.3. A linear pH gradient (5–6). C_a is linear (50 to 20 mM), C_b is calculated (a). Due to higher immobiline concentrations, the buffer capacity level (b) is essentially higher.

expression for the buffer power has as its usual form:

$$\beta = 2.3\left\{\sum_i c_i \frac{K_i(H^+)}{[K_i + (H^+)]^2} + (H^+) + \frac{K_W}{(H^+)}\right\} \tag{5.3}$$

where K_i are the dissociation constants of the dissociating groups (no matter acidic or basic), and 2.3 is the approximate value of ln 10. The last two terms reflect the contribution of water ions, and at the pH extremes, the appropriate contribution to the β-value is essential (Fig. 5.4). Since the pH value depends on concentration, the dependence of β on the concentration is not so evident. In contrast with the amphoteric substances where there is an asymptotic pH approximation to the pI value and the additional contribution of water ions depends on how far the isoelectric point is removed from neutrality (Figs. 5.5 and 5.6), the buffer capacity of a solution of non-ampholeters changes with the concentration in a more complicated way. Strong acids or bases bring the pH to the extreme values where the contribution of water ions to the buffer capacity is essential, but, on the other side, when the pH of a solution is removed from the pK values, the appropriate contribution of the buffering groups cannot be essential (Figs. 5.7 and 5.8). Fig. 5.8 shows the relative contribution of the buffering group (the buffering group contribution divided by the total value of the buffer power) as a function of the concentration for a series of acids with pK values from 0 to 6.

Using a limited set of compounds, there exists always a possibility to provide any desired β-profile for any given pH profile. For example, by means of one pair of immobilines balancing each other (the same concentrations of the opposite charge), one can 'correct' the buffering capacity value while keeping the initial pH profile the same.

It is important to remember that the 'buffering capacity' is a local characteristic; such a value is useful, if we are interested in prediction of some pH changes due to a small concentration variation (see Chapter 8).

A characteristic feature of the conductivity of the media carrying immobilised charges is the absence of mobile ions (the charge centres are grafted onto the matrix

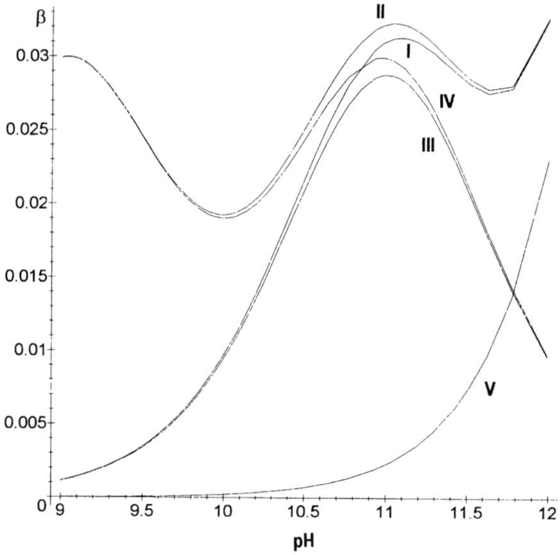

Fig. 5.4. Buffering capacity as a function of pH for non amphotere (pK 11) and amphotere (pK_1 9, pK_2 11; curves I and II, respectively. Curves III and IV correspond to the buffer power without water ion (hydroxyl) contribution, the latter given with curve V.

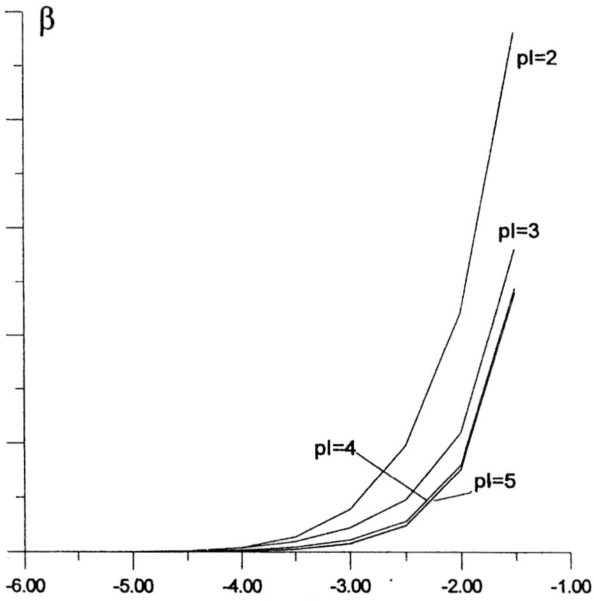

Fig. 5.5. Buffer capacity in a wide region of ampholyte concentration at different ampholyte pIs.

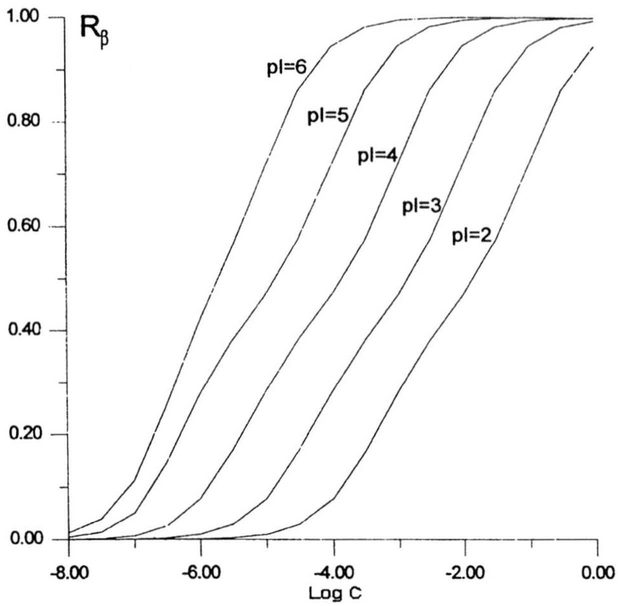

Fig. 5.6. Relative ampholyte buffer power contribution as a function of concentration.

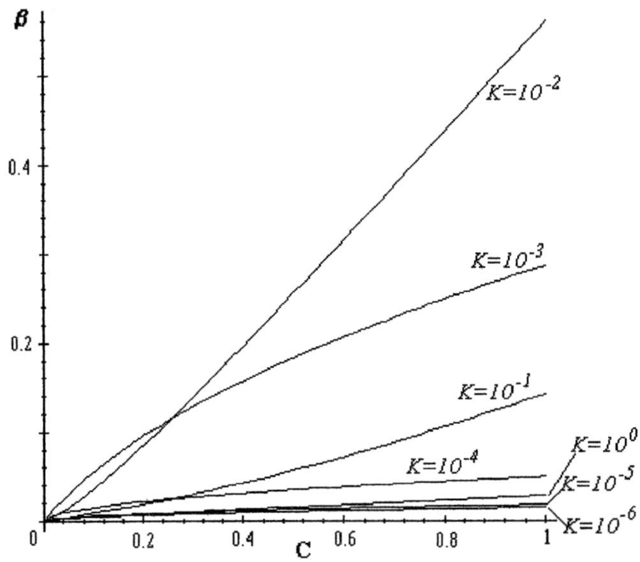

Fig. 5.7. Buffer capacity increase with concentration for monoacids. p*K*s vary from 0 to 6.

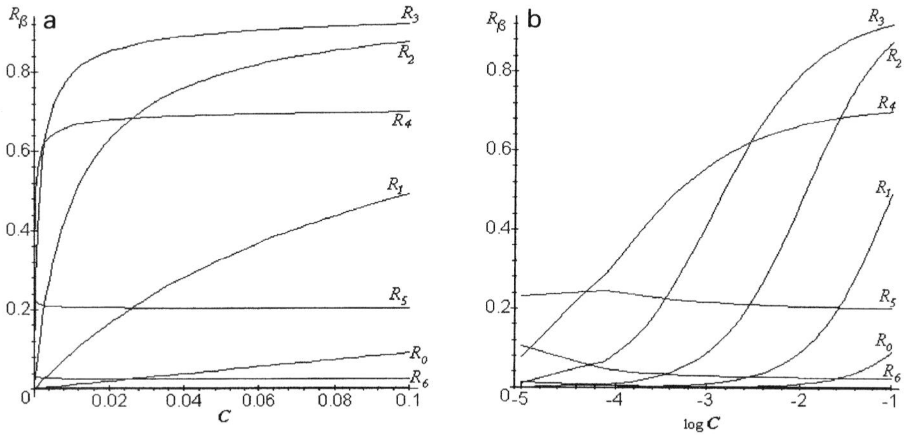

Fig. 5.8. Relative contribution of buffering group (buffering group contribution divided by total value of buffer power) as a function of concentration (pK from 0 to 6); (b) R_β is given versus a logarithm of concentration.

and thus they can not contribute to the conductivity). So the electric current in such media is maintained by the H^+ and OH^- ions only. It would not be quite correct to write the transport equation for these components in the form written previously for the free solution containing hydrogen and hydroxyl ions (Eq. 4.1), since the water ions are involved in mutual equilibria with covalently immobilised ions and their diffusional movement is no longer described with the help of the same diffusion low. By omitting the diffusional term, the appropriate equation for H^+ becomes:

$$\frac{\partial(\mathrm{H}^+)}{\partial t} + \frac{\mathrm{d}[\mu_\mathrm{H}(\mathrm{H}^+)E(x)]}{\mathrm{d}x} = \sigma_\mathrm{H} \tag{5.4}$$

5.4. SOME CHARACTERISTIC FEATURES OF ELECTROPHORESIS IN GEL MEDIA WITH IMMOBILISED ELECTRIC CHARGE

5.4.1. Method of diagonal sample application

As mentioned above, the conductivity of any immobilised pH gradient gel is ideally defined by the concentration of the hydrogen and hydroxyl ions that cause great non-uniformities in conductivity profiles; the latter should be determined by the pH profile and is to be independent of the concrete immobiline composition. On the other hand, a considerable number of molecular mass ionic species is present in the gel, their ratio and total concentrations being dependent on the gel preparation technique. Although an opportunity exists to wash away all these ions prior to the electrophoretic step, their elimination is never complete due to the weak ion-exchange character of the IPG gel. After applying a current, the ions present in the gel create essentially a non-stationary conductivity distribution and, under some conditions, form moving salt fronts [9].

with the electromigrative one and a similar situation is definitely realised when we consider a region sufficiently removed from pI. In the vicinity of the isoelectric point the electrophoretic fluxes are small and thus comparable with diffusional ones. That means, that except for the small region in the vicinity of pI where the sample should be finally condensed, in order to find the solution we can formally assume the diffusional coefficient to be equal to zero. Thus the problem is formulated in the following form:

$$\frac{\partial C_k}{\partial t} + \frac{\partial}{\partial x}\{v_k(x)C_k\} = 0, \qquad k = 1, \ldots, n \tag{7.1}$$

Here k corresponds to the different compounds with analytical concentration C_k. Under the conditions the concentrations c_i are zeroes at the focussing camera ends, at the initial moment of time ($t = 0$), the boundary conditions (flux absence at points $x = 0, L$) are satisfied automatically and the system (7.1) should be supplied by the initial conditions only.

This problem allows the solution [1,2]:

$$c_k(x,t) = c_k^{(0)}(\theta_i(x,t))\frac{v_k(\theta_k(x,t))}{v_k(x)}, \qquad k = 1, \ldots, n \tag{7.2}$$

Where the functions $\theta_i(x,t)$ are defined non-evidently by the relationships

$$t = \int_{\theta_i}^{x} \frac{d\eta}{v_k(\eta)}, \qquad k = 1, \ldots, n \quad (\theta_k(x,0) = 0) \tag{7.3}$$

This method is justified in [2,3] and it is named 'method of characteristics'. The fact that formula (7.2) is the solution of system (7.1) can be verified by direct substitution, taking into account

$$\frac{\partial \theta_i}{\partial t} = v_i(\theta_i(x,t)) \tag{7.4}$$

$$\frac{\partial \theta_i}{\partial x} = \frac{v_i(\theta_i(x,t))}{v_i(x)} \tag{7.5}$$

These latter formulae follow from Eq. (7.3).

The solution (Eqs. (7.3) and (7.4)) possesses clear and evident geometric interpretation. In coordinates (x,t), according to Eq. (7.3) the appropriate curves are drawn with the θ_k value being considered as a parameter. These lines are called 'characteristics' and they possess the following properties: the initial point is situated at $x = \theta_k$ at the x-axis; they do not intersect the line $x = x_k$ ($v(x_k) = 0$); the concentration C_k value at any point (x,t) is defined by Eq. (7.2) provided that θ_k is given. These curves represent, at the (x,t) plane, the trajectories of a small volume of C_k substance which was initially concentrated at the point of $x = \theta_k$ of the x-axis. It is evident, if the substance is absent at point $x = \theta_k$ at the initial moment of time (i.e. $C_k^{(0)}(\theta_k) = 0$), than the concentration will be zero at the line starting from the point of $x = \theta_k$, at any other moment of time. Thus, to study the evolution of the concentration profile, it is sufficient to draw the characteristics for which the concentration is different from zero. The lines $t = t^*$ characterise the concentration distribution at that moment of time; the intersection point of the characteristic with the line of $t = t^*$ gives a new coordinate for the small volume, previously situated at $x = \theta_k$ point (see Fig. 7.1a).

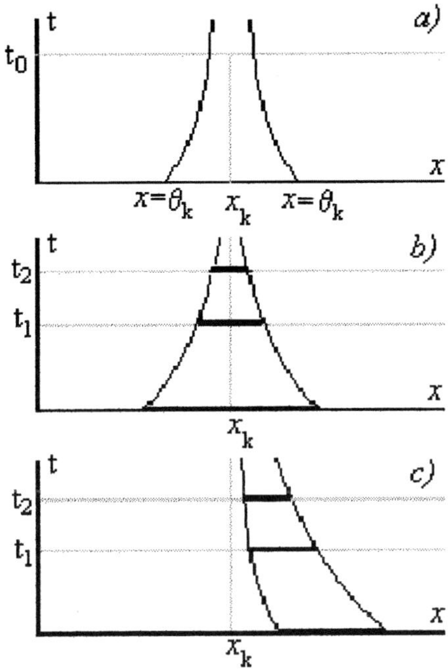

Fig. 7.1. Zone evolution during focussing process at different initial concentration distributions. The explanations are given in the text.

As an example let us consider a linear velocity profile

$$v_k(x) = -G_k(x - x_k), \qquad G_k > 0 \tag{7.6}$$

where G_k is some positive constant.

By substituting Eq. (7.6) for Eqs. (7.4) and (7.5), we obtain

$$\theta_k(x, t) = x_k + (x - x_k)\,\mathrm{e}^{-G_k t}, \qquad C_k(x, t) = C_k^{(0)}(\theta_k(x, t))\,\mathrm{e}^{G_k t} \tag{7.7}$$

Thus, for the example described above, the concentration at each characteristic grows exponentially with time with zone width δ_k being decreased with exponential law also:

$$\delta_k(t) = \delta_k(0)\exp\{-G_k t\} \tag{7.8}$$

Fig. 7.1b,c illustrates the concentration profile at different moments of time, i.e. the solid segments at the lines of t = const. Fig. 7.1b corresponds to the case where the substance was distributed at both sides of the isoelectric point; Fig. 7.1c shows distribution at one side (right) only.

7.3. THE EVALUATION OF FOCUSSING TIME

This approach can be applied for evaluating the time needed to focus the substance for some zone width which is sufficient for the purposes of experimenting by substituting the appropriate values of x in Eq. (7.6). Although one should take into account that the solution derived above is not valid in the vicinity of the focussing point, so the final zone width in such evaluations should be set appropriately.

In the case of a linear velocity profile the exact solution can be easily obtained [2,4]:

$$C_k(x,t) = \frac{e^{G_k t}}{2\sqrt{\pi D_k \eta_k}} \int_{-\infty}^{+\infty} \exp\left(-\frac{(\theta_k - \xi)^2}{4D_k \eta_k}\right) d\xi \tag{7.9}$$

$$\xi_k = x_k + (x - x_k)e^{G_k t}, \qquad \eta_k = (e^{2G_k t} - 1)/2G_k$$

The results given by Eq. (7.8) are in rather good agreement with those obtained with the help of Eq. (7.7) provided the diffusional coefficient values are sufficiently small. Being compared with the experimental results these evaluations proved to be useful [4].

In Ref. [5] the dynamics of sample focussing in the linear pH gradient was analysed by assumptions of the initial Gaussian concentration profile, that allowed to obtain an analytical solution.

Like we have already noted in the section devoted to steady-state electrophoresis, the numerical simulations are a unique tool of obtaining the solution in the case of complex systems which are usually realised in experiments of practical importance. Many examples can be found in [6–9].

7.4. REFERENCES

1. V.G. Babski and M.Yu. Zhukov, Biophysical Methods: Principles of Electrophoresis, Moscow State University Publishers, Moscow, 1990.
2. V.G. Babski, M.Yu. Zhukov and V.I. Yudovich, Mathematical Theory of Electrophoresis, Consultants Bureau, New York, 1989.
3. B.L. Rozhdestvenski and N.N. Yanenko, The Systems of Quasilinear Equations, Nauka, Moscow, 1978.
4. V.G. Babski, M.Yu. Zhukov, A.I. Sazonov and A.V. Stoyanov, Cosmic Science and Engineering, Vol. 4, Naukova Dumka, Kiev, 1989, pp. 15–19.
5. G. Zilbershtein, S. Peltek and L. Frumin, J. Biochem. Biophys. Methods, 45 (2000) 205–209.
6. M. Bier, O.A. Palusinski, R.A. Mosher and D.A. Saville, Science, 219 (1983) 1281–1287.
7. R.A. Mosher, D. Saville and W. Thormann, The Dynamics of Elecrophoresis, 1991.
8. K. Shimao, Jpn. J. Electrophoresis, 38 (1994) 221–225.
9. W. Thorman and R.A. Mosher, Adv. Electrophoresis, 2 (1988) 45–108.

PART I.II

Optimization of the Electrophoretic Separation

CHAPTER 8

Buffering Capacity

CONTENTS

8.1. INTRODUCTION

The 'buffer value' in the form that was introduced by Van Slyke [1], is an important characteristic of any buffer system, and a useful tool for choosing an experimental condition in many different separation techniques. Nevertheless, the 'buffering capacity' is a local value, thus, for practice, when the pH changes considerably during titration, we should operate with the value of the 'buffer resource', the amount of strong titrant which is necessary to provide for the defined pH shift. A similar problem arises when a complex substance, biopolymer for example, is used as a titrating agent.

8.2. BUFFER CAPACITY AND BUFFER RESOURCE

According to definition [1–3], the buffer capacity is the derivative

$$\beta = \frac{dC_b}{d(pH)} \tag{8.1}$$

where C_b represents the concentration of a strong base titrant, that is required to give a pH increase. Since β is a local value, when the pH interval that we analyse is not infinitesimal, we ought to take into account the β variance with pH. Thus to obtain the amount of C that results in the pH shift of $\Delta pH = pH_2 - pH_1$ we have to integrate:

$$C = \int_{pH_1}^{pH_2} \beta(pH)\, d(pH) \tag{8.2}$$

References p. 93

or in another form:

$$C = -\frac{1}{2.303}\int_{(\mathrm{H})_1}^{(\mathrm{H})_2} \frac{\beta(\mathrm{H})}{(\mathrm{H})}\,\mathrm{d(H)} \tag{8.3}$$

Here we used

$$\frac{\mathrm{d}}{\mathrm{d(pH)}} = -2.303\frac{\mathrm{d}}{\mathrm{d(H)}}$$

In many practical tasks we are restricted often by the demand that the pH of the system is to be controlled and not to exceed some defined limit $(\Delta\mathrm{pH})_{\mathrm{max}}$. Eq. (8.2) allows us to calculate the concentration of a strong titrant that causes this change of the pH. We suppose it is suitable to introduce the value of the 'buffer resource', which represents the limit of the titrant concentration when we need to satisfy our restriction on the pH shift. [1]

$$U\beta(\Delta\mathrm{pH}_{\mathrm{max}}) = C_{\mathrm{max}} = \int_{\mathrm{pH}_1}^{\mathrm{pH}_2+\Delta\mathrm{pH}_{\mathrm{max}}} \beta(\mathrm{pH})\,\mathrm{d(pH)} \tag{8.4}$$

Thus the buffer resource is a function of the limit on ΔpH, and any concrete buffer system may be characterised by it.

Let us consider some simple examples. When we have two different buffers, and the values of their buffering capacity are close to each other or even the same, their 'buffer resource', also for a moderate pH limit, may be considerably different (or, in other terms, the same amount of titrant will result in a different pH shift).

In Fig. 8.1A, the behaviour of a buffer power function in the vicinity of the isoelectric point for two different models of biprotic protolytes is shown. If these two substances are ampholytes, the pH of their solutions should be close to their isoelectric points. Let us suppose also that we have moderate pH conditions, so the water ion contribution to the buffer power may be omitted. We see that the *b*-function for an amphotere with $\Delta\mathrm{p}K = 1$ is quasi constant during almost all the pH interval of one unit, Curve I, while for $\Delta\mathrm{p}K = 0$, we can see an appreciated decrease, Curve II. In order to obtain the same β value at the isoelectric point, the protolyte concentration for Curve II is taken lower, $C_{\mathrm{II}}/C_{\mathrm{I}} = 0.73$. One more bright example is given by Fig. 8.1B. Here we have two β-functions with derivatives of the opposite sign: $\Delta\mathrm{p}K = 3$, Curve I; $\Delta\mathrm{p}K = 0$, Curve II ($C_{\mathrm{II}}/C_{\mathrm{I}} = 0.119$). Here we will have an increase in the buffer capacity with titration for Buffer I and a decrease for Buffer II, that obviously will give a considerable difference in the 'buffer resource'.

One should also take into account that a high buffer capacity value of the buffer systems, having their pH at the pH extremes, are always connected with a fast decrease with titration (high negative derivative), that is explained by a relatively high contribution of water ions. Consider an example with two real substances. Iminodiacetic acid (IDA) has two acidic ionogenic groups: $\mathrm{p}K_1 = 1.73$, $\mathrm{p}K_2 = 2.73$ (pI 2.23) and possesses a very high buffer power — for 10 mM solution (pH = 2.56), the calculated β

[1] Certainly, the information we need, may be obtained directly by solving the electroneutrality equation; this form is more suitable for approximate calculations.

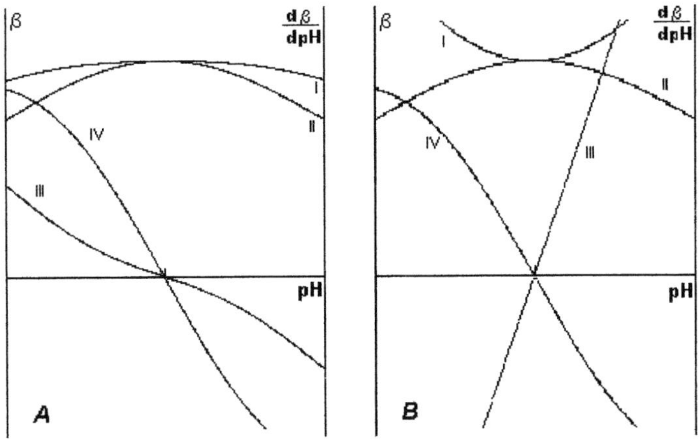

Fig. 8.1. Different behaviour of the buffer power function in the vicinity of the isoelectric point for model biprotic protolytes. (A) Curves I and II are the buffer capacity course for $\Delta pK = 1$ and 0, respectively (in order to obtain the same *b* value at the isoelectric point, the protolyte concentration for Curve II is taken lower — $C_{II}/C_{I} = 0.73$). (B) An example of two β-functions with the derivatives of the opposite sign: $\Delta pK = 3$ — Curve I; $\Delta pK = 0$ — Curve II ($C_{II}/C_{I} = 0.119$). Curves III and IV are the derivatives for Curves I and II, respectively. The total pH interval is one unit. (This figure is from [4].)

value is 6.26×10^{-3} (Mequiv./(L pH)). The solution of glutamic acid ($pK_1 = 2.162$, $pK_2 = 4.324$ (p*I* 3.243) [4]) at the same concentration will provide only $\beta = 1.88 \times 10^{-3}$ (Mequiv./(L pH)). In order to obtain the same level of the buffer power, we should perform a four-fold concentration increase ($C = 40.05$ mM), and the pH will be 3.28.

Now we have the same buffer capacity, but the former is characterised by a high negative derivative (1.45×10^{-2} at the pH of a pure solution). Oppositely, the latter has a small derivative value (2.25×10^{-3}), and it increases as we are performing the titration procedure (Fig. 8.2). If we calculate the 'buffer resource' for these two systems, supposing $\Delta pH_{max} = 0.3$, we will obtain $U_{IDA}(\Delta = 0.3) = 4.5$ mM, and $U_{GLU}(\Delta = 0.3) = 3.64$ mM.

8.3. BUFFER PROPERTIES OF SOLUTIONS OF PROTEINS AND NUCLEIC ACIDS

The buffer power of a biopolymer may be approximated as an additive sum of the ionogenic groups of their monomer units.[2] Each one represents the function $(H)K_d/((H) + K_d)^2$, which decreases rather slowly as the pH goes away from that corresponding to the maximum (Fig. 8.3A,C). In the case of proteins, the main

[2] The expression for the buffering capacity depends on the dissociation scheme we used [4–6]. In frames of 'parallel schemes', the assumption of an independent dissociation leads to the approximation that we used in the present chapter. Nevertheless, one should remember that within monomers some groups may dissociate sequentially, e.g. phosphoric acid, although due to the high ΔpK distance the error is negligible.

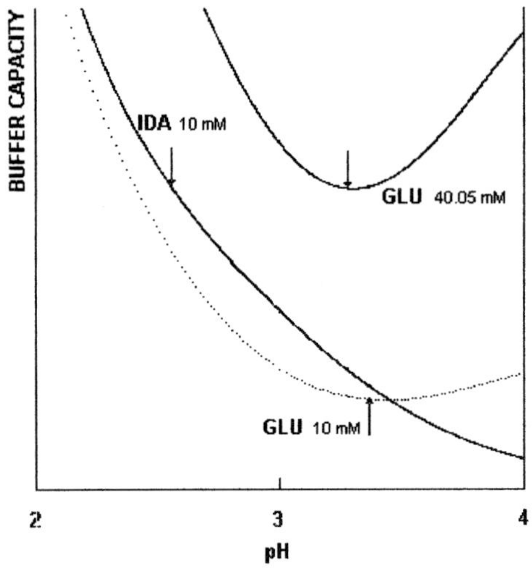

Fig. 8.2. Buffer capacity of iminodiacetic acid (10 mM) and glutamic acid (40.5 mM) solutions in the range of their isoelectric points. These two buffers have the same buffer capacity, but the former is characterised by a high negative derivative. Oppositely, the latter has a small derivative value, and it increases as we are performing the titration procedure. The dotted line shows the β-function of 10 mM solution of glutamic acid. Small arrows indicate the appropriate pH value of the pure solution. (This figure is from [4].)

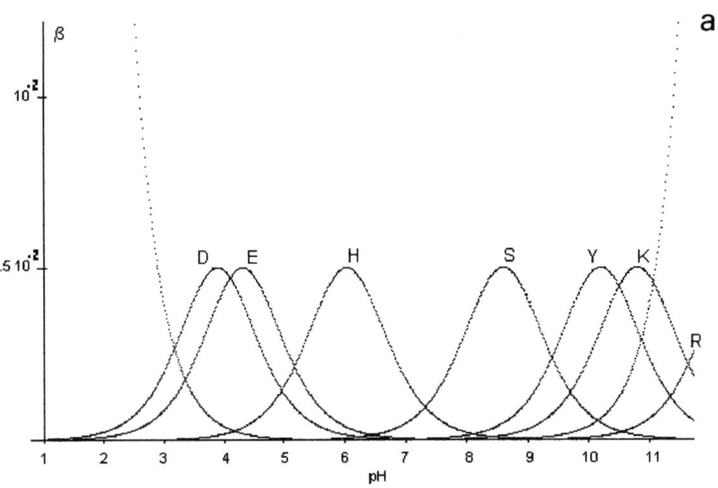

Fig. 8.3. Buffer power of proteins and nucleic acids in a wide pH region. (A) and (C) represent the contribution of monomer units amino acids and nucleotides, respectively. (B) corresponds to the calculated buffer capacity of three different proteins: albumin 10^{-4} mM — Curve I; lisozym 10^{-3} — Curve II; and haemoglobin a-chain 10^{-3} mM — Curve III. (D) corresponds to the solution of 18-mer oligonucleotide (5′-TCTGAAAGTGCTCTACTG-′3). The horizontal line is used to mark the level of 2.303×10^{-3} M/(L pH) (that corresponds to the 1 mM of strong acid). (This figure is from [4].)

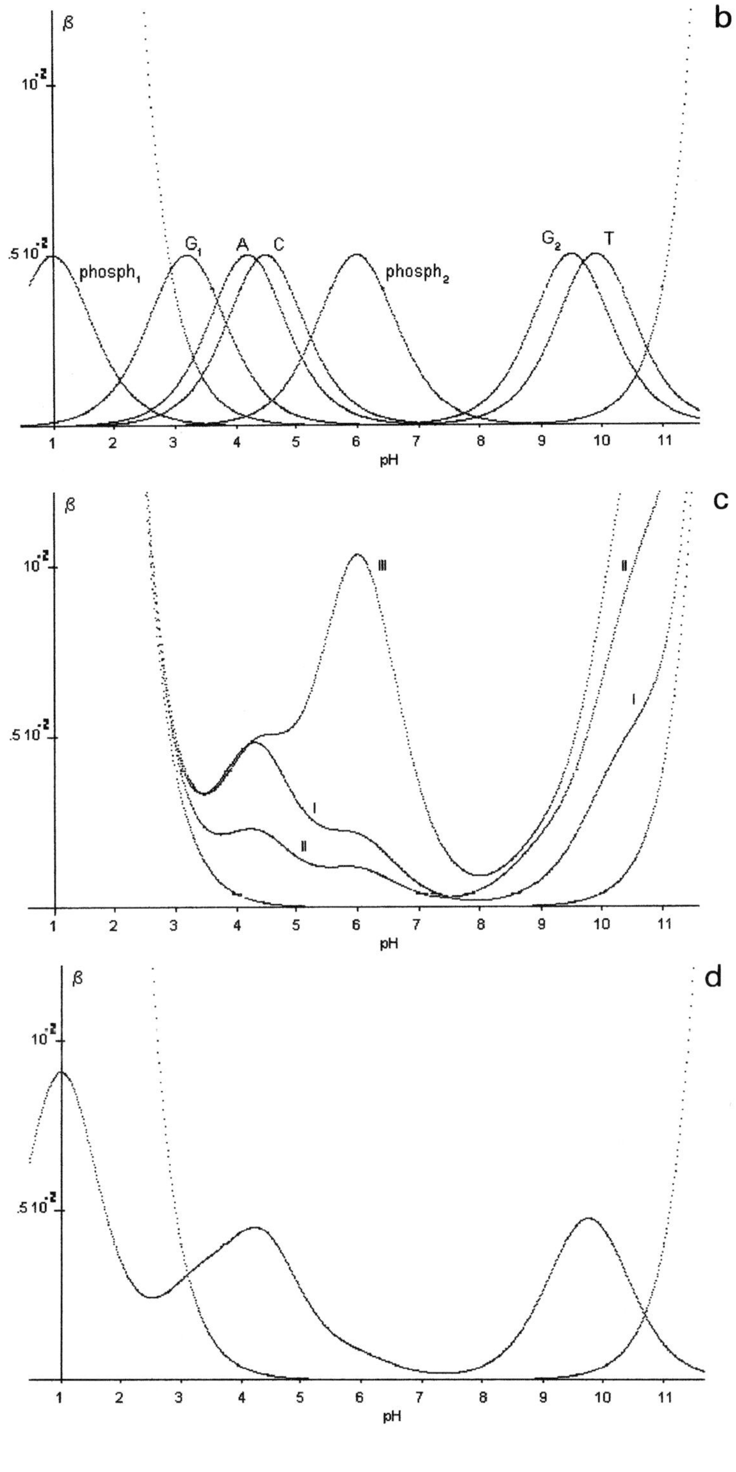
b
ß
10^{-2}
$.5\ 10^{-2}$
phosph$_1$
G_1
A
C
phosph$_2$
G_2
T
1 2 3 4 5 6 7 8 9 10 11
pH
c
ß
10^{-2}
$.5\ 10^{-2}$
III
II
I
I
II
1 2 3 4 5 6 7 8 9 10 11
pH
d
ß
10^{-2}
$.5\ 10^{-2}$
1 2 3 4 5 6 7 8 9 10 11
pH

contribution to the buffering capacity is due to the ionogenic groups of the amino acid side chains. In Fig. 8.1A, the separate contribution of amino acids is shown, from left to right: asp (p*K* 3.9), glu (p*K* 4.324), his (p*K* 6.04), cys (p*K* 8.6), tyr (p*K* 10.189), lys (p*K* 10.79), arg (p*K* 12.48). Here we used the data of Hirokawa [7]. Although the p*K*s of the ionogenic groups are distributed rather uniformly within the pH region, we may suppose that in general for proteins there should exist a minimum in the range of ca. 7–8, since cystein usually does not take part in the dissociation (its relative concentration is also very low). So we may speak about a local minimum in this range (see Fig. 8.3B), although the β value remains still relatively high and, moreover, the proteins, we usually deal with, possess no pH ranges where β approaches zero (in contrast to small amphoteric compounds). We should take into account, also, that the ionogenic group appearance rate may be taken as 0.1, that leads to the solution of any protein providing a relatively moderate buffering power with respect to the mass concentration.

We observe another situation in the case of nucleic acids. First of all, they possess a considerably higher density of ionogenic groups per monomer (and also per mass unit). Fig. 8.3C shows the separate contribution of them, from left to right: phosphate (p*K* 1.0), G_1 (pK_1 2.4), A (p*K* 3.8), C (p*K* 4.5), phosphate (p*K* 6), G_2 (pK_2 9.4), T (p*K* 10.0). The bases incorporated are weak (p*K* 2.4; p*K* 3.8; p*K* 4.5) [8], so the isoelectric points are always in the acidic region, where the buffering power value is very high. For RNAs, it is due to the presence of base groups of G, A, and C; and for DNAs, it is connected with the contribution of the prime phosphate (see Fig. 8.3D).

8.4. BIOPOLYMERS AS TITRATION AGENTS

In the previous section we considered the buffer properties of a solution of biopolymers, but for many practical purposes more important is the influence of a biopolymer sample when added in a buffer. So we are interested in the behaviour of a sample as a 'titrant'.

It is suitable to approximate the action of many dissociating groups in a biopolymer when adding the equivalent amount of a strong titrant. The later is, obviously, coinciding with the value of an additional charge, that the biopolymer molecule will obtain when placed at the pH of the buffer.

Fig. 8.4 provides an equivalent presentation of an 18-mer oligonucleotide as a strong acid (Mol equiv./Mol), when we use a buffer with a different pH. The abscissa axis is the difference between the pH of the buffer that is selected for the experiment and the p*I* value (the calculated p*I* of this nucleotide is 1.28). There is a small difference in the 'titration' properties at the pH range of 6–10, but in the acidic buffers (around pH 3) it acts as an equivalent amount of acid, which is three times lower.

Ideally, one would prefer to take a buffer with an incredible buffer power and then to put it in his sample in a negligible concentration, in order not to pay attention to how great the deviation from the conditions which were initially prescribed may be. If it is impossible to ignore the sample influence on the buffer pH, an opportunity to calculate this value theoretically, starting from the known chemical composition, both for the buffer and sample, always exists. The only thing that we need is to solve the

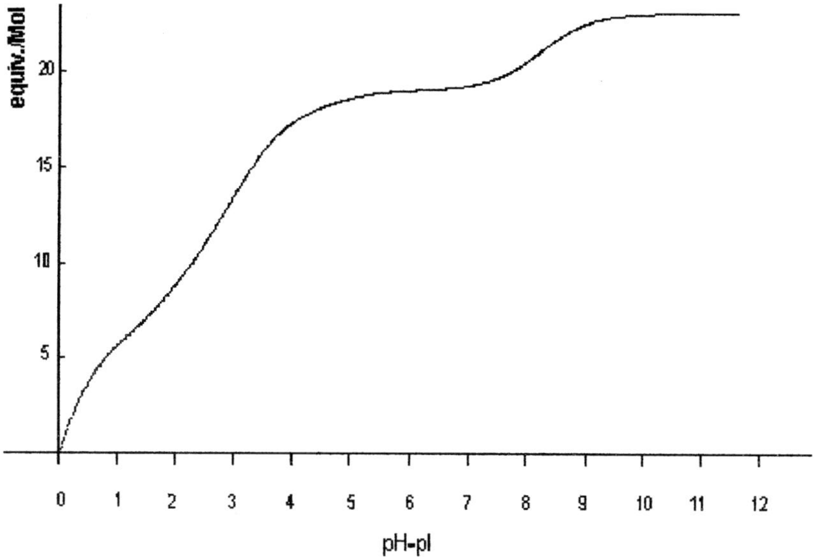

Fig. 8.4. Equivalent presentation of an oligonucleotide as a strong acid (Mol equiv./Mol) when we use a buffer with a different pH. The calculated p*I* for the 18-mer oligonucleotide (see legend to Fig. 8.3) is 1.28. (This figure is from [4].)

electroneutrality equation, although some technical problems may arise. At the same time, the information about the buffering capacity of a buffer (in Eq. (8.1)), calculated or taken from tables, gives a recipe for making our choice, providing we are sure that the pH shift will be rather small.

In the opposite case, we should take into account the b change with the pH. We think it is useful to operate with the term of the 'buffer resource'; the latter may be easily calculated for simple substances (or, at least, be evaluated with the help of the first derivative). When selecting a buffer, it is important to know not only its properties, but also the possible influence of the sample on it. Even when the precise data about the chemical composition are unavailable, some general properties may be useful in order to evaluate the 'equivalent presentation' of a biopolymer as a strong titrant.

8.5. REFERENCES

1. D. Van Slyke, J. Biol. Chem., 52 (1922) 525–536.
2. J.N. Butler, Ionic Equilibrium, Addison-Wesley, New York, 1964.
3. J.E. Ricci, Hydrogen Ion Concentration. New Concept in a Systematic Treatment, Princeton University Press, Princeton, NJ, 1952.
4. A.V. Stoyanov and P.G. Righetti, J. Chromatogr. A, 838 (1999) 11–18.
5. A.V. Stoyanov and P.G. Righetti, Electrophoresis, 19 (1998) 187–191.
6. A.V. Stoyanov and P.G. Righetti, J. Chromatogr. A, 853 (1999) 35–44.
7. T. Hirokawa, M. Nishino, N. Aoki, Y. Kiso, Y. Sawamoto, T. Yagi and J. Akiyama, J. Chromatogr., 273 (1983) D1–D106.
8. H.R. Mahler and E. Cordes, Biological Chemistry, 2nd ed., Harper and Row, New York, 1971.

CHAPTER 9

Optimisation of Electrophoretic Separation

CONTENTS

9.1. OPTIMISATION OF ELECTROPHORETIC SEPARATION USING pH–CHARGE RELATIONSHIP

9.1.1. Calculation of mobility vs. pH

While the first experiments on obtaining the titration curves (acid–base titration) were aimed to study the protein structure organisation [1–6], the 'titration curves' developed later — the mobility vs. pH curves (according to Rozengren et al., [7]), or the theoretical Q vs. pH relationships (e.g. [8–10]) — became used for selecting the proper pH conditions for isolation of proteins from their contaminants with the employment of charge differences.

As discussed earlier in Chapter 2, for the titration curves modelling, a simplified approach can be applied when the consideration is performed in terms of 'reduced' intrinsic constants. By this approach, the appropriate intrinsic constants in polyelectrolyte corresponding to the same ionogenic group are supposed to be equal to each other, and the latter value may be calculated from the experimental data on small compounds. For example, in the case of amino acids, one can consider the macroscopic dissociation constants as intrinsic (microscopic), and these values can be used for calculating the titration curves of proteins. For a protein molecule or other biopolymer with i acidic

References p. 104

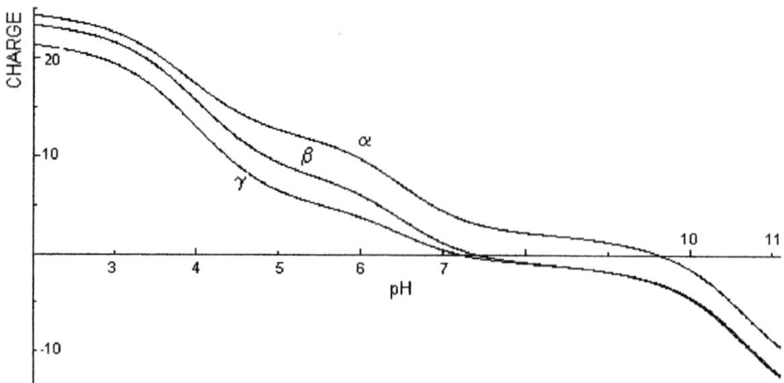

Fig. 9.1. Charge vs. pH for the α, β and γ haem-free globin chains by assuming that all of the −SH groups of Cys are capable of creating an electric charge (free reduced −SH groups). The value of the dissociation constant was taken as pK 8.6. (From Ref. [11].)

groups and j basic groups, net charge Q is calculated according to the equation

$$Q(\mathrm{H}) = \sum_i \frac{n_i(\mathrm{H}^+)}{k_i + (\mathrm{H}^+)} - \sum_j \frac{n_j k_j}{k_j + (\mathrm{H}^+)} \tag{9.1}$$

where k_i and k_j are the ‘intrinsic constants’.

Such an approach, although seeming to be justified to a certain extent, is clearly a crude approximation. Note that the problem is not only in this. The dissociation constants of ionogenic groups in proteins differ from those of free amino acids, but, generally speaking, one can not apply equal constant values even for the same dissociating centre when the number of possible microreactions is analysed. By the way, even in the case of free amino acids, one is to find at least two intrinsic constants corresponding to one dissociation centre (four for a three-dissociating group molecule). Strange, a little bit, but this method provides a reasonable agreement with experimental results; cf., for example, the above theoretical titration curves for haemoglobin chains with the experimental relationships [12].

Good agreement is observed both for the titration curve in general and for pI values. In the above example, by taking into account the fact that the Cys ionogenic groups were blocked, it was possible to achieve a quasi-complete coincidence (see also Figs. 9.1 and 9.2).

9.1.2. Relative charge difference for two components to be separated

When the separation mechanism is not based on sieving, i.e. when the size differences of the biopolymers to be separated are negligible, one will have to resort to differences in the surface charge. In order to find such optima of the resolving power, it is possible to introduce parameter R, reflecting the charge ratio of the two separands:

$$R = [\max\{‘Q'_1, ‘Q'_2\}/\min\{‘Q'_1, ‘Q'_2\}]\,\mathrm{sign}\{Q_1 Q_2\} \tag{9.2}$$

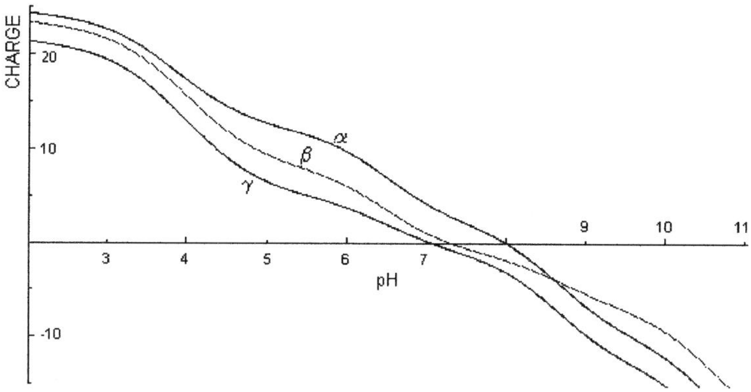

Fig. 9.2. Theoretical charge as a function of pH for the haem-free globin chains (α, β and γ). The influence of −SH groups (Cys) was not taken into account (reduced and alkylated chains). (From Ref. [11].)

where Q_1 and Q_2 are the surface charges of the two oligomers, respectively. It should be borne in mind that, for $R < 0$, the two polymers will have the opposite sign, and thus there will be no opportunity of detecting both the substances in a single electrophoretic run (pH region between two isoelectric points).

Let us consider an example with four 18-mer nucleotides, their compositions and dissociation constant values are given in Tables 9.1 and 9.2. Each one differs from the other by one substitution only, in position of the 9th nucleotide. (The titration curves of the four single nucleotides are visualised in Fig. 9.3.)

TABLE 9.1

SEQUENCE OF THE FOUR 18-MERS UTILISED IN THE PRESENT STUDY

Name	Sequence
T_9 (I)	5′TCTGAAAG**T**GCTCTACTG3′
G_9 (II)	5′TCTGAAAG**G**GCTCTACTG3′
A_9 (III)	5′TCTGAAAG**A**GCTCTACTG3′
C_9 (IV)	5′TCTGAAAG**C**GCTCTACTG3′

TABLE 9.2

p*K* VALUES

Base	Nucleotides	Free bases
	pK_b:	pK_b:
C	4.5	4.5
A	3.8	4.2
G	2.4	3.2
	pK_a:	
G	9.4 (enolate)	
T	10.0 (enolate)	

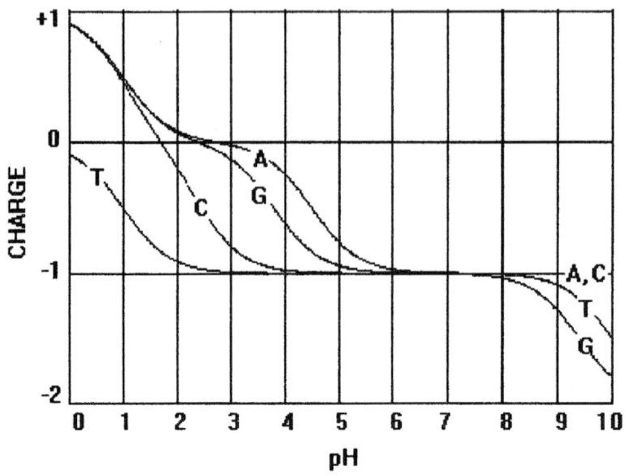

Fig. 9.3. Titration curves of the four nucleotides in DNA, calculated in the pH 0–10 interval. The four curves represent pdT, pdC, pdG and pdA, respectively. The contributions of the dissociation of the second and third groups in phosphate were not taken into account, since in oligonucleotides such groups will form the phosphodiester linkage. The dissociation value for the first phosphate group was taken as 1.0. (From Ref. [13].)

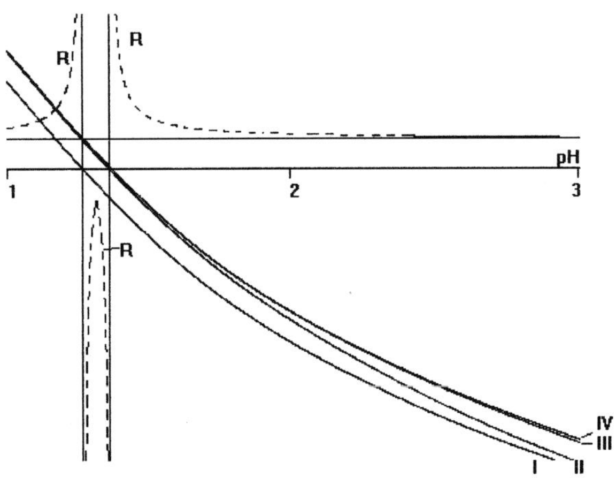

Fig. 9.4. Theoretical charge as a function of pH for the four 18-mers listed in Table 1. The curves are: T_9 (I), $T_9 \rightarrow G_9$ replacement (II); $T_9 \rightarrow A_9$ (III) and $T_9 \rightarrow C_9$ (IV). The dotted lines represent the *R* parameter for 18-mer I (T_9) and II (G_9). The abrupt rise in these two curves below pH 2 and above pH 1 indicates the approach to the p*I* value of each component. (From Ref. [13].)

On the basis of these data, we can now construct the overall titration curves of the four 18-mers of Table 9.1. Fig. 9.4 shows such titration curves in the pH 1–3 region for the original ((T_9) oligomer (Curve I), for a T→G replacement (Curve II), for a T→A substitution (Curve III) and for a T→C replacement (Curve IV). It can be seen that, whereas Curve I remains quite removed from all others in the entire titration interval, Curves II–IV tend to converge below pH 2 and essentially coalesce at pH = p*I* (indicated by the abrupt asymptotic rise of the two upper and two lower *R* curves). It can thus be observed that it should be possible to find experimental pH conditions allowing separa-

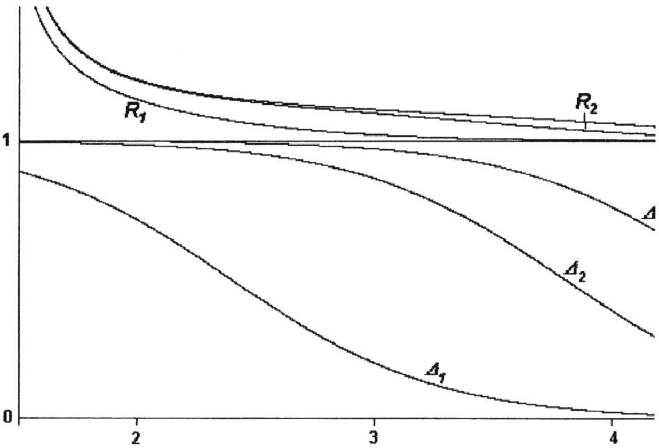

Fig. 9.5. Absolute charge difference (Δ) and charge ratio (R) for the three pairs of oligonucleotides corresponding to the original 18-mer (T_9) and its pairs with the other three substituted oligonucleotides (G_9, C_9, A_9). The subscripts indicate: 1 (Pair I and II); 2 (Pair I and III); and 3 (Pair I and IV). See also Table 1 for the sequences and labelling of various oligomers. (From Ref. [13].)

tion of all the four oligomers without entering 'the dangerous pH regions', i.e. pH values approaching the p*I* (a most unfavourable condition, since it would lead to zero mobility) and in such a strongly acidic environment as to risk rapid depurinisation (see ahead. In Fig. 9.5, we have plotted the absolute charge differences (curves labelled Δ_1 to Δ_3, representing, as above, the charge differences between pairs T→G, T→A and T→C, respectively, in the four sets of the 18-mers of Table 9.1) and charge ratios R_1 to R_3 (upper curves), explored in the pH 1.5–4.5 interval. This graph (especially in regard to curves R_1–R_3) gives an immediate answer to our search: a good pH window for the optimal resolution should exist only in the pH 3–4 interval. In fact, whereas above pH 4, sets R_2 and R_3 would still allow reasonable resolution, doublet R_1 (representing two oligomers T_9 vs. G_9) would be unresolvable, since this ratio approaches unity already at pH 3.7.

Fig. 9.6 shows separation of the mixture of the four oligomers of Table 9.1 by free solution (i.e. in the absence of any sieving liquid polymer) CZE in an acetate buffer at pH 5.7. In agreement with our modelling (see Fig. 9.3), essentially no separation occurs among these four species, although a shoulder is clearly visible to the right of the main peak and the broad peak appearance suggests zone heterogeneity. When, however, the pH is lowered to 4.8 (Fig. 9.7), much better results are obtained: now three out of four oligomers are resolved. By injecting them one by one, it has been possible to identify each peak as a mixture of T_9 and G_9 (first peak to the left), followed by A_9 and C_9. Notice the excellent agreement with the predictions of Fig. 9.3: at this operative pH, curve R_1 predicted in fact complete fusion of doublet T_9 vs. G_9, whereas it still indicated decent resolution for doublets R_2 and R_3. It remained to be seen if one could find a pH value allowing simultaneous resolution of all four oligomers. Previous experience suggested that we can try to explore isoelectric buffers, since the electrophoretic run had to be performed in the shortest possible time. Trying iminodiacetic acid (pH 3.3 with 7 M urea [14]), excellent resolution could be obtained upon simultaneous injection of all four oligomers (Fig. 9.8).

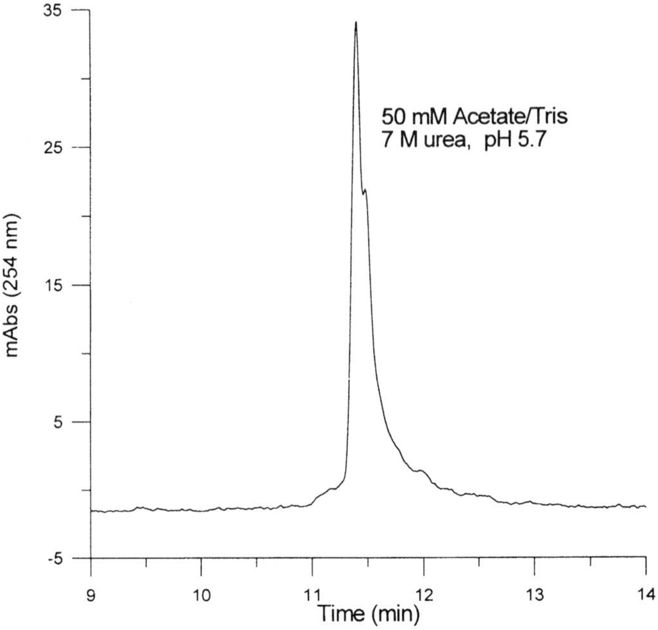

Fig. 9.6. CZE of the mixture of four oligonucleotides of Table 1 in 50 mM acetate, titrated with Tris to pH 5.7, in presence of 7 M urea. Conditions: Bio Rad Bio Focus 2000; injection: 3 s at 10 kV; run: 150 V/cm at 25°C. Notice the lack of resolution of the set of four compounds. (From Ref. [13].)

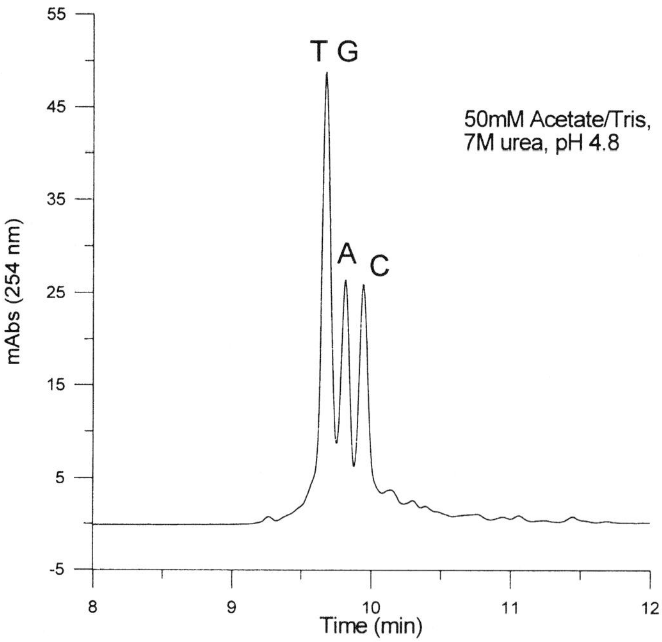

Fig. 9.7. CZE of the mixture of four oligonucleotides of Table 1 in 50 mM acetate buffer, titrated with Tris to pH 4.8, in presence of 7 M urea. Conditions: Bio Rad Bio Focus 2000; injection: 3 s at 10 kV; run: 150 V/cm at 25°C. Notice the good resolution of three out of four compounds. (From Ref. [13].)

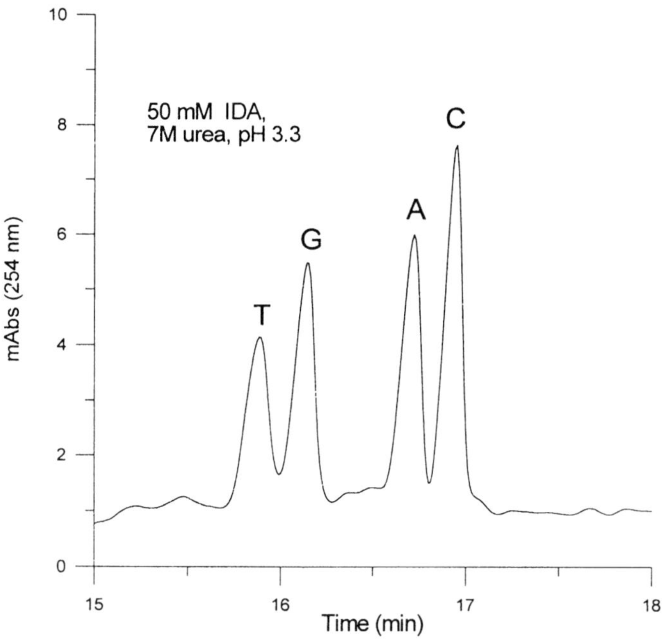

Fig. 9.8. CZE of the mixture of four oligonucleotides of Table 1 in 50 mM iminodiacetic acid (IDA) buffer, in presence of 7 M urea. Conditions: Bio Rad Bio Focus 2000; injection: 3 s at 10 kV; run: 150 V/cm at 25°C. Notice the good resolution of all four compounds. (From Ref. [13].)

9.2. DEPENDENCE OF MOBILITY ON MOLECULAR MASS IN FREE SOLUTION

When titration curves in acrylamide are obtained, one should try to minimise the effect of molecular sieving which discriminates the separated molecules with respect to their size. To achieve this, low acrylamide concentration gels are used, and usually it is sufficient to take the same concentration as for the IEF procedure. Nevertheless, the effect of the mobility dependence on the molecular size even in a free solution still remains, and should be taken into account properly.

There is no good way, probably, to propose any universal correlation between the molecular weight and relative mobility, although there are a lot of works where such attempts are performed [15].

As one simple physically evident function, the cubic root of the molecular mass is used (according to Stokes' law), although this approximation is not always justified

$$\mu(\mathrm{pH}) \approx Q(\mathrm{pH})M^{1/3} \qquad (9.3)$$

Often some other simple correlations, two thirds for example [15], or some more complicated ones [24,16] are used. The main disadvantage of such semi-empirical correlations lies in the fact that they, being constructed to work satisfactory for one defined set as a rule, do not act so well once the initial set is arbitrary extended.

It is important to remember that any reasonable correlation supposes a function of mass as M^{-p}, where index p is lower than one. This leads to the event that the mobility of biopolymers with a constant charge/mass ratio should result in some constant increase of mobility with molecular weight, and such behaviour is observed for 'not too long' DNA chains [17].

Thus, using mobility vs. pH curves of some sample mixture for evaluating pH–charge relationships or performing other tasks of mobility prediction based on theoretical 'titration curves', one should take this effect into account, at least approximately. For the peptides of haemoglobin digest the pattern of peptide mobilities is subjected to an essential transformation when the factor of relative mobility dependence on M_w is taken into account (in this example, as function of 'one third', Figs. 9.9 and 9.10 [18]).

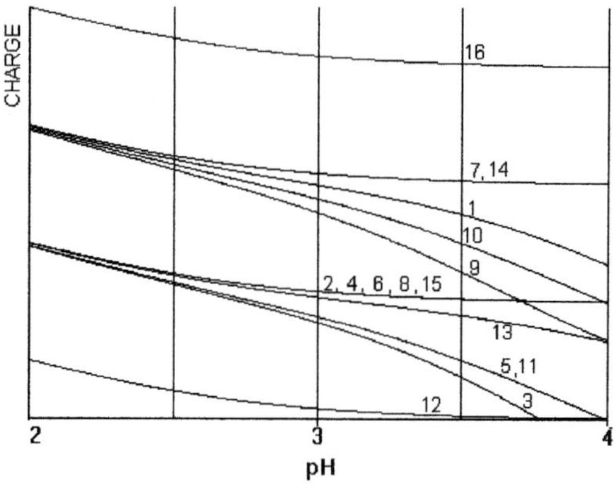

Fig. 9.9. Charge vs. pH curves of the major fragments derived from a tryptic digest of β-globin chains. (From Ref. [18].)

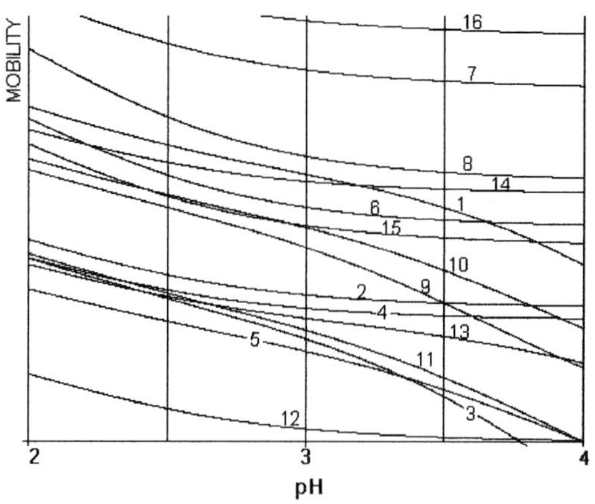

Fig. 9.10. Theoretical titration (pH/mobility) curves of the major fragments derived from a tryptic digest of β-globin chains. Mobility has been calculated as proportional to the fragment charge divided by the cubic root of the molecular mass. Amino acid peptide digest composition is described in [18].

It should be remembered, nevertheless, that these evaluations represent approximations.

9.3. ISOELECTRIC BUFFERS. THE CONCEPT OF 'NORMALISED β/λ RATIO'

After the report of Westermeer and Schickle [19], the use of the so-called isoelectric buffers becomes more and more popular [20,21]. Initially this term meant a solution of one amphoteric compound dissolved in pure water (thus being in the state close to the isoelectric one), and later on an attempt to extend this definition to other buffer systems was made [22]. The fist report on the electrophoretic separation in such media is the one of Mandecki and Hayden [14]. The authors achieved an essential improvement in resolution using the histidine buffer in slab gel electrophoresis. In the above works, the resolving power increase was attributed to the combination of an acceptable buffer capacity and low conductivity level, allowing to increase the applicable field strength.

From this point of view, an 'amphoteric' buffer possesses some evident advantages. Consider the simplest example with only two dissociating centres. If we suppose the pK equivalence for the positive and negative groups, the degree of dissociation (see Chapter 1, Section 1.3) equals one half (see also Fig. 1.6).

Thus even neglecting the relaxating effects (also assuming equal mobilities), one obtains a two times conductivity decrease, while the buffering capacity is expressed as an arithmetic sum of the appropriate contribution of each ionogenic group, no matter if they are free or assembled together (it is supposed that we operate in terms of intrinsic dissociation constants).

The ratio of the buffering capacity to conductivity can be introduced for evaluating the potential usefulness of any buffer. As an appropriate unit, it is suitable to take the value of pure water, as it was proposed in [23], although in this work the simplified approach of a fast interconversion rate was used for some small molecular weight compounds, see discussion in Section 3.3.

This value is:

$$^{\mathrm{Acq}}R_{\beta/\lambda} = 8.55 \times 10^{-3}\ [(\mathrm{ohm\,mol})/(\mathrm{cm}^2\,\mathrm{pH})] \tag{9.4}$$

The latter value was calculated by using absolute mobilities of hydrogen and hydroxyl ions as $\mu_{\mathrm{H}} = 36 \times 10^{-4}$ and $\mu_{\mathrm{OH}} = 19.8 \times 10^{-4}$ [cm^2/(s V)].

By normalising it to the molar concentration, we obtain the unit of [(ohm cm)/pH]:

$$A = \mathrm{Const} = (^{\mathrm{Acq}}R_{\beta/\lambda})/C^{\mathrm{Acq}} = 1.56 \times 10^{-1}\ [(\mathrm{ohm\,cm})/\mathrm{pH}] \tag{9.5}$$

where C^{Acq} denotes the molar concentration of pure water.

For any electrolyte, the 'normalised buffering power/conductivity ratio' can be easily calculated and then used for checking the utility of the buffer in some concentration range.

Such calculations are also an approximation, since it is very difficult to model correctly the conductivity of a multireacting system. Taking into account the discussion (Chapter 3), one should expect for small amphoteric substances the $R_{\beta/\lambda}$ ratio to be close to constant, i.e. independent of concentration in the concentration range practically used (long life-time for each microstate). The opposite model results in a quasi-linear increase of $R_{\beta/\lambda}$ with concentration, and this approach should be more preferable in the case of high M_r species.

Real solutions demonstrate a more complex conductivity dependence on the electrolyte concentration due to ion–ion and ion–dipole interaction, that leads to more and more slow conductivity increase with concentration and further decrease at very high concentrations.

9.4. REFERENCES

1. R.K. Cannan, Chem. Rev., 30 (1942) 395.
2. E.J. Cohn, J.T. Edsall and M.H. Blanchard, J. Biol. Chem., 105 (1934) 319.
3. D.I. Hitchcock, J. Gen. Physiol., 4 (1921) 723.
4. A.V. Kenchington, Biochem. J., 68 (1958) 458–459.
5. C. Tanford and R. Roxby, Biochemistry, 11 (1997) 2192–2198.
6. C. Tanford, S.A. Swanson and W.S. Shore, J. Am. Chem. Soc., 77 (1955) 6416.
7. A. Rozengren, B. Bjellqvist and V. Gasparic, in B. Radola and D. Graesslin (Eds.), Electrofocusing and Isotachophoresis, de Gruyter, Berlin, 1997, pp. 165–171.
8. M. Castagnola, L. Cassiano, I. Messana, G. Nocca, R. Rabino, D.V. Rossetti and B. Giardina, J. Chromatogr. B, 656 (1994) 87–97.
9. B. Skoog and C. Wichman, in P. Peters (Ed.), Protides of the Biological Fluids, Vol. 33, Pergamon Press, Oxford, 1985, pp. 593–596.
10. A.V. Stoyanov, Zh. Fiz. Khim., 68 (1994) 1893–1896.
11. P.G. Righetti, A. Saccomani, A.V. Stoyanov and C. Gelfi, Electrophoresis, 19 (1998) 1733–1737.
12. P.G. Righetti, R. Krishnamoorthy, E. Gianazza and B. Labie, J. Chromatogr., 166 (1978) 455–460.
13. M. Perego, C. Gelfi, A.V. Stoyanov and P.G. Righetti, Electrophoresis, 18 (1997) 2915–2920.
14. W. Mandecki and M. Hayden, DNA, 7 (1998) 57–62.
15. M. Castagnola, L. Cassiano, I.G. Messana, D.V. Rossetti and B. Giardina, in P.G. Righetti (Ed.), Capillary Electrophoresis in Analytical Biotechnology, CRC Press, Boca Raton, FL, 1996.
16. S. Fanali, Electrophoresis, 19 (1998) 3160–3165.
17. N.C. Stellwagen, C. Gelfi and P.G. Righetti, Biopolymers, 42 (1997) 687–703.
18. L. Capelli, A.V. Stoyanov, H. Wajcman and P.G. Righetti, J. Chromatogr. A, 791 (1997) 313–322.
19. R. Westermeer and H. Schickle, Electrophoresis '95, Paris, 1995, Abstr. N3.
20. C. Gelfi, M. Perego and P.G. Righetti, Electrophoresis, 17 (1996) 1470–1475.
21. S. Hjertèn, L. Valtcheva, K. Elenbring and J.L. Liao, Electrophoresis, 16 (1995) 584–594.
22. A.V. Stoyanov and P.G. Righetti, J. Chromatogr. A, 799 (1998) 275–282.
23. A.V. Stoyanov and P.G. Righetti, Electrophoresis, 19 (1998) 1674–1676.
24. L. Cellai, A.M. Onori, C. Desiderio and S. Fanali, Electrophoresis, 19 (1998) 3160–3165.

CHAPTER 10

Two-Dimensional Methods

CONTENTS

10.1. TWO-DIMENSIONAL ELECTROPHORESIS

When a complex mixture is to be analysed, it is often useful to take the advantages of the separation in the second dimension. This can increase essentially the number of the detected peaks. The processes used in the first and in the second separation should be, obviously, different, otherwise there is no necessity of performing the second stage (a gain in resolution can be achieved more easily by a simple increase of the separation path length), although the separation mechanism may be the same during each step. For example, one can perform the first separation by zone electrophoresis and then follow the same technique but at a different pH value.

Usually, as one of the earliest examples, reference is made to Poulik and Smithies [1, 2], but the advantages of this approach became more evident when using polyacrylamide gels [3].

An illustration of some possible combinations which can be used for two-dimensional electrophoresis is given in Table 10.1. Five different techniques are used: electrophoresis of native biopolymers, of biopolymers in a denatured state (e.g. SDS-electrophoresis), electrophoresis in gel with a porosity gradient, with a pH gradient (including IEF), and with a gradient of the denaturing agent. In this way a great number of working combinations can be obtained.

Theoretically, all of the indicated combinations are possible (indicated with a plus sign); more useful combinations are those which employ at least one method connected with separation according to the molecular weight (size) or the isoelectric point. The most important are those combining both the isoelectric point and the

References p. 108

TABLE 10.1

DIFFERENT TECHNIQUES FOR TWO-DIMENSIONAL ELECTROPHORESIS

	Uniform media native conditions (μ_N, electrophoretic mobility)	Uniform media denaturating conditions (μ_D, mobility in denaturated state)	Porosity gradient, grad(C_{A-A})	pH gradient, grad(pH)	Gradient of denaturating agent, grad(C_{denat})
μ_N	+	+	+	+	+
μ_D	+	+	+	++	+
grad(C)	+	+	+	++	+
grad(pH)	+	++	++	+	+
grad(denat)	+	+	+	+	+

See text for explanation.

molecular weight determination. In this case, after the separation by IEF, a 'linear pattern' of a thin polyacrylamide rod is introduced into the gel slab, where the focused proteins are subjected to the SDS-electrophoresis at a constant or increasing acrylamide concentration [4]. Some methodological aspects of the 'two-D maps' are discussed in the next part of this book, where one can find a complete review of recent trends in this field.

Table 10.1 does not include all the possible variants. It serves as an illustration only, and the reader has to use his own inventiveness. A lot of new methods can appear as a result, and perhaps some of them will prove to be useful, indeed.

10.2. OTHER TWO-DIMENSIONAL SEPARATIONS

All the examples considered in the previous section can be treated as 'pure' two-dimensional, i.e. after the first separation, the second one is performed in a perpendicular direction with the sample 'line' belonging to the uniform media at each moment of time (in other words, the gradient, if any, is always collinear to the separation direction, Fig. 10.1A). Some more complicated media for the second stage may be selected, for example, a perpendicular gradient (with respect to migration) may be used. In this case, the environment conditions will vary along the x-axis, and this perpendicular gradient can be coupled or not with the collinear gradient (respectively B and A in Fig. 10.1). Also, two collinear gradients can be used. The interpretation of such patterns becomes rather difficult, and the most important applications are connected with using the uniformly distributed samples. They include electrophoresis in the perpendicular concentration gradients (acrylamid, urea, formamid) and electrophoresis in the perpendicular pH gradient. The latter technique will be analysed in the next section.

A variant of the two-dimensional separation is described for the case in which the two stages are both performed in the same gel [5]. In this work, the sample was applied at the corner of the gel plate with the acrylamid concentration gradient, and the run was performed in the two perpendicular directions.

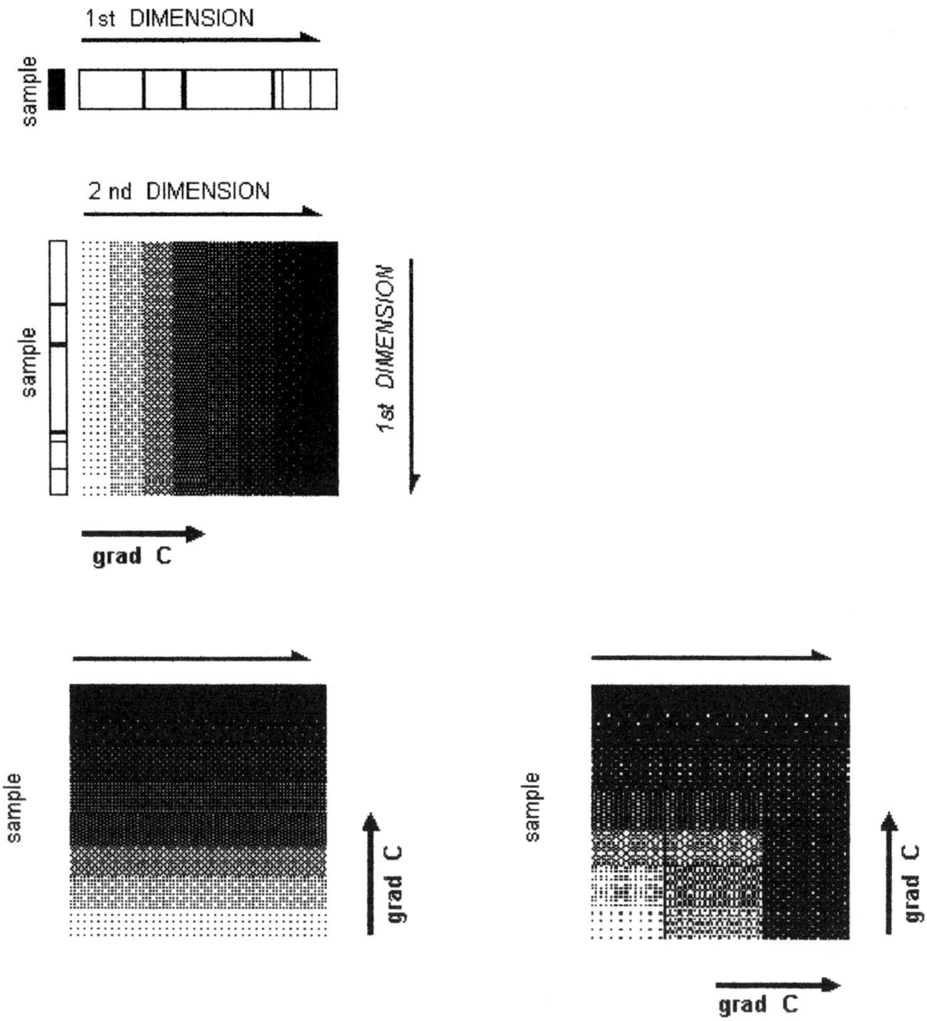

Fig. 10.1. Two-dimensional electrophoresis when, after the separation in the first dimension, a narrow gel band is subjected to a new separation procedure in the perpendicular direction. (A) The concentration gradient is collinear to the migration. (B) Electrophoresis with the perpendicular gradient; the sample is uniform. (C) An example of a more complicated system with two perpendicular gradients (e.g. acrylamide and urea).

10.3. MOBILITY VERSUS pH CURVES

A sample being introduced as a narrow band into a gel where there is already some perpendicular pH gradient, and then subjected to a usual zone electrophoresis procedure, will be distributed according to the component mobilities at the pH values covering the initial pH interval. So, one can easily obtain the curve (or the set of curves in the case of a more than one component sample) representing the mobility versus pH relationship.

References p. 108

This technique was proposed by Rosengren et al. [6] as 'a simple method for choosing optimum conditions for electrophoresis'.

This procedure of the experimental determination of the above dependence is often referred to as 'obtaining titration curves', although one should pay attention to the question of terminology. The true titration curves (acid or base titration) of a biopolymer solution [7–9] can reflect the 'charge versus pH' relationship, provided the concentration of hydrogen (hydroxyl) can be neglected (for an example see Fig. 1.8 in Chapter 1. The 'mobility versus pH' curve is a simple and very effective tool for understanding the 'charge versus pH' behaviour of the biopolymer molecule and for checking the theoretically obtained 'pH–charge' relationships [10]. Here the effect of a different relative mobility for a different molecular weight should be taken into account properly.

10.4. REFERENCES

1. M.D. Poulik and O. Smithies, Biochem. J., 68 (1958) 636.
2. O. Smithies and M.D. Poulik, Nature, 177 (1956) 1033–1034.
3. S. Raymond, Ann. N.Y. Acad. Sci., 121 (1964) 350.
4. P.H. O'Farrell, J. Biol. Chem., 250 (1975) 407.
5. J. Wein, Anal. Biochem., 31 (1969) 405–411.
6. A. Rosengren, B. Bjelqvist and V. Gasparic, in B.J. Radola and D. Graesslin (Eds.), Electrophoresis and Isotachophoresis, de Gruyter, Berlin, 1977, pp. 165–171.
7. R.K. Cannan, Chem. Rev., 30 (1942) 395.
8. E.J. Cohn and J.T. Edsall, Proteins, Amino-Acids and Peptides, Reinhold, New York, 1943.
9. D.I. Hitchcock, J. Gen. Physiol., 15 (1931) 125.
10. C. Tanford, Hydrogen Ion Titration Curves of Proteins, Electrochemistry in Biology and Medicine 248, Wiley, New York, 1995.

CHAPTER 11

Limitations of the Method of Isoelectric Focussing

CONTENTS

11.1. INTRODUCTION

At the present-day level of the IEF art, the immobilised pH gradients [1] provide us an opportunity to achieve the best resolution with respect to the pI difference of the separating components, although the potential of this method is still not completely realised. In addition to the two main ways of the pH gradient creation — practically only these are used widely as standard techniques for a lot of different applications (i.e. 'immobilised' and 'natural', mostly with CA) — there exist some other possible ways of obtaining pH gradients. Some of them will be summarised further, for analysing a possibility of using them both for solving some specific problems and for creating more powerful separation IEF-techniques. However, the isoelectrofocussing has its own intrinsic limits.

11.2. WAYS OF GENERATING pH GRADIENTS

11.2.1. Natural and artificial pH gradients

As it was suggested by Svensson [2], the gradients of pH can be divided into two types: artificial pH gradients and natural pH gradients. The former were developed by Kolin [3–5]. In his works, the pH gradients were generated by allowing diffusion to broaden the interface between the two buffers of different pH, and thus such gradients were destroyed sooner or later by the electric field. Conversely, natural pH gradients are formed as the electric current passes through the solution [2], so they are 'self-organised', indeed. Before Vesterberg proposed his elegant method [6], which enables us to synthesise a mixture of ampholytes suitable for IEF, some peptide mixtures [7,8] were used with moderate success for this purpose.

With respect to the way of gradient formation, immobilised pH gradients should be considered, certainly, as artificial ones. Although they are able to withstand the electric field quite 'naturally', i.e. the electric current does not cause any damage to them, so one should pay attention to the terminology question [9]. It is possible to say, also, that any artificial pH gradient moves in its 'natural form' once the electric field is applied.

There exist other possible ways of pH gradient creation which are not directly connected with the electrolyte concentration gradients; some of them will be analysed in this section.

11.2.2. Thermal pH gradients

11.2.2.1. External temperature field

Dissociation constants of acids and bases change their values with temperature. Thus, the solution of one more electrolyte, when subjected to applying a temperature gradient, should exhibit some pH gradient, even in the case of a uniform electrolyte concentration. This approach was realised in the works of [10–13]. The external temperature gradient was maintained by appropriate thermostatting, the opposite ends of the electrophoretic camera were kept at different temperatures with the help of stationary water jackets. To keep the stability of the system, an auxiliary density gradient was also used.

An absolute increase of pK obviously should cause an effect that the base becomes stronger (see Figs. 11.1 and 11.2) and the solution pH increases correspondingly. (For acids the situation is opposite.) Temperature coefficients $\mathrm{d}\,\mathrm{p}K/\mathrm{d}t$ are rather small, they usually do not exceed 0.01 pH unit/grad. As it was pointed out in [11], this is one of the highest values of such a coefficient ($\mathrm{d}\,\mathrm{p}K/\mathrm{d}t$ is around 0.03 in absolute values).

To illustrate the pH behaviour depending on the buffer concentration, temperature, temperature coefficient, and pK of ordinary buffers, two three-dimensional patterns are given (Figs. 11.3 and 11.4).

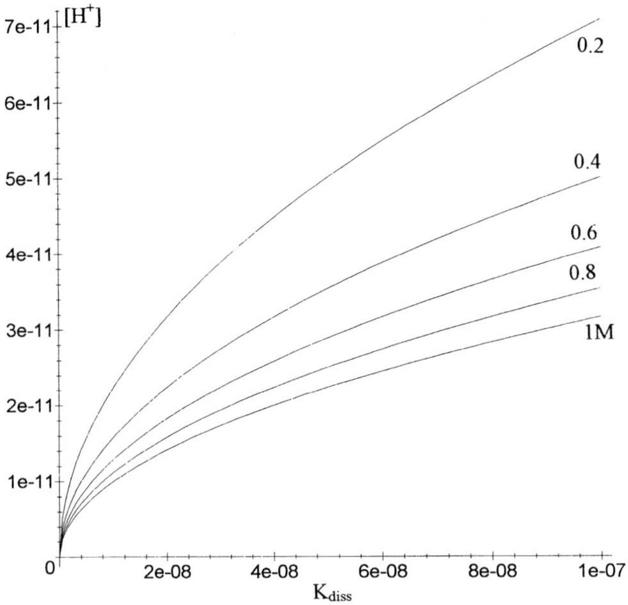

Fig. 11.1. Hydrogen ion concentration ($[H^+]$) of the solution of an ordinary electrolyte (base) vs. K_{diss} value (dissociation constant value range from 10^{-7} to 10^{-8}) at different concentrations (0.2, 0.4, 0.6, 0.8 and 1 M).

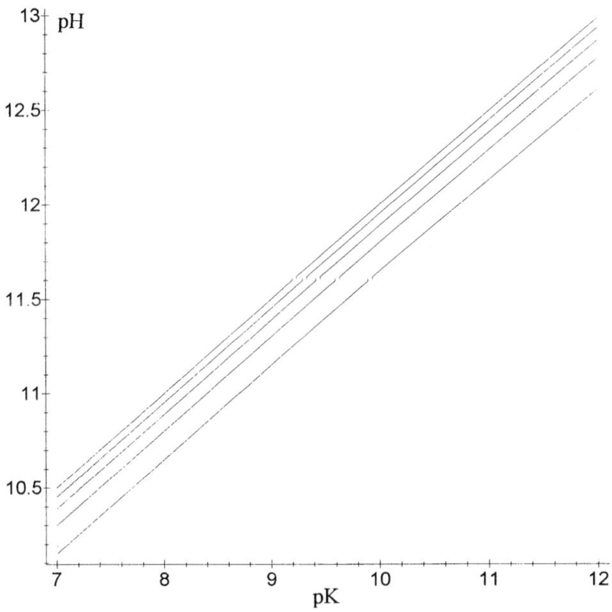

Fig. 11.2. pH vs. pK. The lines correspond to the same concentration as on Fig. 11.1.

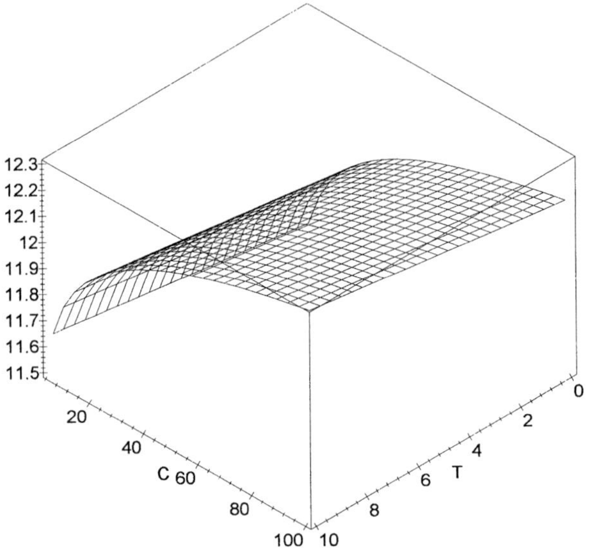

Fig. 11.3. Three-dimensional pH surface as a function of concentration and temperature. The buffer concentration is 0.1 M.

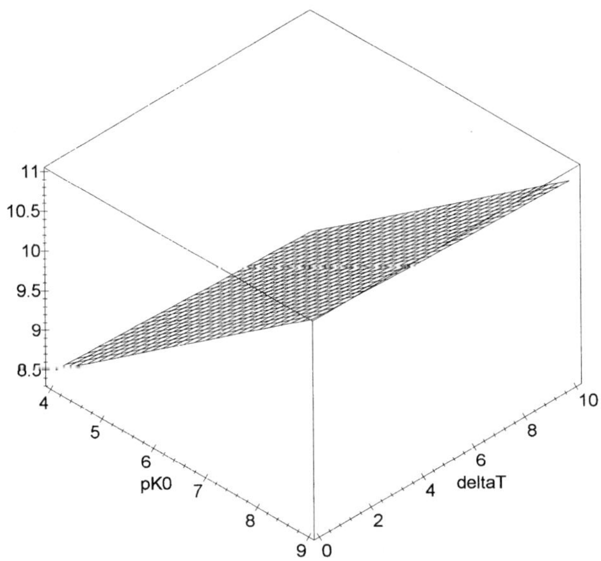

Fig. 11.4. Three-dimensional pH dependence on temperature and pK_0. The temperature coefficient is supposed to be constant ($\mathrm{d\,p}K/\mathrm{d}t = -0.03$), p$K_0$ is taken as 8.0.

For different compounds, one can see different signs of the temperature coefficient. Moreover, for the same compound, this coefficient changes its sign with temperature (e.g. acetic acid); and working with a buffer mixture, one can take into account the effect of the mutual suppression of the pH-temperature dependence.

11.2.2.2. Thermal pH gradients created by Joule heat dissipation in an electrophoretic chamber with a non-constant cross-section

A quite different approach was proposed by Pawliszyn [14]. He suggested that a cone-shaped capillary filled with a non-carrier ampholyte buffer should be used.

Generally, by varying the form of the electrophoretic chamber and the boundary conditions, the temperature profile can be changed in a wide range, the appropriate temperature distribution is obtained by solving the equation of thermoconductivity. For the purpose of crude evaluation, a simple approximation can be used. Let us consider an electrophoretic chamber with a non-constant cross-section. When $S(x)$ is not constant, current density j is not constant either, while I is always constant.

Electric power (W) per volume $\mathrm{d}V = S(x)\mathrm{d}x$ is given by

$$w = \frac{W}{\Delta x S(x)} = \frac{I^2}{\sigma(x) S^2(x)} \tag{11.1}$$

By supposing w to be a linear function of x,

$$w = \alpha x + \beta \tag{11.2}$$

$S(x)$ should be expressed as

$$S(x) = \frac{I}{\sqrt{\sigma(x)(\alpha x + \beta)}} \tag{11.3}$$

This relationship can be used for generating the temperature gradient close to linear.

Due to rather small $\mathrm{d}pK/\mathrm{d}T$ values, this method is considered to be useful for the creation of some narrow pH gradients only. In this connection, the idea of using some special buffers, e.g., which are considered in Section 2.4 seems to be very promising. Such buffers could be more temperature sensitive, thus, with an acceptable temperature variance, an essentially wider pH range can be covered.

11.2.3. Gradient in dielectric constant

11.2.3.1. Dielectric constant influence on buffer dissociation constant

An essentially different approach to the problem of stable pH gradient creation was proposed by Troitzki and co-workers [15,16]. They suggested that the dielectric properties of aqueous media can be modified by a gradient of organic solvents, with the dielectric constant variance causing the appropriate pK deviation.

The authors came to the conclusion that, although many organic solvents can be used for the pH gradient creation by an appropriate concentration gradient, the mechanism of the pH gradient formation may be different for different buffer mixtures. Further, this group developed the so-called borate–polyol systems (which includes boric acid and some polyoxy compound, glycerol, for example). Such systems have demonstrated a rather good efficiency and, to a certain extent, can be considered as an alternative to CA, at least for preparative purposes [17,18].

11.2.3.2. Gradient of electric field coupled with dielectric constant gradient

Giddings and Dahlgren put forward the idea of employing the dielectrical force [19]. In a strongly non-uniform electric field, uncharged species with a high dielectric constant are to migrate to the high field regions. A secondary gradient in the dielectric constant being imposed results in each species trying to reach its own equilibrium point where the dielectric constant of the medium equals its own. Thus one obtains separation according to the differences in the dielectric constant. As it was correctly pointed out by the authors, this method would be applicable only to rather large particles.

11.3. INTRINSIC LIMITS OF IEF

At the present state of the art, immobilised pH gradients are the very tool which can provide maximum resolution with respect to the pI difference of the components to be separated. Natural pH gradients do not possess such a potential. In addition to those disadvantages usually referred to in manuals, there is some other principal restriction. Namely, the introduction of a sample may completely reorganise the initial concentration distribution of the species forming the gradient (carrier-ampholytes), and the final separating properties of any CA-system are defined, unfortunately, by the concrete chemical composition, rather than by the resulting pH profile.

This problem no longer exists with the technology of the immobilised pH-gradient invention; moreover, it was supposed that with time 'all limits will be broken', since one can create theoretically as narrow a pH gradient as is desired. At first glance, this gives us a possibility of separating anything we want. In reality, it proved that any narrow immobilised gradient by itself is unable to provide an essential progress in resolution (although a higher voltage gradient is applied), since individual peaks become wider and wider.

One can put a question: 'Is there any limit in the resolving capability for IEF, and if any, what is the minimum pI difference that can be detected?' Expression of the standard deviation (see Section 6.3 in Chapter 6) does not put any restrictions, provided we take as narrow a gradient as we want, and we can apply any high voltage also.

Such equations are derived from two strong assumptions (as well as some others but not so essential). The first one is connected with the fact that pH is presumed to be some continuous function of distance x.

Obviously, at very small distances, the pH monotony can be violated. Indeed, in the vicinity of charge centres, the concentration of counterions should be higher. Taking into account that the number of immobilised charges per unit volume is:

$$N = CN_{\mathrm{A}}, \tag{11.4}$$

one can easily evaluate the average volumes per one charge by inverting the value, and taking the cube root we come to the characteristic distance (r) between the two nearest charges

$$r = \frac{1}{\sqrt[3]{CN_{\mathrm{A}}}} \tag{11.5}$$

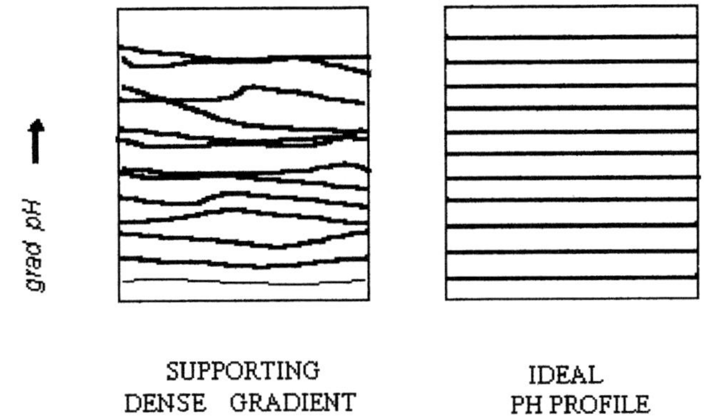

Fig. 11.5. Non-uniform pH profile in a real IPG gel. Lines in the left panel are the ones with a constant pH value (schematically). An ideal pH profile is shown in the right panel.

Here, N_A is the Avogadro number, C is the immobiline concentration.

For micromole concentrations, the r value approaches 10^{-6} cm (10^{-2} μm), and thus the value of at least 0.1 μm should be taken as one natural limit.

The other strong assumption is the way of mobility approximation; namely, by deriving Eq. (6.1) the mobility of the focussing sample was supposed to be unique, that is, the reactions of mutual interconversion were essentially ignored [20]. In reality, any amphotere compound being dissolved results in some number of microscopic states, each having its own mobility (μ) and lifetime (τ). This consideration gives us the second limit on the final zone width; the latter should be comparable with:

$$r \approx \mu E \tau \tag{11.6}$$

Since mobility values have an order of 10^{-4} cm^2 V^{-1} s^{-1} at high voltage (1000 V/cm), the velocity should be around 0.1 cm/s. Depending on the lifetime (τ), in some cases r can reach rather high values, as it follows from Eq. (11.6), although this expression does not take into account the degree of dissociation.

Coming back to the immobilised pH gradients, it is possible to say that, in reality, it has been proved that any narrow immobilised gradient by itself is unable to provide an essential progress in resolution (although a higher voltage gradient is applied), since the individual peaks become wider and wider. It is easy to understand the origin, if we remember the technology of such gradient preparation. For this purpose, a gradient mixture is used (with two solutions: one light and one dense). Thus a concentration gradient of the so-called immobilines is created with the help of an accessorial gradient of a neutral substance. When the glass form is filling with such a device, a set of hydrodynamic instabilities arises; so the final pH gradient profile is far from the ideal one, see Fig. 11.5, left panel. These are the lines of the equal pH value, with the fixed delta-pH step. So the potential of the IEF method has not been completely realised at present, and let us hope that this task of the resolving possibility increasing up to the 'natural limits' will be solved in the near future.

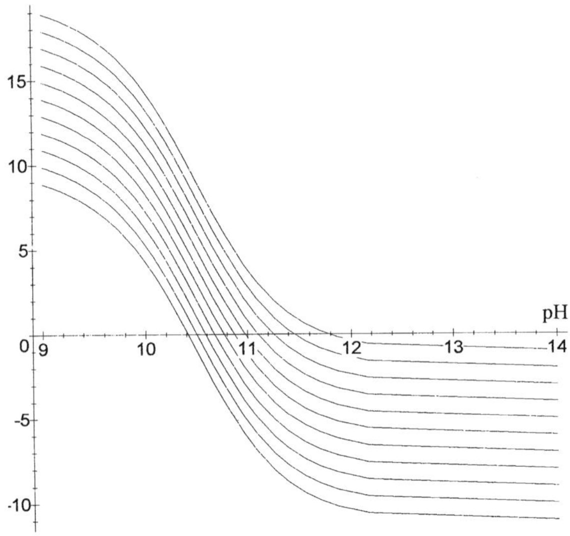

Fig. 11.6. Titration curves for the polypeptide with twenty basic groups (pK = 10) after progressive addition of a negative charge unit. The upper curve corresponds to the unmodified polypeptide (pI = 11.78).

11.4. MICROHETEROGENEITY OF PROTEINS AND OTHER BIOPOLYMERS

The question of microheterogeneity is connected with the resolving capacity limits. Many proteins, for example, show microheterogeneity when analysed by isoelectric focussing; there are many reasons for this (see, e.g. Williamson [21]). The term microheterogeneity is often used to emphasise the impossibility of detecting the molecular weight differences of the appropriate components of the IEF-spectrum (by SDS-electrophoresis).

The single component of the IEF-spectrum (band or spot), in its turn, may not always correspond to an individual substance. So, there exists a problem of interpreting the IEF-pattern obtained. On the other hand, one can put a question: 'What could we see, should the resolving capacity of the IEF method increase?'

11.4.1. pH shift due to single modifications of ionogenic groups

In frames of a simplified approach which assumes an independent dissociation process (Chapters 1 and 2) for any modification of the protein changing the number of ionogenic groups in the molecule (e.g. a single positive charge can be eliminated due to deamidation of asparagine or glutamine, or a negative charge can be removed with the loss of the charged sugar residue in glycoprotein), the appropriate pH-shift may be easily calculated. In Fig. 11.6, the fragments of the appropriate titration curves for the polypeptide are given (20 basic groups with pK = 10.5, plus N- and C-terminative

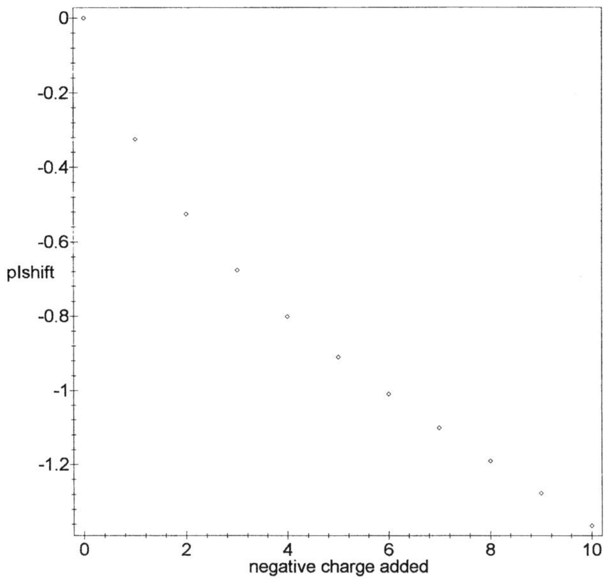

Fig. 11.7. pH-shift vs. the introduced charge for the system given in Fig. 11.5.

groups) before (upper line) and after a progressive addition of a negative-charge unit (up to ten). The initial pI value of this polypeptide is rather high (around 11.8); each negative-charge addition causes its appropriate negative shift (toward neutrality). Such a shift is higher farther from the pK value of the buffering groups (Fig. 11.7).

In order to obtain an approximate value of the pH shift, instead of solving directly the electroneutrality equation, one can use some simplified expression taking into account the influence of those ionogenic groups which are situated closer to the isoelectric point:

$$\theta = \frac{\alpha[n + q(1 + \alpha)]}{\alpha n - q(1 + \alpha)} \tag{11.7}$$

Here θ is used to designate the hydrogen concentration ratio after and before the modifications corresponding to the pI values (so $\theta = [H^+]_1/[H^+]_0$, thus $\Delta pH = -\log\theta$); n is the number of 'buffering' groups, $\alpha = K/[H^+]_0$, q is the introduced charge. The value of the pI shift, obviously, depends on the position of buffering groups pK and the number of these groups, Fig. 11.8. For rather big biopolymers, the pI becomes small, but the latter is still considerable, provided the value of the introduced charge is comparable with a unit.

11.4.2. Multiple one-type modifications

One can imagine two proteins with the same number of the modified groups, but with a positional difference of some of them (at multiple postsynthetic modification of one type). For the protein having P groups which are capable of being modified, we can observe the IEF-pattern, which contains a number of bands coinciding with the number

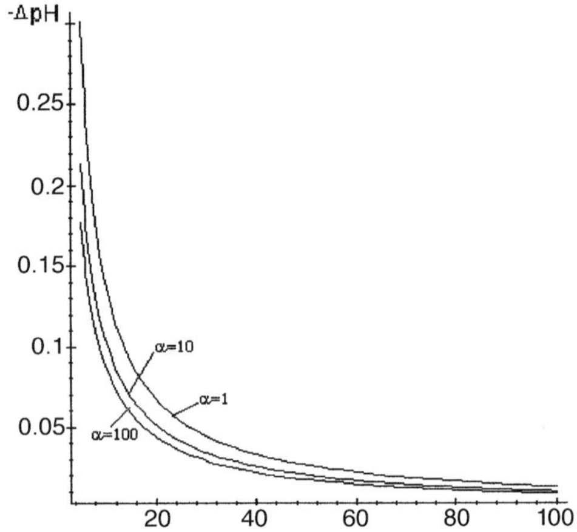

Fig. 11.8. pH-shift vs. the number of the buffering groups, at different α-parameter values.

of all possible values of the modified groups differing from each other sequentially in one unit (e.g. this is the situation with the pI standards used in 2D electrophoresis [22]). Nevertheless, there exists a possibility of distributing F 'charges' between P centres by the $Cn^k = P!/[F!(P - F)!]$ ways. F is the number of modified groups. Thus there are reasons to suppose that some of the detected bands may, in their turn, be the result of the superposition of several protein isoforms. In frames of the approach of an independent dissociation, one can speak about some differences in pKs for the same amino acids in a different surrounding. Such pK difference (several tenths of pH or even smaller [23]) can provide very small additional charge values. Thus the appropriate pI-shift should be out of the present limits of the IEF technique.

In reality, the situation is considerably more complex. Not only the approach of an independent dissociation is an approximation, but the microscopic forms corresponding to the molecule are in mutual kinetic equilibrium as well (Chapter 3). In their turn, these complexes may stay in mutual interconversion. The latter processes can result in additional conformational changes. An increase in resolution will provide, also, an opportunity of studying the kinetics of such processes.

11.5. REFERENCES

1. B. Bjellqvist, K. Ek, P.G. Righetti, E. Gianazza, A. Gorg, R. Westermeier and W. Postel, J. Biochem. Biophys. Methods, 6 (1982) 317–339.
2. H. Svensson, Acta Chem. Scand., 15 (1961) 325–341.
3. A. Kolin, J. Chem. Phys., 22 (1954) 1628–1629.
4. A. Kolin, J. Chem. Phys., 23 (1955) 407–410.
5. A. Kolin, Proc. Natl. Acad. Sci., 41 (1955) 101–110.
6. O. Vesterberg, US Patent 3485736, 1969.
7. H. Svensson, Arch. Biochem. Biophys. Suppl., 1 (1962) 132–140.
8. O. Vesterberg and H. Svensson, Acta Chem. Scand., 20 (1966) 220–824.
9. V.G. Babskii, M.Y. Zhukov and V.I. Yudovich, Mathematical Theory of Electrophoresis, Consultants

Bureau, New York, 1989.
10. S. Hjerten, Chromatogr. Rev., 9 (1965) 122–219.
11. P. Lundal and S. Hjerten, Ann. N.Y. Acad. Sci., 209 (1973) 94–111.
12. S. Luner and A. Kolin, US Patent 3664939, 1972.
13. S. Luner and A. Kolin, Proc. Natl. Acad. Sci., 66 (1970) 889–893.
14. J.B. Pawliszyn, US Patent 5759370.
15. G.V. Troitzki, B.P. Zavialov and V.M. Abramov, Dokl. Akad. Nauk, 214 (1974) 955–958.
16. G.V. Troitzki, B.P. Zavialov and I.F. Kiriukhin, Bull. Exp. Biol. Med., 75 (1974) 118–120.
17. Prof. S. Peltek, personal communication.
18. G.V. Troitzki and G.Y. Agitzki, Isoelectric Focusing in Natural and Artificial pH Gradients, Naukova Dumka, Kiev, 1984.
19. J.K. Giddings and K. Dahlgren, Sep. Sci., 6 (1971) 345–356.
20. A. Stoyanov and P.G. Righetti, J. Biochem. Biophys. Methods, 46 (2000) 21–30.
21. A.R. Williamson, M.R. Salaman and H.W. Kreth, Ann. N.Y. Acad. Sci., 209 (1973) 210–222.
22. N.L. Anderson and B.J. Hickman, Anal. Biochem., 93 (1979) 312–320.
23. C. Cantor and P. Shimmel, Biophysical Chemistry, Part I, W.H. Freeman and Company, San Francisco, CA.

PART II

Methodology

CHAPTER 12

Conventional Isoelectric Focussing in Gel Slabs and Capillaries and Immobilised pH Gradients

CONTENTS

12.1. INTRODUCTION

This part of the book is organised as follows: we will first treat conventional isoelectric focussing in soluble, carrier ampholyte buffers, as originally envisaged by

Svensson–Rilbe [1]. In this section, we will give a brief historical survey of its evolution, from a preparative-scale approach, to the extremely popular gel-slab version. This part will also contain an ample description of the properties of CA buffers, especially in regard to their buffering capacity (β) and conductivity, since these basic concepts have been found to be fundamental in capillary zone electrophoresis (CZE). We will then proceed into the description of immobilised pH gradients (IPG), the novel version of focussing launched in 1982 [2]. In a third part, we will describe IEF in a capillary format, the rising star in Separation Science [3–5]. As a final part, we will describe CZE separations exploiting isoelectric buffers as a background electrolyte. Although this is not a focussing technique per se, it is the natural evolution of the know-how developed in IEF, and it appears to hold a unique separation potential in CZE. Since this zonal electrophoresis in isoelectric buffers relies heavily on the properties of CAs, the importance of the theoretical treatment offered at the beginning on the physico-chemical properties of such buffers will be here appreciated.

12.1.1. A brief historical survey

In 1960–1962 as we, as freshmen in the University, went down to the beach to the tune of *Surfin' Safari* (Beach Boys), others spent their time trying to surf the rough seas of Separation Science, in the hope of some major developments. A major breakthrough was indeed the discovery of conventional isoelectric focussing (IEF) in carrier ampholytes (CA, soluble, amphoteric buffers) reported by Svensson–Rilbe in a series of now classical articles [6–8]. At just about the same time, Meselson et al. [9] described isopycnic centrifugation (IPC), a related, high-resolution technique, in fact another member of a family called by Kolin [10] isoperichoric focussing. Unlike conventional chromatographic and electrophoretic techniques available up to those times, which could not provide any means for avoiding peak decay during the transport process, IEF and IPC had built-in mechanisms opposing entropic forces trying to dissipate the zone. As the analyte reached an environment (the 'perichoron', in Greek) in which its physico-chemical parameters were equal (iso) with those of the surroundings, it focussed, or condensed, in an ultrathin zone, kept stable and sharp in time by two opposing force fields: diffusion (tending to dissipate the zone) and external fields (voltage gradients, in IEF, or centrifugal fields, in IPC) forcing the 'escaping' analyte back into its 'focussing' zone. It was truly a magic event in Separation Science, by which all of us, general practitioners in the field, and humble Clark Kents in the lab, all at once felt (and acted) like Superman.

According to the original idea of Svensson–Rilbe, CA-IEF was born as a preparative technique, in vertical, hollow columns exploiting two co-linear gradients: a sucrose density gradient acting as an anticonvective medium and a pH gradient generated by the current during the focussing step [11]. A typical experiment took several days to complete and the subsequent analyses of fractions eluted from the gradients were tedious and laborious. It was the advent of gel IEF [12–14] that rendered the technique so popular. The evolution of CA-IEF can be followed in a series of meetings, starting with the cornerstone symposium in New York [15] soon followed by Glasgow [16],

Milan [17], Hamburg [18] and Cambridge (MA, USA) [19]. With the latter meeting, the specific interest in IEF and isotachophoresis (ITP, note that ITP is also a steady-state technique with this proviso: in IEF all zones have identical mobilities all equal to zero, whereas in ITP all zones acquire the same, non-zero, velocity) moved to the general field of electrophoresis (in which, however, IEF still played the role of Prima Donna); there followed a series of meetings in Munich [20], in Charleston [21], in Athens [22], in Tokyo [23], in Göttingen [24], in London [25] and in Copenhagen [26]. With the latter meeting, the publication of Proceedings ceased, since meanwhile, the official journal of the International Electrophoresis Society was launched, i.e. the journal Electrophoresis, which appeared to be the proper forum for such events. Nevertheless, this collection of Proceedings is an interesting means for following the evolution of IEF and, later on, of general electrophoretic techniques. Some ideas were far-reaching, like the one of a 'rain-box', proposed by Hinckley [27]. To quote him verbatim: "at its simplest, the rain-box principle is a droplet-forming disperser or nozzle, with travel of the droplet from the donor buffer to the open recipient buffer pool through an insulating fluid". The problems with the rain-box? "If ionization or tracking breakdown does occur... then the rain-box will become a lightning-box"! Some more ideas from the same author [27]: "this discussion has left out of account all kinds of bucket fountains, archimedean screws, and other gadgets which await the ingenuity of a latter-day Hero of Alexandria, and are not necessarily the worse for a debt to the limited means of archaic or amateur engineering". That reminds us that the first inventor, perhaps, of the rain-box was Jupiter, who mutated himself into 'golden rain' to be able to visit Danae in her prison. As a result of this 'heavy rain', Perseus was born; it is not known to the present authors, however, if Hinckley's brain-child ever came to see the light. The same Hinckley [28], at another of these meetings, proposed a revolutionary, modular apparatus design, thus described in his terse and crystal-clear English: "starting with a suitable vocabulary of interfacing modules, apparatus for any method can be assembled. By interpolating column-holding separation modules, and optional separand-introduction, in-run fraction-elution or detection modules, any type or length of interpolated module stack is possible. Non-shunting vessel plumbing allows discontinuous buffer methods (electrofocussing, isotachophoresis), and separate optional flow circuits with individual removal of dense and light electrode liquors, otherwise segregated by the field-shaped removable labyrinths".

It was in 1975 that IEF took another, important turn: in that year O'Farrell [29] introduced two-dimensional electrophoresis (2-DE) and demonstrated that, by sequentially running IEF in the first dimension, followed by SDS electrophoresis at right angles, >1100 individual polypeptides could be resolved in an *Escherichia coli* lysate. These 2-D maps became the nightmare of chromatographers: they still had to accept the hard reality that, even in the best gas-chromatographic separation, it was hard to resolve barely 100 distinct peaks [30]. Soon 2-DE excited grandiose projects, like the Human Protein Index System of Anderson and Anderson [31], who started the far-reaching goal of mapping all possible phenotypes expressed by any and all different cells in our organism. An Herculean task, if one considers that there might be close to 75,000 such phenotypes expressed (assuming a total of ca. 250 differentiated cells in our body, each expressing >3000 polypeptides, of which 300 specific to a given cell; J. Celis, personal communication in Seattle, March 1997). This started a series of meetings of the 2-DE group, particularly strong in the field of clinical chemistry [32,33]. A number of books were devoted to this 2-DE issue [34,35] and a series of meetings were started on this special topic [36,37]. Today, 2-DE has found a proper forum in the journal Electrophoresis, which began hosting individual papers dealing with variegate topics in 2-D maps. Starting with 1988, Electrophoresis launched special issues devoted to 2-D maps, not only in clinical chemistry and human molecular anatomy, but in fact in every possible living organism and tissue; the first one, in reality, being dedicated to plant proteins [38]. Soon, a host of such 'Paper Symposia' appeared, collecting databases on any new spots of which a sequence and a function could be elucidated [37,39–58]. This deluge seems now to have subsided a bit, although there is no way to predict if and when such prolific 'guest editors' as Julio Celis will strike again. Also this collection of 'Paper Symposia' is worth a perusal, since it is a gold mine on new information and on novel evolutionary steps on the IEF technique (such as new solubilisers, new staining procedures, sample pre-treatment before IEF and the like).

At the end of this excursus, one might ask how vast is the literature on IEF: we posed the same question to the World Wide Web by interrogating Med Line (search from 1966 to present-day). Here is the answer: plain IEF, a total of 19,663 articles; just 2-D maps, another 3968 articles. The combination of the two (excluding duplicates) comes to a total of 23,631 papers. Considering that perhaps Med Line does not cover so thoroughly all possible journals in the scientific arena, it is not unreasonable to think that the total should be close to 30,000 articles. So, it appears that the followers of Svensson have grown from a trickle to an army.

12.2. CONVENTIONAL ISOELECTRIC FOCUSSING IN AMPHOTERIC BUFFERS

12.2.1. General considerations

All fractionations which rely on differential rates of migration of sample molecules, for example along the axis of a chromatographic column or along the electric field lines in electrophoresis, generally lead to concentration bands or zones which are essentially always out of equilibrium. The narrower the band or zone the steeper the concentration gradients and the greater the tendency of these gradients to dissipate spontaneously. This dissipative transport is thermodynamically driven; it relates to the tendency of entropy to break down all gradients, to maximise dilution and, during this process, to thoroughly mix all components [30]. Most frequently, entropy exerts its effects via diffusion, which causes molecules to move down concentration gradients and so produces band broadening and component intermixing.

The process of isoelectric focussing (IEF) in carrier ampholytes (CA) [6–8] and in immobilised pH gradients (IPG) [59] provides an additional force which counteracts CAs diffusion and so maximises the ratio of separative to dissipative transports. This substantially increases the resolution of the fractionation method. The sample focusses towards its isoelectric point (p*I*) driven by the voltage gradient and by the shape of

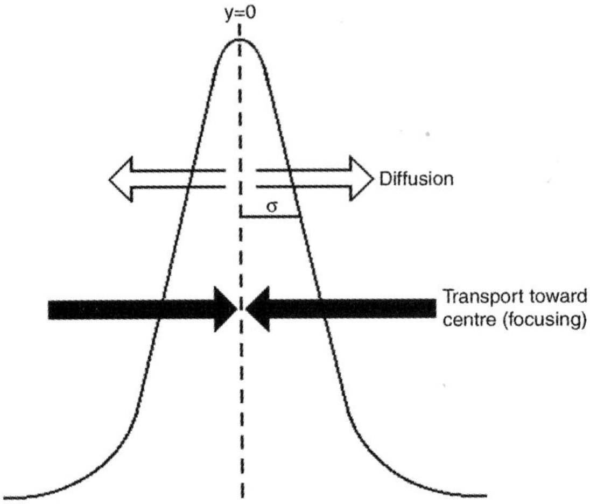

Fig. 12.1. Illustration of the forces acting on a condensed zone in isoelectric focussing (IEF). The focussed zone is represented as a symmetric Gaussian peak about its focussing point (pI; $y = 0$). Migration of sample towards the pI position is driven by the voltage gradient and by the slope of the pH gradient. σ is the standard deviation of the peak. (By courtesy of Dr. O. Vesterberg.)

the pH gradient along the separation axis (Fig. 12.1). The separation can be optimised by using thin or ultrathin matrices (0.5 mm or less in thickness; [60,61]) and by applying very low sample loads (as permitted by high-sensitivity detection techniques, such as silver and gold staining, radioactive labelling, immuno-precipitation followed by amplification with peroxidase- or alkaline phosphatase-linked secondary antibodies).

12.2.1.1. The basic method

IEF is an electrophoretic technique by which amphoteric compounds are fractionated according to their pIs along a continuous pH gradient [62]. Contrary to zone electrophoresis, where the constant (buffered) pH of the separation medium establishes a constant charge density at the surface of the molecule and causes it to migrate with constant mobility (in the absence of molecular sieving), the surface charge of an amphoteric compound in IEF keeps changing, and decreasing, according to its titration curve, as it moves along a pH gradient until it reaches its equilibrium position, i.e. the region where the pH matches its pI. There, its mobility equals zero and the molecule comes to a stop.

The gradient is created, and maintained, by the passage of an electric current through a solution of amphoteric compounds which have closely spaced pIs, encompassing a given pH range. The electrophoretic transport causes these carrier ampholytes (CA) to stack according to their pIs, and a pH gradient, increasing from anode to cathode, is established. At the beginning of the run, the medium has a uniform pH, which equals the average pI of the CAs. Thus most ampholytes have a net charge and a net mobility. The most acidic CA moves toward the anode, where it concentrates in a zone whose pH equals its pI, while the more basic CAs are driven toward the cathode. A less acidic ampholyte migrates adjacent and just cathodal to the previous one and so on, until all the components of the system reach a steady state. After this stacking process is completed,

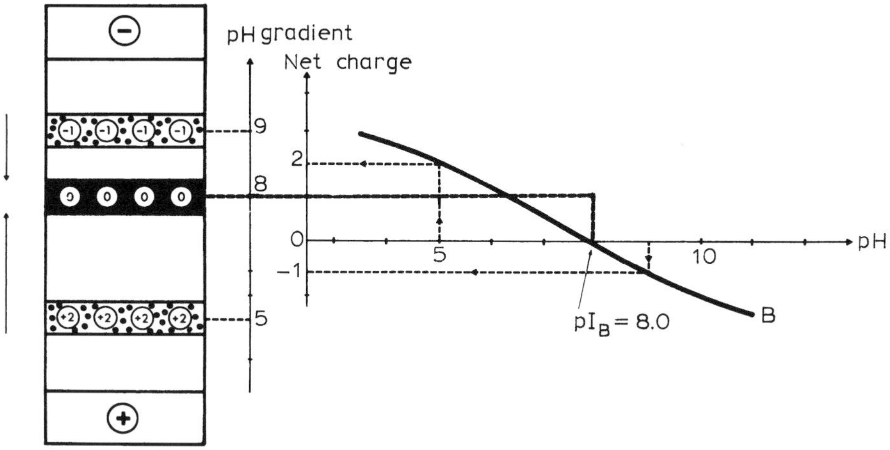

Fig. 12.2. The principle of IEF in carrier ampholyte (CA) buffers. When a protein macroion is applied to a pre-focussed gel slab, both above and below its p*I* (here p*I* = 8.0), it acquires negative and positive charges, according to its titration curve (shown on the right). The two fronts then merge at pH = p*I*, where the net charge is zero. (By courtesy of LKb Produkter AB.)

some CAs still enter zones of higher, or lower, pH by diffusion where they are not any longer in isoelectric equilibrium. But as soon as they enter these zones, the CAs become charged and the applied voltage forces them back to their equilibrium position. This pendulum movement, diffusion versus electrophoresis, is the primary cause of the residual current observed under isoelectric steady-state conditions. Finally, as time progresses, the sample protein molecules also reach their isoelectric point. Fig. 12.2 shows this focussing process for a protein with a p*I* = 8.0. If this macroion is applied simultaneously above and below its p*I*, its negatively and positively charged species (whose net charge is defined by the titration curve shown at the right), respectively, migrate towards each other till they fuse, or merge (or focus, tout court) at the p*I* zone, having zero net charge.

12.2.1.2. Applications and limitations

The technique only applies to amphoteric compounds and more precisely to good ampholytes with a steep titration curve around their p*I*, conditio sine qua non for any compound to focus in a narrow band. This is very seldom a problem with proteins but it may be so for short peptides, that need to contain at least one acidic, or basic, amino acid residue, in addition to the $-NH_2$ and –COOH termini. Peptides which have only these terminal charges are isoelectric over the entire range of approximately pH 4 and pH 8 and so do not focus. Another limitation with short peptides is encountered at the level of the detection methods: CAs are reactive to most peptide stains. This problem may be circumvented by using specific stains, when appropriate [63,64], or by resorting to immobilised pH gradients (IPG) which do not give background reactivity to ninhydrin and other common stains for primary amino groups (e.g. dansyl chloride, fluorescamine) [65].

In practice, notwithstanding the availability of CAs covering the pH 2.5–11 range, the practical limit of CA-IEF is in the pH 3.5–10 interval. Since most protein p*I*s cluster between pH 4 and 6 [66], this may pose a major problem only for specific applications.

When a restrictive support like polyacrylamide (PAA) is used, a size limit is also imposed for sample proteins. This can be defined as the size of the largest molecules which retain an acceptable mobility through the gel. A conservative evaluation sets an upper molecular mass limit of about 750,000 when using standard techniques. The molecular form in which the proteins are separated strongly depends upon the presence of additives, such as urea and/or detergents. Moreover, supramolecular aggregates or complexes with charged ligands can be focussed only if their K_d is lower than 1 μM and if the complex is stable at pH = p*I* [67]. An aggregate with a higher K_d is easily split by the pulling force of the current.

12.2.1.3. Specific advantages

(a) IEF is an equilibrium technique; therefore the results do not depend (within reasonable limits) upon the mode of sample application, the total protein load or the time of operation.

(b) An intrinsic physico-chemical parameter of the protein (its p*I*) may be measured.

(c) IEF requires only a limited number of chemicals, is completed within a few hours, is less sensitive than most other techniques to the skill (or lack of it) of the operator.

(d) IEF allows excellent resolution of proteins whose p*I*s differ by only 0.01 pH units (with immobilised pH gradients, up to about 0.001 pH units); the protein bands are very sharp due to the focussing effect.

12.2.1.4. Carrier ampholytes

Table 12.1 lists the general properties of carrier ampholytes, i.e. of the amphoteric buffers used to generate and stabilise the pH gradient in IEF. The fundamental and performance properties listed in this table are usually required for a well-behaved IEF system, whereas the 'phenomena' properties are in fact the drawbacks or failures

TABLE 12.1

PROPERTIES OF CARRIER AMPHOLYTES

Fundamental 'classical' properties	
1	Buffering ion has mobility of zero at p*I*.
2	Good conductance.
3	Good buffering capacity.
Performance properties	
1	Good solubility.
2	No influence on detection systems.
3	No influence on sample.
4	Separable from sample.
'Phenomena' properties	
1	'Plateau' effect (i.e. drift of the pH gradient).
2	Chemical change in sample.
3	Complex formation.

inherent to the technique. For instance, the 'plateau effect' or 'cathodic drift' is a slow decay of the pH gradient with time, whereby, upon prolonged focussing at high voltages, the pH gradient with the focussed proteins drifts towards the cathode and is eventually lost in the cathodic compartment. There seems to be no remedy to this problem (except from abandoning CA-IEF in favour of the IPG technique), since there are complex physico-chemical causes underlying it, including a strong electroosmotic flow generated by the covalently bound negative charges of the matrix (carboxyls and sulphate in both polyacrylamide and agarose) (as reviewed in [68]). In addition, it appears that basic CAs may bind to hydrophobic proteins, such as membrane proteins, by hydrophobic interaction. This cannot be prevented during electrophoresis, whereas ionic CA–protein complexes are easily split by the voltage gradient [69].

In chemical terms, CAs are oligo-amino, oligo-carboxylic acids, available from different suppliers under different trade names. There are three basic synthetic approaches: Vesterberg's approach, which involves reacting different oligoamines (tetra-, penta- and hexa-amines) with acrylic acid [70]; the Söderberg et al. [71] synthetic process, which involves the copolymerisation of amines, amino acids and dipeptides with epichlorohydrin, and the Pogacar–Grubhofer approach, which utilises ethyleneimine and propylenediamine for subsequent reaction with propanesultone, sodium vinylsulphonate and sodium chloromethyl phosphonate ([72,73]). Accordingly, there are 3 types of products on the market: the Amersham-Pharmacia Biotech Ampholines (formerly from LKB-Produkter AB) and Bio-Rad Biolytes which belong to the first class; the Amersham-Pharmacia Biotech Pharmalytes which should be listed in the second class, and the Novex-Servalyt Ampholytes and Genomic Solutions pH 3–10 Ampholytes, which should be classified into the third category.

The wide-range synthetic mixtures (pH 3–10) contain hundreds, possibly thousands, of different amphoteric chemicals having p*I*s evenly distributed along the pH scale. Since they are obtained by different synthetic approaches, CAs from different manufacturers are bound to have somewhat different p*I*s. Thus, if higher resolution is needed, particularly for two-dimensional maps of complex samples, we suggest using blends of the different commercially available CAs. A useful blend is 50% Pharmalyte, 30% Ampholine and 20% Biolyte, by volume).

CAs from any source should have an average molecular mass of about 750 (size interval 600–900, the higher M_r referring to the more acidic CA species) [74]. Thus CAs should be readily separable (unless they are hydrophobically complexed to proteins) from macromolecules by gel filtration. Dialysis is not recommended due to the tendency of CAs to aggregate. Salting out of proteins with ammonium sulphate seems to completely eliminate any contaminating CAs.

A further complication arises from the chelating effect of acidic CAs, especially towards Cu^{2+} ions, which may inactivate some metallo-enzymes [75]. In addition, focussed CAs represent a medium of very low ionic strength (less than 1 mequiv./l at the steady state) [76]. Since the isoelectric state involves a minimum of solvation, and thus of solubility, for the protein macroion, there is a tendency, for some proteins (e.g. globulins) to precipitate during the IEF run near their p*I* position. This is a severe problem in preparative runs. In analytical procedures it can be minimised by reducing the total amount of sample applied.

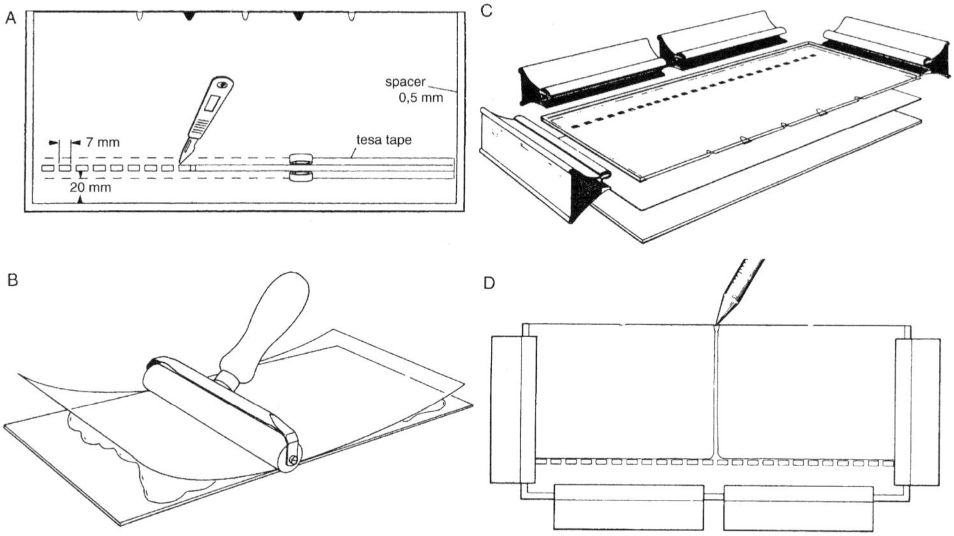

Fig. 12.4. Preparation of the gel cassette. (A) Preparation of the slot former: onto the cover plate (bearing the rubber gasket U-frame) is glued a strip of Tesa tape out of which rectangular tabs are cut with a scalpel. (B) Application of the Gel Bond PAG film to the supporting glass plate. (C) Assembling the gel cassette. (D) Pouring the gelling solution in the vertically standing cassette using a pipette. (By courtesy of LKB Produkter AB.)

with γ-methacryl-oxypropyl-trimethoxy-silane is the most reliable reactive substratum and is the most suitable for autoradiographic procedures [78] (see Protocol 1 for the procedure). It is also the cheapest of such supports: dried-out gels can be removed with a blade and then a brush with some scrubbing powder. Unreacted silane may be hydrolysed by keeping the plates in Clorox for a few days. This step is unnecessary, however, if they have to go through successive cycles of silanisation. The glass plates used as a support should not be thicker than 1 to 1.2 mm. On the other hand, thin plastic sheets designed to bind polyacrylamide gels firmly (e.g. Gel Bond PAG by Marine Colloids, PAG foils by Amersham-Pharmacia, Gel Fix by Serva) are more practical if the records of a large number of experiments have to be filed, or when different parts of the gel need to be processed independently (e.g. the first step of a two-dimensional separation or a comparison between different stains). The plastic sheet is applied to a supporting glass plate and the gel is cast onto this. The binding of the polyacrylamide matrix to these substrata, however, is not always stable and so care should be taken in using them, especially for detergent-containing gels and when using aqueous staining solutions. For good adherence, the best procedure is to cast 'empty' gels (i.e. polyacrylamide gel lacking CAs), wash and dry them, and then reswell with the solvent of choice (see Section 12.3.2.2).

12.2.2.2.2. Spacer U-gaskets of any thickness, between 0.2 and 5 mm can be cut from rubber sheets (para-, silicone- nitrile-rubber). For thin gels, a few layers of Parafilm (each about 120 μm thick) can be stacked and cut with a razor blade. The width of such

Protocol 1. Siliconising glass plates

Equipment and reagents
- Binding silane (γ-methacryloxy-propyl-trimethoxy-silane) prepared by Union Carbide (available through Pharmacia, Serva or LKB) or Repel-silane (dimethyl-dichloro-silane) available from suppliers Merck, Serva, etc.
- IEF gel supporting plates

Method
(1) Binding silane
Two alternative procedures are available.
Either
1. Add 4 ml of silane to 1 litre of distilled water adjusted to pH 3.5 with acetic acid.
2. Leave the plates in this solution for 30 min.
3. Rinse with distilled water and dry in air.

or
1. Dip the plates for 30 sec in a 0.2% solution of silane in anhydrous acetone.
2. Thoroughly evaporate the solvent using a hair drier.
3. Rinse with ethanol if required.

In either case, store the siliconised plates away from untreated glass.

(2) Repel-Silane
1. Swab the glass plates with a wad impregnated with a 2% (w/v) solution of Repel-silane in 1,1,1-trichloroethane.
2. Dry the plates in a stream of air and rinse with distilled water.

U-gaskets should be about 4 mm. In addition, cover plates with a permanent frame are commercially available (# 2117-901 from Amersham-Pharmacia). It is the same for a plastic tray with two lateral ridges for horizontal polymerisation (Bio-Rad # 170-4261 and 170-4258; see Fig. 12.5A). A similar device may be home-made using Dymo tape strips, which are 250 μm thick to form the permanent spacer frame. Mylar foil strips or self-adhesive tape may be used as spacers for 50–100 μm thick gels. Rubber- or tape-gaskets should never be left to soak in soap (which they absorb) but just rinsed and dried promptly.

12.2.2.2.3. Cover plate Clean glass, glass coated with dimethyl dichloro silane (Repel Silane; see Protocol 1) or a thick Perspex sheet are all suitable materials for the cover plate. If you wish to mould sample application pockets into the gel slab during preparation, attach Dymo tape pieces to the plate, or glue small Perspex blocks to the plate with drops of chloroform. Perspex should never be exposed unevenly to high temperatures (for example by being rinsed in running hot water) because it bends even if cut in thick slabs.

12.2.2.2.4. Clamps Clamps of adequate size and strength should be chosen for any gel thickness. Insufficient pressure may result in leakage of the polymerising solution. The pressure of the clamps must be applied on the gasket, never inside it.

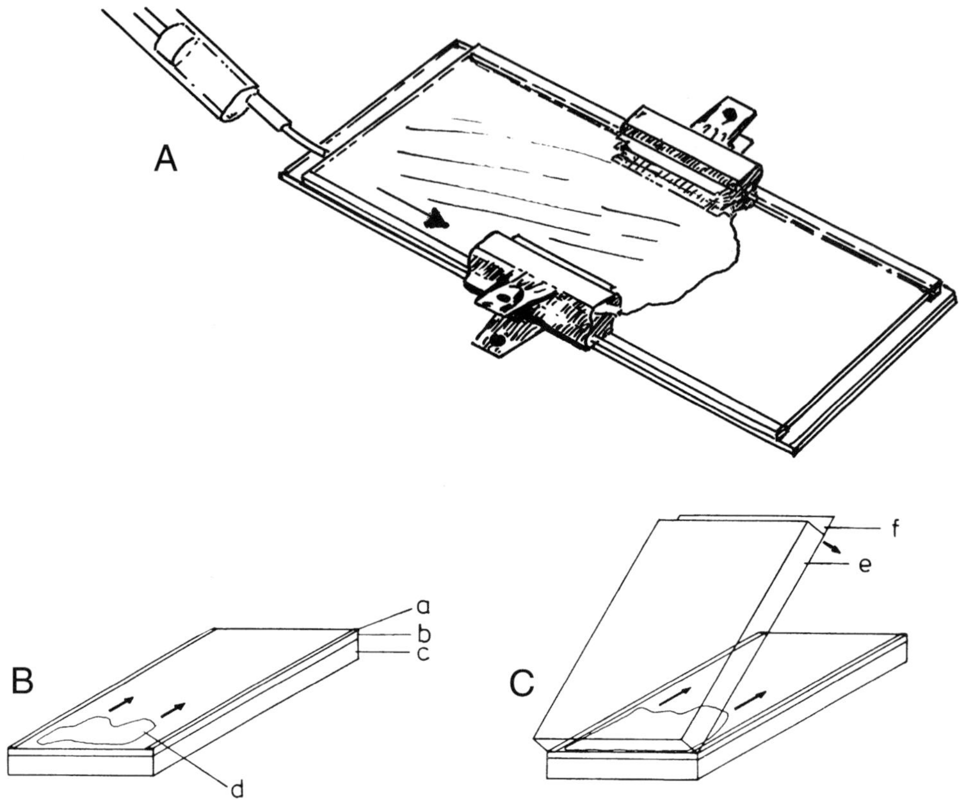

Fig. 12.5. Casting of thin gel plates. (A) By capillary filling of an horizontally placed cassette. (B and C) By the 'flap' technique. a = spacer strips; b = silanised glass plate or polyester film; c = glass base plate; d = polymerisation mixture; e = glass cover plate; f = cover film. A: by courtesy of LKB; B and C: with permission from Ref. [20].

12.2.3. The Polyacrylamide gel matrix

12.2.3.1. Reagents

Stocks of dry chemicals (acrylamide, Bis, ammonium persulphate) may be kept at room temperature, provided they are protected from moisture by being stored in air-tight containers. Very large stocks are better sealed into plastic bags, together with Drierite (Mercks), and stored in a freezer. TEMED stocks should also ideally be kept in a freezer, in an air-tight bottle or better, under nitrogen. Avoid contaminating acrylamide solutions with heavy metals, which can initiate its polymerisation.

Acrylamide and Bis for IEF must be of the highest purity to avoid poor polymerisation and strong electroosmosis resulting from acrylic acid. Bis is more hydrophobic and more difficult to dissolve than acrylamide, so start by stirring it in a little amount of luke-warm distilled water (the solution process is endothermic), then add acrylamide and water as required. Recently, novel monomers, endowed with extreme resistance to alka-

TABLE 12.2

STOCK SOLUTIONS FOR POLYACRYLAMIDE GEL PREPARATION

Monomer solution [a,b]	
30% *T*, 2.5% *C*	Mix 29.25 g of acrylamide and 0.75 g of bisacrylamide. Add water to 100 ml.
30%*T*, 3% *C*	Mix 29.10 g of acrylamide and 0.90 g of bisacrylamide. Add water to 100 ml.
30% *T*, 4% *C*	Mix 28.80 g of acrylamide and 1.20 g of bisacrylamide. Add water to 100 ml.
Initiator solution	
40% (w/v) ammonium persulphate	This reagent is stable at 4°C for no more than 1 week.
Catalyst	
TEMED	This reagent is used undiluted. It is stable at 4°C for several months.

[a] Monomer solutions can be stored at 4°C for 1 month. Gels of ≥10% *C* (Table 12.5) require two monomer stock solutions, 30% acrylamide and 2% bisacrylamide.
[b] %*T* = g monomers per 100 ml; %*C* = g cross-linker per 100 g monomers.

TABLE 12.3

COMMON ADDITIVES FOR IEF

Additive	Purpose	Concentration	Limitations
Sucrose, Glycerol	To improve the mechanical properties of low %*T* gels and to reduce water transport and drift.	5–20%	The increased viscosity slightly slows the focussing process.
Glycine, Taurine	To change the dielectric constant of the medium. This increases the solubility of some proteins (e.g. globulins) and reduces ionic interactions.	0.1–0.5 M	Glycine is zwitterionic between pH 4 and 8, taurine between pH 3 and 7. Their presence somewhat slows the focussing process and shifts the resulting gradient.
Urea	Disaggregation of supramolecular complexes.	2–4 M	Unstable in solution especially at alkaline pH.
	Solubilisation of water-insoluble proteins, denaturation of hydrophilic proteins.	6–8 M	Urea is soluble at ≥10°C; it accelerates polyacrylamide polymerisation, so reduce the amount of TEMED added.
Non-ionic and zwitterionic detergents	Solubilisation of amphiphilic proteins.	0.1–1%	To be added to the polymerising solutions just before the catalysts to avoid foaming; they interfere with polyacrylamide binding to reactive substrata; they are precipitated by TCA and require a specific staining protocol.

line hydrolysis and with higher hydrophilicity, have been reported: they are *N*-acryloyl amino ethoxyethanol [79] and *N*-acryloyl amino propanol [80,81]. Table 12.2 gives the composition and general storage conditions for monomers and catalyst solutions, while Table 12.3 lists the most commonly used additives in IEF.

TABLE 12.4
CHOICE OF POLYACRYLAMIDE GEL CONCENTRATION

Protein M_r (upper limit for a given %T)	Gel composition	
40,000	%$T = 7$	%$C = 5$
75,000	%$T = 6$	%$C = 4$
150,000	%$T = 5$	%$C = 3$
800,000	%$T = 4$	%$C = 2.5$

12.2.3.2. Gel formulations

In order to allow all the sample components to reach their equilibrium position at essentially the same rate, and the experiment to be terminated before the pH gradient decay process adversely affects the quality of the separation, it is best to choose a non-restrictive anti-convective support. There are virtually no theoretical but only practical lower limits for the gel concentration (the minimum being about 2.2% *T*, 2% *C*). Large pore sizes can be obtained both by decreasing %*T* and by either decreasing or increasing %*C* from the critical value of 5%. Although the pore size of polyacrylamide can be enormously enlarged by increasing the percentage of cross-linker, two undesirable effects also occur in parallel, namely increasing gel turbidity and proneness to syneresis [82,83]. In this respect, *N,N′*-(1,2-dihydroxyethylene)bisacrylamide (DHEBA), with its superior hydrophilic properties, appears superior to bisacrylamide. In contrast, *N,N′*-diallyltartardiamide (DATD) inhibits the polymerisation process and so gives porous gels just by reducing the actual %*T* of the matrix. Because unpolymerised acryloyl monomers may react with $-NH_2$ and react readily with –SH groups on proteins [84–86] and, once absorbed through the skin, act as neurotoxins, the use of DATD should be avoided altogether. Table 12.4 gives the upper size of proteins that will focus easily in different %*T* gels. Because of these limitations, agarose is more suitable than polyacrylamide for the separation of very large proteins (M_r approx. 1,000,000), or supramolecular complexes. Table 12.5 gives details of preparation for 4–7.5% polyacrylamide gels containing 2–8 M urea for total gel volumes ranging from 4 to 30 ml, together with the recommended amounts of catalysts.

12.2.3.3. Choice of carrier ampholytes

A simple way to extend and stabilise the extremes of a wide (pH 3–10) gradient is to add acidic and basic (natural) amino acids. Thus lysine, arginine, aspartic acid and glutamic acid are prepared as individual stock solutions containing 0.004% sodium azide and stored at 0–4°C. They are added in volumes sufficient to give 2–5 mM final concentration. To cover ranges spanning between 3 and 5 pH units, a few narrow cuts of CA need to be blended, with the proviso that the resulting slope of the gradient will be (over each segment of the pH interval) inversely proportional to the amount of ampholytes isoelectric in that region.

Shallow pH gradients are often used to increase the resolution of sample components. However, longer focussing times and more diffuse bands will result, unless the gels

TABLE 12.5

RECIPES FOR IEF GELS (4–7.5% *T* and 2–8 M UREA)

Gel vol. (ml)	30% *T* acrylamide–bisacrylamide monomer solution (ml)								2% Carrier ampholytes[a]		Urea (g)				TEMED (μl)	40% ammonium persulphate[b]
	4% *T*	4.5% *T*	5%*T*	5.5% *T*	6% *T*	6.5% *T*	7% *T*	7.5% *T*	A	B	2 M	4 M	6 M	8 M		(μl)
30	4.00	4.50	5.00	5.50	6.00	6.50	7.00	7.50	1.50	1.88	3.60	7.20	10.40	14.40	9.0	30
25	2.34	3.75	4.17	4.58	5.00	5.42	5.83	6.25	1.25	1.56	3.00	6.00	9.00	12.00	7.5	25
20	2.66	3.00	3.34	3.66	4.00	4.33	4.66	5.00	1.00	1.25	2.40	4.80	7.20	9.60	6.0	20
15	2.00	2.27	2.50	2.75	3.00	3.25	3.50	3.75	0.75	0.94	1.80	3.60	5.40	7.20	4.5	15
10	1.33	1.50	1.66	1.83	2.00	2.16	2.33	2.50	0.50	0.63	1.20	2.40	3.60	4.80	3.0	10
8	1.06	1.21	1.33	1.46	1.60	1.73	1.86	2.00	0.40	0.50	0.96	1.92	2.88	3.84	2.4	8
7	0.93	1.05	1.17	1.28	1.40	1.51	1.63	1.75	0.35	0.44	0.84	1.68	2.52	3.36	2.1	7
6	0.80	0.90	1.00	1.10	1.20	1.30	1.40	1.50	0.30	0.38	0.72	1.44	2.16	2.88	1.8	6
5	0.66	0.75	0.83	0.91	1.00	1.08	1.16	1.25	0.25	0.31	0.60	1.20	1.80	2.40	1.5	5
4	0.53	0.60	0.66	0.73	0.80	0.86	0.93	1.00	0.20	0.25	0.48	0.96	1.44	1.92	1.2	4

[a] Use A for 40% solution (Ampholine, Servalyte, Resolyte); B for Pharmalyte.

[b] To be added after degassing the solution and just before pouring it into the mould.

TABLE 12.6

GOOD CARRIER AMPHOLYTES ACTING AS SPACERS

Carrier ampholyte	p*I*	Carrier ampholyte	p*I*	Carrier ampholyte	p*I*
Aspartic acid	2.77	*p*-Aminobenzoic acid	3.62	Lysyl–glutamic acid	6.10
Glutathione	2.82	Glycyl–aspartic acid	3.63	Histidyl–glycine	6.81
Aspartyl–tyrosine	2.85	*m*-Aminobenzoic acid	3.93	Histidyl–histidine	7.30
o-Aminophenylarsonic acid	3.00	Diiodotyrosine	4.29	Histidine	7.47
Aspartyl–aspartic acid	3.04	Cysteinyl–diglycine	4.74	L-Methylhistidine	7.67
p-Aminophenylarsonic acid	3.15	α-Hydroxyasparagine	4.74	Carnosine	8.17
Picolinic acid	3.16	α-Aspartyl–histidine	4.92	α,β-Diaminopropionic acid	8.20
Glutamic acid	3.22	β-Aspartyl–histidine	4.94	Anserine	8.27
β-Hydroxyglutamic acid	3.29	Cysteinyl–cysteine	4.96	Tyrosyl–arginine	8.38–8.68
Aspartyl–glycine	3.31	Pentaglycine	5.32	L-Ornithine	9.70
Isonicotinic acid	3.44	Tetraglycine	5.40	Lysine	9.74
Nicotinic acid	3.44	Triglycine	5.59	Lysyl–lysine	10.04
Anthranilic acid	3.51	Tyrosyl–tyrosine	5.60	Arginine	10.76
		Isoglutamine	5.85		

are electrophoresed at higher field strengths. Shallow pH gradients (shallower than the commercial 2-pH unit cuts) can be obtained in different ways.

(a) By subfractionating the relevant commercial carrier ampholyte blend. This can be done by focussing the CAs at high concentration in a multicompartment electrolyser [87].

(b) By allowing trace amounts of acrylic acid to induce a controlled cathodic drift during prolonged runs. This is effective in the acidic pH region, but is rather difficult to obtain reproducible results from run to run.

(c) By preparing gels containing different concentrations of carrier ampholytes in adjacent strips [88] or with different thickness along the separation path [89].

(d) By adding specific amphoteric compounds (spacers) at high concentration [90].

In the last case, two kinds of ampholytes may be used for locally flattening the pH gradient: 'good' and 'poor'. Good CAs, those with a small $\mathrm{p}I - \mathrm{p}K_1$ (i.e. possessing good conductivity and buffering capacity at the pI) are able to focus in narrow zones. Low concentrations (5–50 mM), are sufficient to induce a pronounced flattening of the pH curve around their pIs. A list of these CAs is given in Table 12.6. Poor carrier ampholytes, on the other hand, form broad plateaux in the region of their pI, and should be used at high concentrations (0.2–1.0 M). Their presence usually slows down the focussing process. Some of them are listed in Table 12.7.

A note of caution: in an IEF system, the distribution of acids and bases is according to their dissociation curve, in a pattern that may be defined as protonation (or deprotonation) stacking. If large amounts of these compounds originate from the samples (in the form of buffers), the limits of the pH gradient shift from the expected values. For example, β-mercaptoethanol, as added to denatured samples for two-dimensional PAGE analysis, lowers the alkaline end from pH 10 to approx. pH 7.5 [91]. This effect, however, may sometimes be usefully exploited. For example, high levels of TEMED in the gel mixture appear to stabilise alkaline pH gradients [92]. In addition, TEMED is

TABLE 12.7

POOR CARRIER AMPHOLYTES ACTING AS SPACERS

Carrier ampholyte	pK_1	pK_2	pI
Carrier ampholytes with pIs 7–8			
β-Alanine	3.55	10.24	6.90
γ-Aminobutyric acid	4.03	10.56	7.30
δ-Aminovaleric acid	4.26	10.77	7.52
ε-Aminocaproic acid	4.42	11.66	8.04
'Good' buffers with acidic pIs			
Mes	1.3	6.1	3.70
Pipes	1.3	6.8	4.05
Aces	1.3	6.8	4.05
Bes	1.3	7.1	4.20
Mops	1.3	7.2	4.25
Tes	1.3	7.5	4.40
Hepes	1.3	7.5	4.40
Epps	1.3	8.0	4.65
Taps	1.3	8.4	4.85

utilised in c-IEF for blocking the capillary region after the detection point, so that basic proteins would focus in the region prior to the detector [93].

12.2.4. Gel preparation and electrophoresis

Protocol 2 outlines the series of steps required for an IEF run. The key steps are described in more detail below.

Protocol 2. CA-IEF flow sheet

1. Assemble the gel mould.
2. Mix all components of the polymerising mixture, except ammonium persulphate.
3. Degas the mixture for a few minutes, and re-equilibrate (if possible) with nitrogen.
4. Add the required amount of ammonium persulphate stock solution and mix.
5. Transfer the mixture to the gel mould and overlay it with water.
6. Leave the mixture to polymerise (at least 1 h at room temperature or 30 min at 37°C).
7. Open the gel mould and blot any moisture from the gel edges and surface.
8. Lay the gel on the cooling block of the electrophoretic chamber.
9. Apply the electrodic strips.
10. Pre-run the gel, if appropriate.
11. Apply the samples.
12. Run the gel.
13. Measure the pH gradient.
14. Reveal the protein bands using a suitable detection procedure.

12.2.4.1. Assembling the gel mould

Assembling the gel mould is visualised in Fig. 12.4A–C. Fig. 12.4A shows the preparation of the slot former which will give 20 sample application slots in the final gel. In the method shown, the slot former is prepared by gluing a strip of embossing tape onto the cover plate and cutting rectangular tabs, with the dimensions shown, using a scalpel. A rubber U-gasket covering three edges of the cover plate is also glued to the plate. In Fig. 12.4B, a sheet of Gel Bond PAG film is applied to the supporting glass plate in a thin film of water. When employing reactive polyester foils as the gel backing, use glycerol rather than water. Avoid leaving airpockets behind the gel backing sheet since this will produce gels of uneven thickness, which will create distortions in the pH gradient. Also ensure that the backing sheet is cut flush with the glass support since any overhang easily bends. Finally, the gel mould is assembled using clamps (Fig. 12.4C). Assembly is usually made easier by wetting and blotting the rubber gasket just before use. Note that the cover plate has three V-shaped indentations on one edge, which allow insertion of a pipette or syringe tip into the narrow gap between the two plates of the mould to facilitate opening the gel.

12.2.4.2. Filling the mould

One of the three methods may be chosen: by gravity, capillarity or the flap technique.

(i) *By gravity*. This procedure uses a vertical cassette with a rubber gasket U-frame glued to the cover plate (Fig. 12.4A). If the cover plate has V-indentations along its free edge as shown in Fig. 12.4A, the gel mixture can be transferred simply by using a pipette or a syringe with its tip resting on one of these indentations (Fig. 12.4D).

Avoid filling the mould too fast, which will create turbulence and trap air bubbles. If an air bubble appears, stop pouring the solution and try to remove the bubble by tilting and knocking the mould. If this manoeuvre is unsuccessful, displace the bubble with a 1-cm wide strip of polyester foil.

(ii) *By capillarity*. This method is mainly used for casting reasonably thin gels (0.2–0.5 mm). It requires a horizontal sandwich with two lateral spacers (see Fig. 12.5A). The solution is fed either from a pipette or from a syringe fitted with a short piece of fine-bore tubing. It is essential that the solution flows evenly across the whole width of the mould during casting. If an air bubble appears, do not stop pumping in the solution or you will produce more bubbles. Remove all the air bubbles at the end, using a strip of polyester foil. A level table is not mandatory but the mould should be left laying flat until the gel is completely polymerised.

(iii) *The flap technique*. This procedure is mainly used for preparing ultrathin gels. A 20–50% excess of gel mixture is poured along one edge of the cover (with spacers on both sides) (Fig. 12.5B) and the support plate is slowly lowered on it (Fig. 12.5C) ([60]). Air bubbles can be avoided by using clean plates. If bubbles do get trapped, they can also be removed by lifting and lowering the cover plate once more. Since this method may lead to spilled unpolymerised acrylamide, take precautions for its containment (wear gloves and use absorbent towels to mop up excess).

Protocol 3. Polymerisation of the gel for IEF

Equipment and reagents
- 30% *T* acrylamide–bisacrylamide solution (see Table 12.5)
- 2% carrier ampholytes (Ampholine, Servalyte, Resolyte or Pharmalyte)
- Urea
- TEMED
- 40% ammonium persulphate (prepared fresh)
- Detergents (as desired, for the gel mixture)
- Graduated glass measuring cylinder sufficient to hold the gel mixture during preparation
- Vacuum flask
- Mechanical vacuum pump
- Gel mould

Method

1. Mix all the components of the gel formulation (from Table 12.5, except TEMED, ammonium persulphate and detergents, when used) in a cylinder, add distilled water to the required volume and transfer the mixture to a vacuum flask.

2. Degas the solution using the suction from a water pump; the operation should be continued as long as gas bubbles form. Manually swirl the mixture or use a magnetic stirrer during degassing. The use of a mechanical vacuum pump is desirable for the preparation of very soft gels but is unnecessarily cumbersome for urea gels (urea would crystallise). At the beginning of the degassing step the solution should be at, or above, room temperature, to decrease oxygen solubility. Its cooling during degassing is then useful in slowing down the onset of polymerisation.

3. If possible, it is beneficial to re-equilibrate the degassed solution against nitrogen instead of air (even better with argon, which is denser than air). From this step on, the processing of the gel mixture should be as prompt as possible.

4. Add detergents, if required. If they are viscous liquids, prepare a stock solution beforehand (e.g. 30%) but do not allow detergent to take up more than 5% of the total volume. Mix briefly with a magnetic stirrer.

5. Add the volumes of TEMED and then ammonium persulphate as specified in Table 12.5. Immediately mix by swirling, and then transfer the mixture to the gel mould. Carefully overlay with water or butanol.

6. Leave to polymerise for at least 1 h at room temperature or 30 min at 37°C. Never tilt the mould to check whether the gel has polymerised; if it has not polymerised when you tilt it, then the top never will, because of the mixing caused by tilting. Instead, the differential refractive index between liquid (at the top and usually around the gasket) and gel phase is an effective, and safe, index of polymerisation and is shown by the appearance of a distinct line after polymerisation.

7. The gels may be stored in their moulds for a couple of days in a refrigerator. For longer storage (up to two weeks for neutral and acidic pH ranges), it is better to disassemble the moulds and (after covering the gels with Parafilm and wrapping with Saran foil) store them in a moist box.

8. Before opening the mould, allow the gel to cool at room temperature, if polymerised at a higher temperature. Laying one face of the mould onto the cooling block of the electrophoresis unit may facilitate opening the cassette. Carefully remove the overlay by blotting and also remove the clamps.

Protocol 3. Continued

9. (a) If the gel is cast against a plastic foil, remove its glass support, then carefully peel the gel from the cover plate.

(b) If the gel is polymerised on silanised glass, simply force the two plates apart, with a spatula or a blade.

(c) If the gel is not bound to its support, lay the mould on the bench, with the plate to be removed uppermost and one side protruding a few cm from the edge of the table. Insert a spatula at one corner and force against the upper plate. Turn the spatula gently until a few air bubbles form between the gel and the plate then use the spatula as a lever to open the mould. Wipe any liquid from the gel surface with a moistened Kleenex but be careful; keep moving the swab to avoid it sticking to the surface. If bubbles form on both gel sides of the gel, try at the next corner of the mould. If the gel separates from both glass plates and folds up, you may still be able to salvage it, provided the gel thickness is at least 400 μm and the gel concentration is at least 5% *T*. Using a microsyringe, force a small volume of water below the gel, then carefully make the gel lay flat again by manoeuvring it with a gloved hand. Remove any remaining air bubbles using the needle of the microsyringe. Cover the gel with Parafilm, and gently roll it flat with a rubber roller (take care not to damage it with excess pressure). Carefully remove any residual liquid by blotting.

12.2.4.3. Gel polymerisation

Protocol 3 gives a step by step procedure for casting a CA-IEF gel which might include, if needed, additives such as urea and surfactants. As a variant of this protocol, one can polymerise 'empty' gels (i.e. devoid of CAs and of any leachable additive), wash and dry them and reswell them in the appropriate CA solution (including any additive, as needed). This is a direct application of the IPG technology [2]. After polymerisation (as above, but in the absence of CAs) wash the gel 3 times, in 300 ml distilled water each time, in order to remove catalysts and unreacted monomers. Equilibrate the washed gel (20 min with shaking) in 1.5% glycerol and finally dry it onto Gel Bond PAG foil. It is essential that the gel does not bend so, before drying, the foil should be made to adhere to a supporting glass plate (taking care to remove all air bubbles in between and fastening it in position with clean, rust-proof clamps). Drying must be at room temperature, in front of a fan. Finally, mount the dried gel back in the polymerisation cassette and allow it to reswell in the appropriate CA (and suitable additives) solution. This can be conveniently done by using a reswelling cassette for dry IPG gels (refer to Fig. 12.22 for more details).

In ultrathin gels, pH gradients are sensitive to the presence of salts, including TEMED and persulphate. Moreover, unreacted monomers are toxic and noxious to proteins. The preparation, washing and re-equilibration of 'empty' gels removes these components from the gel and so avoids these problems.

12.2.4.4. Sample loading and electrophoresis

The electrophoresis procedure is described in Protocol 4. The gel is placed on the cooling plate of the electrophoresis chamber. It is necessary to perform the electrophoresis at a constant temperature and with well-defined conditions, since the

Protocol 4. Electrophoretic procedure

Equipment and reagents

- Electrophoresis apparatus (see Section 12.2.2.1)
- Gel polymerised in the gel mould (from Protocol 3)
- Whatmann #17 filter paper
- Electrode solutions (see Table 12.8)
- Protein samples
- Filter paper (e.g. Whatmann #1, Whatmann 3 mm, Paratex II) if the sample is to be applied via filter paper tabs (see step 6 below)
- 10 mM KCl
- Narrow-bore combination pH electrode (e.g. from Radiometer)

Method

1. Set the cooling unit at 2–4°C for normal gels, at 8–10°C for 6 M urea and at 10–12°C for 8 M urea gels.
2. To enable rapid heat transfer, pour a few ml of a non-conductive liquid (distilled water, 1% non-ionic detergent or light kerosene), onto the cooling block of the electrophoretic chamber. Form a continuous liquid layer between the gel support and the apparatus and gently lower the plate into place, avoiding trapping air bubbles and splashing water onto or around the gel. Should this happen, remove all liquid by careful blotting. When the gel is narrower than the cooling block, apply the plate on its middle. If this is not possible, or if the electrode lid is too heavy to be supported by just a strip of gel, insert a wedge (e.g. several layers of Parafilm) between the electrodes and cooling plate.
3. Cut electrode strips (e.g. from Whatmann #17 filter paper); note that most paper exposed to alkaline solutions becomes swollen and fragile, so we use only Whatmann #17 approx. 5 mm wide and about 3 mm shorter than the gel width. Saturate them with electrode solutions (see Table 12.8). However, they should not be dripping; blot them on paper towels if required.
4. Wearing disposable gloves, transfer the electrode strips onto the gel. They must be parallel and aligned with the electrodes. The cathodic strip firmly adheres to the gel: do not try to change its position once applied. Avoid cross-contaminating the electrode strips (including with your fingers). If any electrode solution spills over the gel, blot it off immediately, rinse with a few drops of water and blot again. Check that the wet electrode strips do not exceed the size of the gel and cut away any excess pieces. Be sure to apply the most alkaline solution at the cathode and the most acidic at the anode (if you fear you have misplaced them: the colour of the NaOH soaked paper is yellowish but, of course, you may also check this with litmus paper). If you discover a mistake at this point, simply turn your plate around or change the electrode polarity. However, there is no remedy after the current has been on for a while.
5. The salt content of the samples should be kept as low as possible. If necessary dialyse them against glycine (any suitable concentration) or dissolve in diluted CAs. When buffers are required, use low-molarity buffers composed of weak acids and bases.
6. The samples are best loaded into precast pockets. These should not be deeper than 50% of the gel thickness. If they are longer than a couple of millimetres and it is necessary to pre-electrophorese the gel (see step 7), the pockets should be filled with dilute CAs. After the pre-electrophoresis (see Table 12.9 for conditions) remove this solution by blotting. Then apply the samples. The amount of sample applied should fill the pockets. Try to equalise the volumes and the salt content among different samples. After about 30 min of electrophoresis at high voltage (see Table 12.9 for conditions), the content of the pockets can be removed by blotting

Protocol 4. Continued

---

and new aliquots of the same samples loaded. The procedure can be repeated a third time. Alternatively, the samples can be applied to the gel surface, absorbed into pieces of filter paper. Different sizes and material of different absorbing power are used to accommodate various volumes of liquid (e.g. a 5 × 10 mm tab of Whatman #1 can retain about 5 μl, of Whatman 3MM about 10 μl, of Paratex II more than 15 μl). Up to three layers of paper may be stacked. If the exact amount of sample loaded has little importance, the simplest procedure is to dip the tabs into it, then blot them to remove excess sample. Otherwise, align dry tabs on the gel and apply measured volumes of each sample with a micropipette. For stacks of paper pieces, feed the solution slowly from one side rather than from the top. Do not allow a pool of sample to drag around its pad, but stop feeding liquid to add an extra tab when required. This method of application of samples is not suitable for samples containing alcohol.

7. Most samples may be applied to the gel near the cathode without pre-electrophoresis. However, pre-running is advisable if the proteins are sensitive to oxidation or unstable at the average pH of the gel before the run. Pre-running is not suitable for those proteins with a tendency to aggregate upon concentration, or whose solubility is increased by high ionic strength and dielectric constant, or which are very sensitive to pH extremes. Anodic application should be excluded for high-salt samples and for proteins (e.g. a host of serum components) denatured at acidic pH. Besides this guideline, however, as a rule the optimal conditions for sample loading should be experimentally determined, together with the minimum focussing time, with a pilot run, in which the sample of interest is applied in different positions of the gel and at different times. Smears, or lack of confluence of the bands after long focussing time, denote improper sample handling and protein alteration.

8. After electrophoresis the pH gradient in the focussed gel may be read using a contact electrode. The most general approach, however, is to cut a strip along the focussing path (0.5–2 cm wide, with an inverse relation to the gel thickness). Then cut segments between 3 and 10 mm long from this and elute each for about 15 min in 0.3–0.5 ml of 10 mM KCl (or with the same urea concentration as present in the gel). For alkaline pH gradients, this processing of the gel should be carried out as quickly as possible, the elution medium should be thoroughly degassed and air-tight vials flushed with nitrogen should be used. Measure the pH of the eluted medium using a narrow-bore combination electrode. For most purposes, it is sufficient to note just the temperature of the coolant and the pH measurement in order to define an operational pH. For a proper physico-chemical characterisation, the temperature differences should be corrected for as suggested in Table 12.10 (which gives also corrections for the presence of urea).

---

temperature influences the pH gradient and consequently the separation positions of the proteins. Also the presence of additives (e.g. urea) strongly affects the separation positions by changing the physico-chemical parameters of the solution trapped inside the gel matrix; some compensation factors are listed in Table 12.8. Electrodic strips filled with electrode solutions (see Table 12.9) are placed on the surface of the gel (at anodic and cathodic extremes). Whenever possible (and compatible with the width of the pH gradient to be developed) strong acids and bases as anolytes and catholytes, respectively, should be avoided. For example, 1 M NaOH at the cathode could hydrolyse the amido groups in the underlying polyacrylamide gel, thus inducing a strong electroosmotic flow. Also amphoteric compounds could be used for this purpose, e.g. at the anode aspartic

TABLE 12.8

COMPENSATION FACTORS

Effect of temperature		Effect of urea	
pH	Δ25-4°C	pH	Δ [a]
–	–	3.0	+0.09
3.5	+0.0	3.5	+0.08
4.0	+0.06	4.0	+0.08
4.5	+0.10	4.5	+0.07
5.0	+0.14	5.0	+0.06
5.5	+0.20	5.5	+0.06
6.0	+0.23	6.0	+0.05
6.5	+0.28	6.5	+0.05
7.0	+0.36	7.0	+0.05
7.5	+0.39	7.5	+0.06
8.0	+0.45	8.0	+0.06
8.5	+0.52	8.5	+0.06
9.0	+0.53	9.0	+0.06
9.5	+0.54	9.5	+0.05
10.0	+0.55	–	–

[a] For 1 M urea. The effect is proportional to the urea concentration.

TABLE 12.9

IEF ELECTRODE SOLUTIONS

Solution	Application	Concentration
H_3PO_4	Anolyte for all pH ranges	1.0 M
H_2SO_4	Anolyte for very acidic pH ranges ($pH_a < 4$) [a]	0.1 M
CH_3COOH	Anolyte for alkaline pH ranges ($pH_a > 7$) [a]	0.5 M
NaOH	Catholyte for all pH ranges	1 M [b]
Histidine	Catholyte for acidic pH ranges ($pH_c < 5$) [a]	0.2 M
Tris	Catholyte for acidic and neutral pH ranges ($pH_c < 5$) [a]	0.5 M

[a] pH_a, pH_c represent the lower and higher extremes of the pH range, respectively.
[b] Store in air-tight plastic bottles.

acid (p*I* 2.77 at 50 mM concentration) or iminodiacetic acid (p*I* 2.33 for a 100 mM solution) and at the cathode lysine free-base (p*I* 9.74).

The electrophoretic procedure is divided into two steps. The first is a pre-electrophoresis (characterised by the application of low voltages; see Table 12.10) which allows pre-focussing of carrier ampholytes and the elimination of all contaminants or unreacted catalysts from the gel matrix. Then the protein sample is loaded. In IEF gels, the sample should be applied along the whole pH gradient in order to determine its optimum application point. The second electrophoretic step is then started (see Table 12.10, 'electrophoresis of samples'). This is carried out at constant power; low voltages are used for the sample entrance, followed by higher voltages during the separation and bands sharpening. The actual voltage used in each step depends on gel thickness (see Table 12.10).

TABLE 12.10
ELECTROPHORESIS CONDITIONS

	Voltage	Time (h)
Pre-electrophoresis		
Thick gels (>1.5 mm)	200	0.5
Thin gels (<1.5 mm)	400	0.5
Electrophoresis [a] *of samples*		
Thick gels (2 mm)	200	5
Thick gels (1 mm)	400	3
Thin gels (0.5 mm)	600	2
Thin gels (0.25 mm)	800	1.5

[a] Electrophoresis is carried out at constant power so as to give an initial voltage as indicated.

12.2.5. General protein staining

Table 12.11 gives a list, with pertinent references, of some of the most common protein stains used in IEF. Detailed recipes are given below. Extensive reviews covering general staining methods in gel electrophoresis have recently appeared [101,102]. Merril and Washart [103] have also given a very extensive bibliography list covering all aspects of polypeptide detection methods, including enzyme localisation protocols (>350 citations). Additionally, although not specifically reported for IEF, some recent developments include: use of Eosin B dye [104]; a mixed-dye technique comprising Coomassie Blue R-250 and Bismark Brown R [105]; Stains-All for highly acidic molecules [106]; fluorescent dyes for proteins, such as SYPRO Orange and SYPRO Red [107,108]; a two-minute Nile Red staining [109].

Before entering in detail in all methods listed below, a general comment seems appropriate. This stems from a recent review by Rabilloud [110] who evaluated all possible stains adopted for IEF and 2-D maps according to key issues such as sensitivity (detection threshold), linearity, homogeneity (i.e. variation from one protein to another) and reproducibility. Accordingly, four steps have been considered: (a) affixing the label before IEF; (b) labelling in between IEF and SDS-PAGE; (c) tagging after SDS-PAGE;

TABLE 12.11
PROTEIN STAINING METHODS

Stains	Application	Sensitivity	Reference
Coomassie Blue G-250 (micellar)	General use	Low	[94]
Coomassie Blue G-250/$CuSO_4$	General use	Medium	[95]
Coomassie Blue R-250/sulphosalicylic acid	General use	High	[96]
Silver stain	General use	Very high	[97]
Coomassie Blue G-250/$CuSO_4$	In presence of detergents	Medium	[95]
Light Green SF	General use	Medium	[98]
Fast Green	General use	Low	[99]
Coomassie Blue G-250 at 60°C	In presence of detergents	High	[100]

and (d) labelling after blotting (i.e. transfer of the separated protein onto an inert membrane). We will thus briefly review here these four steps, with the understanding that they apply not only to plain IEF and IPG, as in this chapter, but also to the various electrophoretic steps reported in the following chapters.

(a) *Affixing the label before IEF*. In principle, this approach would be very convenient, since it could be done in a minimal volume with a high concentration of reactants; however, any non-covalently bound tag will be disrupted. On the other hand, any pre-labelling method that alters the charge state of the protein, either by removing a charged group or by adding a spurious one, is not compatible with 2-D maps. Thus, amine or thiol alkylation with reactive acidic dyes such as Remazol [111] must be avoided because the proteins will gain negative charges. Covalent fluorescent tagging would appear to be superior, provided that fluorophors with intense light absorption, high quantum yield, large Stokes shifts and limited fading can be found. Since fluorescein derivatives carry negative charges, pre-IEF labelling has been limited for a long time to thiol alkylation with electrically neutral probes [112], although these species offer limited sensitivity. Use of cyanine-based probes [113], with attachment via amine acylation, improves the process and does not alter the protein charge (except for rather basic proteins) since the positive charge lost upon covalent bonding is replaced by the quaternary ammonium on the cyanine fluorophore. Radioactive labelling, followed by fluorography [114] or by phosphor storage [115] appears to be the most sensitive for IEF- and 2-D-compatible methods. The latter technique, in which the β-radiation induces an energy change in a europium salt, which, via laser excitation, is converted into visible light, appears to offer a 20- to 100-fold higher sensitivity than autoradiography, coupled to a linear dynamic range or 4 orders of magnitude. Another modern detection method is based on amplification detectors similar to those used in high-energy physics, and is compatible with dual-isotope detection [116]. The conclusions: radioactive detection probably offers the best S/N ratio and the best ultimate sensitivity.

(b) *Labelling in between IEF and SDS-PAGE*. This stage offers much greater flexibility, because the charge-state of the proteins can be altered provided that the apparent M_r is kept constant. Covalent labelling dominates at this point and thus fluorescent tags are preferred. However, since protein nucleophiles (amino and thiol groups) are the preferred targets for such a grafting, care to remove carrier ampholytes should be exerted, since these compounds could consume most of the label. This process is usually achieved with a few aqueous–acid alcohol baths or with 10% TCA [117]. When tagging with fluorophores, it should be remembered that detection is an important point. In 2-D maps, a scanning device should be used for acquiring all the spots in the gel. Since laser-induced fluorescence offers the highest sensitivity, fluorophors having an excitation maximum close to an available laser source (e.g. fluorescein and rhodamine) are preferred. In another set-up, illumination is done in the gel plane, preferably by UV excitation, and the fluorescent light is collected by a CCD camera [118].

(c) *Tagging after SDS-PAGE*. After this step, which can be either mono- or bi-dimensional (2-D maps), any solution can be adopted, according to the sensitivity desired. Non-covalent detection with organic dyes, the most common being the Coomassie Brilliant Blue G and R, is quite popular, due to a simple protocol, although the sensitivity is of the order of ca. 1 μg [119]. These kinds of stains are of the regressive type, i.e. the gel is first saturated with dye and then destained to remove the dye unbound to the protein. Alternatively, in dilute dye solutions, the gel can be stained to the end point, which however requires very long staining times [120]. As an additional option, as described below, micellar or colloidal stainings can be adopted, with sensitivities of about 100 ng. Non-covalent, fluorescent probes are also very popular after this stage. Generally, such probes are non-fluorescent in water but highly fluorescent in apolar media, such as detergent, thus they take advantage of SDS binding to proteins to create a micro-environment promoting fluorescence at the protein spot. Typical labels of this type are naphthalene derivatives [121], SYPRO dyes [108] and Nile Red [122]. Detection by differential salt binding is also an attractive procedure, although negative stains are problematic if photographic documentation is required. All negative stains use divalent cations (Cu, Zn) for forming a precipitate with SDS. Submicrogram sensitivities are claimed, the staining is very rapid and destaining is achieved by simple metal chelators such as EDTA. Increased sensitivities are obtained when the precipitate in the gel is no longer Zn-SDS, but a complex salt of zinc and imidazole [123,124]. In this last case, nanogram sensitivities can be obtained. Perhaps one of the most popular stains, however, is metal ion reduction, i.e. silver staining, although it is one of the most complex detection procedures. What makes a silver stain highly sensitive is the strong autocatalytic character of silver reduction [125]. This condition is achieved by using a very weak developer (e.g. dilute formaldehyde) and sensitisers between fixation and silver impregnation. Silver staining protocols are divided into two families. In one, the silvering agent is silver nitrate and the developer is formaldehyde in an alkaline carbonate solution. In the second, the silvering agent is a silver–ammonia complex and the developer formaldehyde in dilute citric acid. The sensitivity with silver staining in modern protocols is in the low nanogram range for both procedures. This is 100-fold better than classical Brilliant Blue staining, 10-fold better than colloidal Brilliant Blue tags and about 2-fold better than zinc staining, with the extra benefit of a much better contrast.

(d) *Labelling after blotting.* Staining on blots with organic dyes is done almost exclusively by regressive protocols. Because most dyes bind weakly to neutral membranes (nitrocellulose and polyvinylidene difluoride), and due to the concentration effect afforded by blotting, high-nanogram sensitivities are easily reached [126]. Non-covalent fluorescence detection is also feasible on blots. In this case, lanthanide complexes are used [127], since they have the advantage of providing time-lapse detection with a very good S/N ratio. In case of PVDF membranes, not wettable with water, fluorescein or rhodamine can be used, since they will bind only to the wetted protein spots [128]. Under these conditions, the detection limit on blots appears to be around 10 ng protein.

At the end of this long excursus, some interesting conclusions have been drawn by Rabilloud [110]. According to this author, most staining methods have reached a plateau close to their theoretical maximum. For organic dye staining, the maximum occurs when the dye is bound to all available sites on the protein, as is the case with colloidal Brilliant Blue staining. A maximum also seems to have been reached for silver staining, since no increase of sensitivity has been reported in the last five years. The only method left with most potential for improvement is fluorescence, as we hope it will be demonstrated by some very recent literature.

Protocol 5. Coomassie Blue G-250 staining procedure

Reagents

• Staining solution: mix 2 g Coomassie Blue G-250 with 400 ml 2 M H_2SO_4, dilute the suspension with 400 ml distilled water and stir for at least 3 h; filter through Whatmann #1 paper; then add 89 ml 10 KOH and 120 ml of 100% (w/v) TCA while stirring.

Method

1. Immerse the gel in staining solution until the required stain intensity is obtained.
2. To remove all salts and to increase the colour contrast, rinse the gel extensively with water.

12.2.5.1. Micellar Coomassie Blue G-250

The advantages of this procedure (see Protocol 5) are that only one step is required (i.e. no protein fixation, no destain), peptides down to approx. 1500 molecular mass can be detected, there is little interference from CAs and, finally, the staining mixture has a long shelf life [94]. The small amount of dye that may precipitate with time can be removed by filtration or washed from the surface of the gels with liquid soap. Note that the dye, as prepared, is in a leuco form (i.e. it is almost colourless, pale greenish instead of deep blue) and in a micellar state. When adsorbed by the peptide/protein zone, it stains the latter in the usual blue colour while leaving the gel slab colourless. The process is rather slow, since the dye has to be extracted from the micelle and slowly be adsorbed by the protein zone; it is best to leave the gel in this solution for at least a day, under gentle agitation.

12.2.5.2. Coomassie Blue R-250/$CuSO_4$

Protocol 6 describes this staining procedure, which is highly recommended; it is easily carried out and has good sensitivity [95]. Apparently, divalent copper becomes coordinated around the peptide bond, thus enhancing the colour yield.

Protocol 6. Coomassie Blue R-250/$CuSO_4$ staining procedure

Reagents

• Staining solution: dissolve 1.09 g $CuSO_4$ in 650 ml water and then add 190 ml of acetic acid; mix this solution with 250 ml of ethanol containing 0.545 g of Coomassie R-250.
• Destaining solution: mix 600 ml ethanol, 140 ml acetic acid, 1260 ml water.

Method

1. Stain the gel (without previous fixation) in staining solution for 30 min to a few hours, depending on its thickness. During immersion in the staining solution, unsupported gels shrink and their surfaces become sticky. Therefore avoid any contact with dry surfaces.
2. Destain the gel in several changes of destaining solution.

12.2.5.3. Coomassie Blue R-250/sulphosalicylic acid

Heat the gels for 15 min at 60°C in a solution of 1 g Coomassie R-250 in 280 ml methanol and 730 ml water, containing 110 g TCA and 35 g sulphosalicylic acid (SSA). Destain at 60°C in 500 ml ethanol, 160 ml acetic acid, 1340 ml water. Precipitation of the dye at the gel surface is a common problem (remove it with alkaline liquid soap) [96]. In a milder approach, Neuhoff et al. [129] have substituted strong acids, such as sulphosalicylic acid, with an acidic medium containing ammonium sulphate. The micro-precipitates of Coomassie G act as a reservoir of dye molecules, so that enough dye is available to occupy all the binding sites on all the proteins, provided the staining is long enough to reach steady state (48 h). The presence of ammonium sulphate in the colloidal dye suspension increases the strength of hydrophobic interaction with the protein moiety, resulting in better sensitivity.

Warning: TCA, often present in many colloidal Coomassie formulations, often leads to esterification of glutamic acid residues, which might hinder identification of proteins using MALDI-TOF MS, as now routinely done after 2-D mapping [130–132]. In addition, the aspartate–proline bond is susceptible to hydrolysis in acidic staining and destaining solutions, resulting in cleavage of some proteins [133].

12.2.5.4. Silver stain

A typical method for silver staining is described in Protocol 7 [97]. A myriad of silvering procedures exists. A good review with guidelines can be found in Rabilloud [134].

Protocol 7. Silver staining procedure

Reagents
The reagents required are listed below.

Method
The silver staining procedure involves exposing the gel to a series of reagents in a strict sequence, as described in the following steps.

1. 12% TCA, 500 ml, 30 min
2. 50% methanol, 12% acetic acid, 1 l, 30 min
3. 1% HJO_4, 250 ml, 30 min
4. 10% ethanol, 5% acetic acid, 200 ml, 10 min (repeat 3×)
5. 3.4 mM potassium dichromate, 3.2 mM nitric acid, 200 ml, 5 min
6. Water, 200 ml, 30 s (repeat 4×)
7. 12 mM silver nitrate, 200 ml, 5 min in the light, 25 min in the dark
8. 0.28 M sodium carbonate, 300 ml, 30 s (repeat 2×); 0.05% formaldehyde, 300 ml, several minutes
9. 10% acetic acid, 100 ml, 2 min
10. Photographic fixative (e.g. Kodak Rapid Fix), 200 ml, 10 min
11. Water, several litres, extensive washes
12. Finally, remove any silver precipitate on the gel surface using a swab

---

Protocol 8. Coomassie Blue G-250/urea/perchloric acid staining procedure [a]

Reagents

- Fixative: dissolve 147 g TCA and 44 g sulphosalicylic acid in 910 ml of water.
- Staining solution: dissolve 0.4 g Coomassie Blue G-250 and 39 g urea in ca. 800 ml of water; immediately before use, add 29 ml of 70% $HClO_4$ with vigorous stirring; bring the volume to 1 l with distilled water.
- Destaining solution: prepare this by mixing 100 ml acetic acid, 140 ml ethanol, 200 ml ethyl acetate, 1560 ml water.

Method

1. Fix the protein bands for 30 min at 60°C in fixative.
2. Stain the gel for 30 min at 60°C in staining solution.
3. Destaining for 4–22 h in destaining solution.

---

[a] This staining protocol may be applied in presence as well as in absence of detergents.

Warning: all silver staining protocols which utilise aldehydes are in general not compatible with subsequent MS analysis. Formaldehyde in general leads to alkylation of α- and ε-amino groups [130–132], which hinders identification of proteins using MS analysis. When fixation is avoided, good results are often obtained with methods in which the proteins are digested, such as in peptide mass fingerprinting. In this last case, protocols that include silver nitrate [135] or silver ammonia [136] have been reported to be successful in peptide mass identification. Also, destaining silver-stained gels with ferricyanide and thiosulphate seems to greatly improve subsequent analysis by mass spectrometry [137].

12.2.5.5. Coomassie Blue G-250/urea/perchloric acid

This staining procedure is described in Protocol 8 [100].

12.2.5.6. Fluorescence protein detection

At the beginning of this section, Rabilloud [110] concluded that, whereas most common staining techniques had reached a plateau, fluorescence detection was the only method with the most likely potential for improvement. Such a method might have now appeared: it is an ultrasensitive protocol utilising SYPRO Ruby IEF Protein Gel Stain (Molecular Probes, Eugene, OR, USA). This luminescent dye can be excited with 302 nm or 470 nm light and it has been optimised for protein detection in IEF (and IPG) gels. Proteins are stained in a ruthenium-containing metal complex overnight and then simply rinsed in distilled water for 2 h. Stained proteins can be excited by UV light at ca. 302 nm (UV-B transilluminator) or with visible wavelengths at ca. 470 nm. Fluorescent emission of the dye is maximal at 610 nm. The sensitivity of SYPRO Ruby is superior to colloidal Coomassie stains and the best silver staining protocols by a factor of 3–30 times. SYPRO Ruby is suitable for staining proteins in both non-denaturing and denaturing carrier ampholyte IEF and IPGs. The stain is compatible with *N,N'*-methylenebisacrylamide or bisacrylyl piperazine cross-linked polyacrylamide gels as well as with agarose gels. The unique advantage of this stain is that it does not contain extraneous chemicals (formaldehyde, glutaraldehyde, Tween-20) that frequently interfere with peptide identification in mass spectrometry. In fact, successful identification of stained proteins by peptide mass profiling was demonstrated, rendering this stain procedure a most promising tool for 2-D mapping [138]. It was demonstrated that SYPRO Ruby interacts strongly with Lys, Arg and His residues in proteins, and as such it closely resembles Coomassie stains. On the contrary, acidic silver nitrate stain primarily interacts only with Lys. Although it has been reported that Nile red [139], as well as SYPRO Red and SYPRO Orange, can also stain IEF gels, the gels must be first incubated in SDS, since all three of these lipophilic dyes bind to proteins indirectly through the anionic detergent. Moreover, with these three dyes, the sensitivity is often poorer than with standard Coomassie staining; on all these accounts, it would appear that this novel SYPRO Ruby stain could represent a major revolution in detection techniques.

12.2.6. Specific protein detection methods

Table 12.12 lists some of the most common specific detection techniques: it is given as an example and with appropriate references, but it is not meant to be exhaustive.

12.2.7. Quantitation of the focussed bands

Some general rules about densitometry are worth repeating.

(a) The scanning photometer should have a spatial resolution of the same order as the fractionation technique (for correct analysis, a 50- to 100-μm resolution is required).

(b) The stoichiometry of the protein–dye complex varies among different proteins.

(c) This same relationship is linear only over a limited range of protein concentration.

(d) At high absorbance, the limitations of the spectrophotometer interfere with the photometric measurement.

TABLE 12.12

SPECIFIC PROTEIN DETECTION TECHNIQUES

Protein detected	Technique	Reference
Glycoproteins	PAS (periodic acid-Schiff) stain	[140]
Lipoproteins	Sudan Black stain	[141]
Radioactive proteins	Autoradiography	[142]
	Fluorography	[143]
Enzymes	Zymograms [a]	[144]
Antigens	Immunoprecipitation in situ	[145]
	Print-immunofixation	[146]
	Blotting	[147]

[a] The concentration of the buffer in the assay medium usually needs to be increased in comparison with zone electrophoresis to counteract the buffering action by carrier ampholytes.

12.2.8. Troubleshooting

12.2.8.1. Waviness of bands near the anode

This may be caused by (a) carbonation of the catholyte: in this case, prepare fresh NaOH with degassed distilled water and store properly; (b) excess catalysts: reduce the amount of ammonium persulphate; (c) too long sample slots: fill them with dilute CAs; (d) too low concentrations of CAs: check the gel formulation.

To alleviate the problem, it is usually beneficial to add low concentrations of sucrose, glycine or urea, and to apply the sample near the cathode. To salvage a gel during the run, as soon as the waves appear, apply a new anodic strip soaked with a weaker acid (e.g. acetic acid vs. phosphoric acid) inside the original one, and move the electrodes closer to one another.

12.2.8.2. Burning along the cathodic strip

This may be caused by (a) the formation of a zone of pure water at pH = 7: add to your acidic pH range a 10% solution of either the 3–10 or the 6–8 range ampholytes; (b) the hydrolysis of the acrylamide matrix after prolonged exposure to alkaline pH: choose a weaker base, if adequate, and, unless a pre-run of the gel is strictly required, apply the electrodic strips after loading the samples.

12.2.8.3. pH gradients different from expected

(a) For acidic and alkaline pH ranges, the problem is alleviated by the choice of anolytes and catholytes whose pH is close to the extremes of the pH gradient.

(b) Alkaline pH ranges should be protected from carbon dioxide by flushing the electrophoretic chamber with moisture-saturated N_2 (or better with argon) and by surrounding the plate with pads soaked in NaOH. It is worth remembering that pH readings on unprotected alkaline solutions become meaningless within half an hour or so.

(c) A large amount of a weak acid or base, supplied as sample buffer, may shift the pH range (β-mercaptoethanol is one of such bases). The typical effect of the addition of urea is to increase the apparent p*I*s of the CAs (see Table 12.8).

(d) It may be due to the cathodic drift; to counteract this (1) reduce the running time to the required minimum (as experimentally determined for the protein of interest, or for a coloured marker of similar M_r), (2) increase the viscosity of the medium (with sucrose, glycerol or urea), (3) reduce the amount of ammonium persulphate, (4) remove acrylic acid impurities by recrystallising acrylamide and Bis, and by treating the monomer solution with mixed-bed ion-exchange resin, (5) for a final cure, incorporate into the gel matrix a reactive base, such as 2-dimethylamino propyl methacrylamide (Polyscience) [148]: its optimal concentration (of the order of 1 μM) should be experimentally determined for the system being used.

12.2.8.4. Sample precipitation at the application point

If large amounts of material precipitate at the application point, even when the M_r of the sample proteins is well below the limits recommended in Table 12.4, the trouble is usually caused by protein aggregation. Some remedies:

(a) Try applying the sample in different positions on the gel, with and without pre-running: some proteins might be altered only by a given pH.

(b) If you have evidence that the sample contains high M_r components, reduce the value of $\%T$ of the polyacrylamide gel.

(c) If you suspect protein aggregation brought about by the high concentration of the sample (for example when the problem is reproduced by disc electrophoresis runs), do not pre-run and set a low voltage (100–200 V) for several hours to avoid the concentrating effect of an established pH gradient at the beginning of the run. Also consider decreasing the protein load and switching to a more sensitive detection technique. The addition of surfactants and/or urea is usually beneficial.

(d) If the proteins are only sensitive to the ionic strength and/or the dielectric constant of the medium (in this case they perform well in disc electrophoresis and are precipitated if dialysed against distilled water), increasing the CA concentration, adding glycine or taurine, and sample application without pre-running may overcome the problem.

(e) The direct choice of denaturing conditions (8 M urea, detergents, β-mercaptoethanol) very often minimises these solubility problems, dissociating proteins (and macromolecular aggregates) to polypeptide chains.

12.2.9. Some typical applications of IEF

IEF is a fine-tuned analytical tool for investigating post-translational processing and chemical modification of proteins. Protein processing may be grouped into two kinds. The first covers changes in primary structure, with proteolysis at peptide bonds, e.g. for removal of intervening sequences or of leader peptides, as in intracellular processing. Such modifications are typically identified by size fractionation techniques, such as SDS-PAGE. A second class covers another mode of processing, in which the size is only marginally affected and no peptide bond is cleaved, but the surface charge is modified. Such post-translational processing is the typical realm for IEF analysis. It includes attachment of oligosaccharide chains, such as in glycoproteins, or of individual sugar moieties, such as in glycated proteins (e.g. haemoglobin A_{1c}), or phosphorylation, methylation, ADP-ribosylation, ubiquitination, to name just a few. But there are also a number of modifications occurring at the NH_2 groups (e.g. acetylation), at the COOH groups (e.g. deamidation) at the SH groups as well as a number of chemical modifications occurring in vitro, such as accidental carbamylation due to the presence of urea. All of these modifications can be properly investigated by IEF, as amply discussed in a broad review dedicated to this topic [149]. Of course, IEF can only give a pI shift, and cannot tell much more than that in assessing the modification. However, IEF, as a fractionation tool, coupled to mass spectrometry, can be a formidable hyphenated method for assessing the extent of modification involved, the stoichiometry of the reaction and the reaction site too. From this point of view, MALDI-TOF appears to be one of the finest probes available. IEF can also be used as a probe for interacting systems, such as protein–ligand interactions. If the interacting species are stable in the electric field, IEF enables determination of intrinsic ligand-binding constants for statistical binding of a charged ligand, binding to heterogeneous sites and cooperative binding. The interacting systems can be a protein with a small ligand or a protein with a macromolecular ligand. In extreme cases, such interactions could be engendered by isomerisation (pH-dependent conformational transition) or by interaction with the carrier ampholyte buffers. An excellent review on all these aspects is now available [150].

12.2.10. Examples of some fractionations

After so much methodological tips, it is refreshing to look at some results obtained with this technique. Since its inception, especially in the gel format (preparative IEF in sucrose density gradients was a quite demanding technique and never became popular), it was immediately realised that IEF represented a turning point in separation techniques, unrivalled by any other fractionation method available in the sixties. At its inception, however, IEF did not have an easy life and had to fight its way through scientific journals to gain acceptance. The old establishment, sitting in the editorial board of prestigious journals, had rejection crises when data, demonstrating microheterogeneity of the pet proteins in which they had invested a life-time of efforts in purifying, where brought to the lime-light and tried hard to 'suffocate the scandals'. The defense was adamantine: 'artefacts' and it was on these grounds that most papers were rejected, without a right to appeal. A case in point was the haemoglobin (Hb) separation shown in Fig. 12.6. It is known that Hb, in vitro, can slowly undergo spontaneous auto-oxidation; in vivo, however, in healthy individuals, Hb is usually in its reduced form, since, in order to fulfil its physiological role, the haem iron must remain in a ferrous state. Haem that has oxidised to the ferric form (met Hb), is unable to bind oxygen. Despite the significant difference in charge between ferro Hb ($\alpha_2\beta_2$) and its oxidised forms ($\alpha_2^+\beta_2$, $\alpha_2\beta_2^+$, $\alpha_2^+\beta_2^+$) these species had never been isolated before by conventional electrophoresis. In contrast, gel IEF permitted clear-cut separations of fully and partially oxidised intermediates. After Hb A was partially oxidised with an agent such as ferricyanide, three new bands could be visualised (Fig. 12.6). The fully oxidised Hb ($\alpha_2^+\beta_2^+$) appears as a brown band with a p*I* of 7.20, 0.25 pH units above that of Hb A [151]. Equidistant between the fully oxidised Hb and reduced Hb ($\alpha_2\beta_2$), two closely spaced bands, designated IB_I and IB_{II}, could be detected, which later on were shown by Park [152] to be $\alpha_2\beta_2^+$ and $\alpha_2^+\beta_2$, respectively. The fact that these two components could be separated in this high-resolution system added to the large body of evidence indicating differences in the haem environments of the α- and β-chains, inducing a slightly different net surface charge (conformational transitions) in otherwise identical total charge proteins. Despite this massive experimental evidence, this paper visited many journals and witnessed a large number of rejections before appearing in *Biochim. Biophys. Acta*! As it turned out, this story of the valence hybrids, which stirred so much concern in the establishment of the epoch, was in a way trivial compared to the real facts of life. In theory, there should be ten valence hybrids obtainable by partial oxidation of $\alpha_2\beta_2$. Since only four were found, it means that these were the only species stable under those experimental conditions. By quenching a Hb solution partially saturated with CO into a

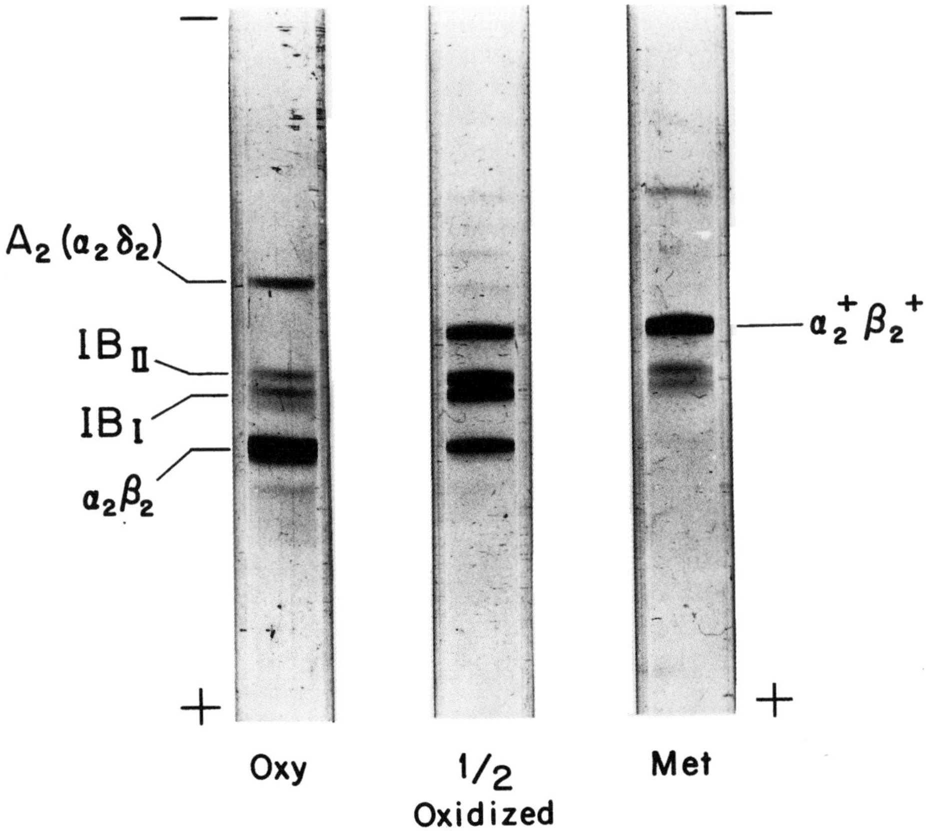

Fig. 12.6. Separation of oxidised forms of human haemoglobin. IEF fractionations were performed on haemolysates in 50 mM phosphate buffer, pH 7.1. Left, no oxidant added; middle, Hb partially oxidised by the addition of a half equivalent amount of ferricyanide; right, Hb fully oxidised by the addition of an excess of ferricyanide. Spectra of these specimens showed the presence of 0%, 50% and 100% ferrihaemoglobin, respectively. (From Ref. [151], with permission.)

hydro-organic solvent containing ferricyanide, Perrella et al. [153] were able to produce a population of partially oxidised and CO-bound Hb molecules. When these valence hybrids were analysed by IEF at temperatures as low as −23°C, all the intermediate species were resolved. This was an excellent example of the use of IEF as a structural probe and vindicated the poor fate of the previous results published 10 years earlier.

IEF, in a gel-slab format, was soon adopted also in clinical chemistry. A nice example is the screening of thalassaemia syndromes, which had to be performed on a large scale in regions of high incidence, such as the Island of Sardinia. In order to assess the extent of the syndrome, the amount of γ-globin (belonging to fetal Hbs) chains in newborns had to be evaluated and the ratio β/γ determined. In the past, this was accomplished by the three-step procedure of Clegg et al. [154]: incubation of red blood cells with [^{3}H]leucine, removal of haem and chromatographic separation of globin chains in CM-cellulose. This procedure was very complex and placed a heavy burden on laboratories where large populations at risk had to be screened. In addition, huge amounts of eluent (containing 8 M urea and plenty of β-mercaptoethanol, ME, for keeping in a reduced state the highly-reactive –SH groups of globin chains, in order to avoid mixed disulphide bridges among different chains) were consumed, further adding to the costs and contributing to the unpleasant smell of these laboratories, due to the volatile fumes of ME impregnating benches and walls. Aware of these shortcomings, we proposed a novel method for globin chain separation, consisting in an IEF step in conventionally thick [155,156] or in ultrathin gel layers [157], in presence of urea and detergents, the latter greatly improving the β/γ separation and even inducing the splitting of two γ-chain phenotypes, called Aγ and Gγ, due to a replacement of a Ala with a Gly residue in position 136 of the γ-chains. Even this remarkable improvement fell short of expectations, since it was still labour-intensive, due to the need of preparing haem-free globin chains and to the lengthy staining/destaining steps, caused by the high levels of surfactant in the gel. A breakthrough finally came with the work of Cossu et al. [158], as shown in Fig. 12.7, who proposed focussing of the intact components of umbilical cord blood: Hb F, Hb A and Hb F_{ac} (acetylated Hb F). Nobody had attempted that before, due to the very poor resolution between the Hb A and Hb F_{ac} bands, caused by minute p*I* differences. The authors succeeded in that by using non-linear pH 6–8 gradients, flattened in the region around pH 7 by addition of 0.2 M β-Ala and 0.2 M 6-amino caproic acids, a technique called 'separator IEF'. It can be seen, by examination of Fig. 12.7A and by a close-up of it (Fig. 12.7B; here a homozygous patient can be spotted at a glance due to the absence of Hb A), that the resolution is now excellent. The technique was user-friendly, did not require any sample manipulation steps prior to IEF, was fast (ca. 2 h focussing time) and adaptable to large sample numbers (up to 30 tracks could be run simultaneously). Even the staining was rapid: it consisted in dipping the gels into 0.2% bromophenol blue in 50% ethanol and 5% acetic acid and in a quick destaining step in 30% ethanol, 5% acetic acid, a method first reported by Awdeh [159] and long since then forgotten. Curiously, the Hb bands were stained in deep red–purple colour and the lab of Dr. Cossu, in Ozieri, who performed >10,000 analyses/year, was full of stained and dried gels, which could be easily stored for further evaluation, if needed. As a result of this massive screening (which soon spread to other regions at risk, such as Sicily, Calabria, etc.), and of genetic counselling to couples at risk, homozygous thalassaemic conditions in Italy have been essentially eradicated today, to the point that, a few years ago, a hospital for thalassaemic kids built on the sea shores of Calabria had to be turned over to the Faculty of Pharmacy of the local University for lack of patients.

Another interesting aspect of the power of IEF comes from Fig. 12.8, which shows the analysis of active fragments of the human growth hormone (hGH), synthesised by the solid-phase method of Merrifield [160]. In the early days of IEF it was thought that IEF of peptides would not be feasible, first of all because they have a higher diffusion coefficient than proteins, and secondly because they would not be precipitated and fixed into the gel matrix by the common protein stains, like alcoholic solutions of Coomassie Blue. Fig. 12.8 dispelled both myths and proved that IEF of peptides was highly feasible and would indeed produce very sharp bands [94]. The problem of fixation was solved by adopting a leuco-stain, a micellar suspension of Coomassie Brilliant Blue G-250 (G stays for its greenish hue, which cannot be appreciated in monomeric solutions, but is

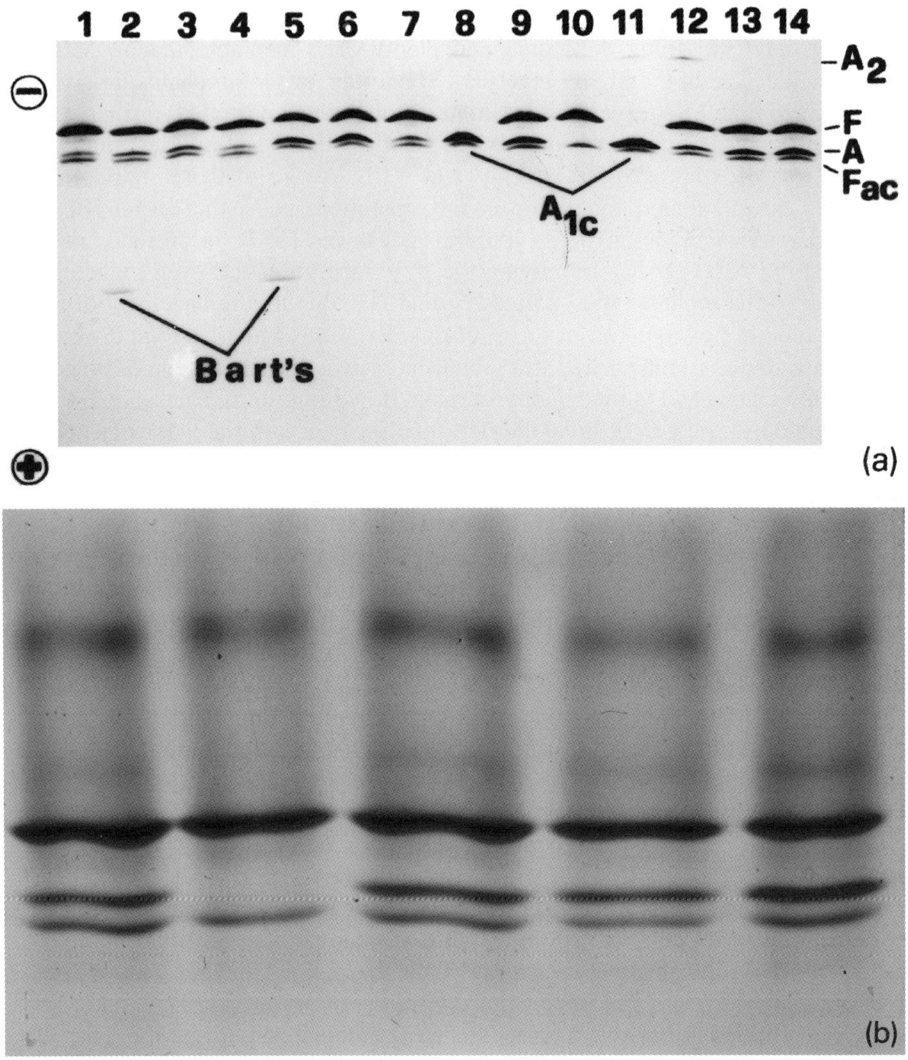

Fig. 12.7. (A) IEF of cord blood haemolysates in a pH 6–8 gradient added with 0.2 M β-Ala and 0.2 M 6-amino caproic acid. Experimental: 6% *T*, 4% *C* gel; run: 15 min at 400 V, 90 min at 15 W – 1500 Vmax; protein load: 20 μg. Levels of Hb A in samples: 1, 9.1%; 2, 11.4%; 3, 14.7%; 4, 7.9%; 5, 31.1%; 6, 28.9%; 7, 18.9%; 9, 31.3%; 10, zero; 13, 28%; 14, 34.3%. Samples 8 and 11: normal human adult lysates (notice the bands of Hb A_{1c}). Note, in samples 2 and 5, the presence of Hb Bart (the homotetramer γ_4), denoting α-thalassaemic conditions. (B) Close-up of a few sample tracks. Note that one of them is a homozygous β thalassaemic, since there is a complete absence of adult haemoglobin. (From Ref. [158], with permission.)

clearly visible in the micellar state) in TCA: as the focussed gel is bathed directly in this solution, the peptides adsorb the dye from the micelle and are fixed by the dye molecules, which probably act as cross-links over the different peptide chains, thus forming a macromolecular aggregate which is trapped in the gel matrix fibres. This

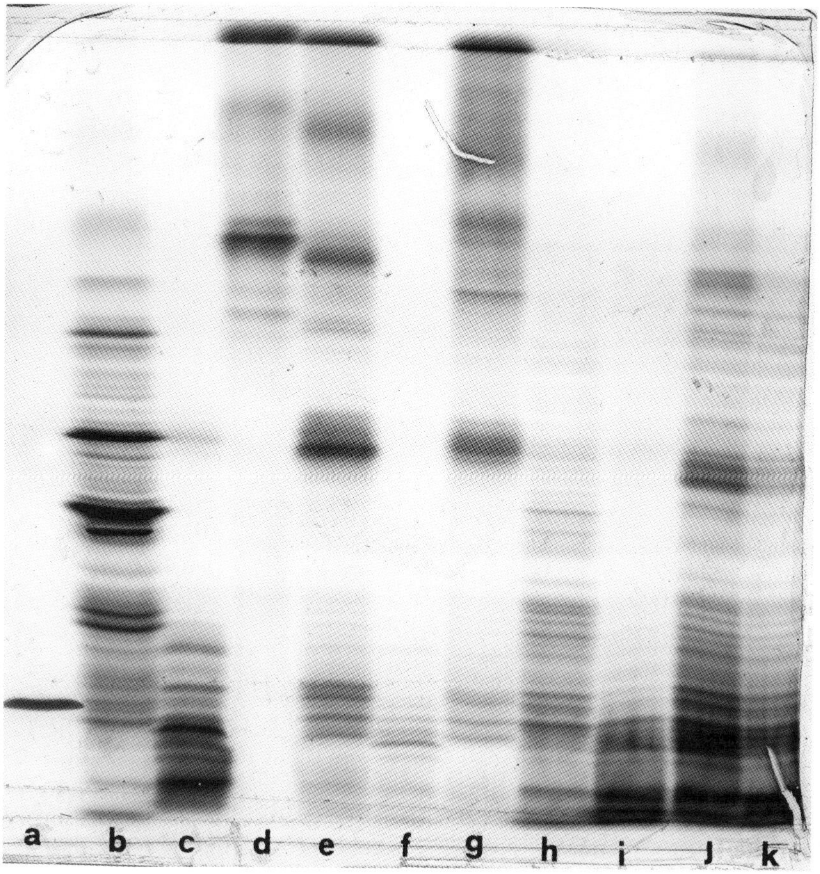

Fig. 12.8. IEF of peptides in the length range 8–54 amino acids. The gel slab was 0.7 mm thick and contained 7% acrylamide, 2% Ampholine pH 3.5–10 and 8 M urea. Ten to fifteen μl of sample (10 mg/ml) were applied in filter paper strips at the anode after 1 h prefocussing. Total running time, 4 h at 10 W (1000 V at equilibrium). The gel was then dipped in a colloidal dispersion of Coomassie Brilliant Blue G-250 in 12% TCA. The samples were the following synthetic fragments of the human growth hormone (hGH): a = hGH 31–44; b = hGH 15–36; c = hGH 111–134; d = hGH 1–24; e = hGH 166–191; f = hGH 25–51; g = hGH 157–191; h = hGH 1–36; i = hGH 96–134; j = hGH 115–156; and k = hGH 103–156. Shorter fragments (octa-, nona- and deca-peptides) were neither fixed in the gel nor stained. (From Ref. [94], with permission.)

experiment was very important for peptide chemists, since it helped redirecting their synthetic strategies. As it turned out from Fig. 12.8, when medium length peptides were produced via the Merrifield approach, the amount of failed and truncated sequences were in large excess over the desired product, to the point that the latter could not be recognised any longer. Although this had been theoretically predicted, it had not been experimentally verified up to that time, due to lack of high-resolution techniques. As news spread out, in vitro synthesis of peptides took an important turn: it was done only in short sequences (5–6 amino acids at most), which were at the end joined together by splicing.

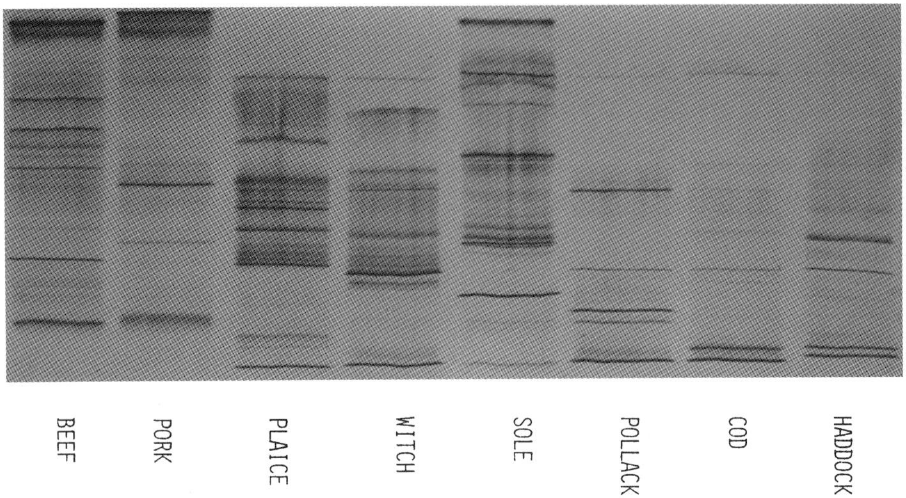

Fig. 12.9. Application of IEF to food analysis. The gel slab was 0.7 mm thick and contained 7% acrylamide, 2% Ampholine pH 3.5–10 and 8 M urea. Total running time, 4 h at 10 W (1000 V at equilibrium). The gel was then dipped in a colloidal dispersion of Coomassie Brilliant Blue G-250 in 12% TCA. The samples are (from left to right): beef, pork and the following fish tissue extracts: plaice, witch, sole, pollack, code, haddock (unpublished results).

IEF in soluble CAs soon became very popular also in food analysis and for detecting frauds. A case in point is shown in Fig. 12.9, in which samples from frozen meats and fishes were analysed in a pH 3–10 gradient. It was often not even necessary to grind and clarify the homogenate by centrifugation: a thin slice of frozen tissue could be deposited close to the anode or cathode: the local pH would lyse the cells and the soluble proteins would just migrate out and focus at their p*I* position. A curious story developed out of this analysis: in those days (late seventies) in Italy it was very popular to buy frozen sole (*Sogliola limanda*) from a well-known distributor of frozen food. We found out that these soles were not the expensive variety, but a cheaper one, known as *Pleuronectes platessa* (rondine di mare in Italian, plaice in English). The company was brought to court and lost the proceedings. The results? A unique case of modern democracy: the price of this food package went up, since they had to change the label (the content remained the same!) and this was a costly operation! Others (especially our German colleagues) went as far as to map all possible cultivars of potatoes, simply by focussing their soluble proteins: as shown in Fig. 12.10, in a collection of 16 different varieties, the various cultivars can be recognised easily from the quantitative/qualitative pattern differences. Once again, these examples show how IEF helped solving important problems in the life sciences and finding new solutions to old problems [161].

12.2.11. Artefacts or not?

We have just discussed at length how the weapon of 'artefacts' would be used by referees in the early days to reject IEF manuscripts. But how is it in real life? This

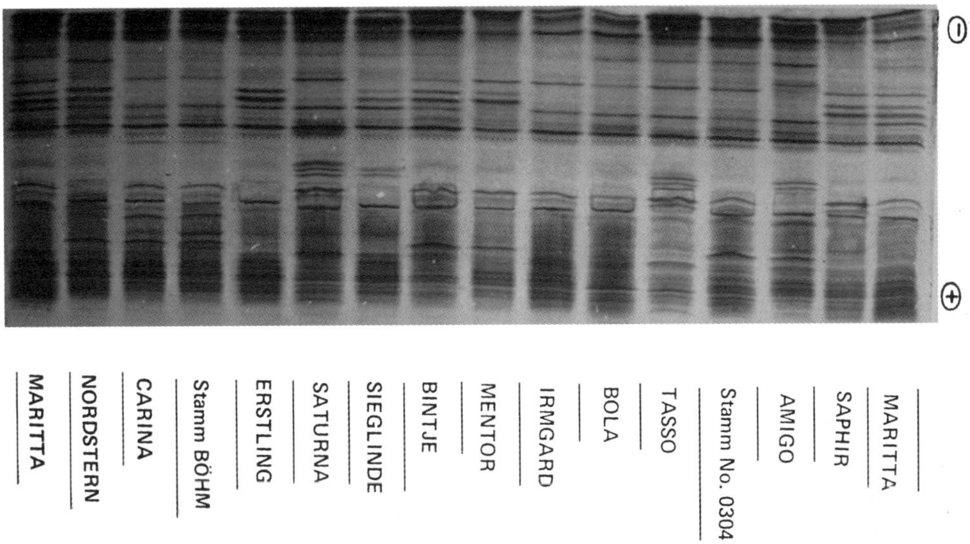

Fig. 12.10. IEF of 16 extracts of soluble potato proteins from different cultivars. The gel slab was 0.7 mm thick and contained 7% acrylamide, 2% Ampholine pH 3.5–10 and 8 M urea. Ten to fifteen μl of sample (10 mg/ml) were applied in filter paper strips at the anode after 1 h prefocussing. Total running time, 4 h at 10 W (1000 V at equilibrium). The gel was then dipped in a colloidal dispersion of Coomassie Brilliant Blue G-250 in 12% TCA.

question has been the object of hot debate, at the inception of the technique, and there were just as many reports on artefacts as additional ones denying them. The field was so controversial that in 1983 Righetti wrote a chapter of his book on IEF by the title: 'artefacts: a unified view' [1]. The final consensus was that artefacts, when reported, occurred only in extreme cases and with 'extreme' structures. One of the most glorious examples was the suspicious report that heparin gave a large number (up to 21) of bands focussing in the pH 4–5 range [162]. Now, how could a pure polyanion, bearing no positive counterions on its polymeric backbone, exhibit an 'isoelectric point' remained a mystery to anybody with a minimum knowledge of chemistry. It took the work of Righetti and Gianazza [163] and Righetti et al. [164] to find out that these 21 fractions indeed represented 21 different complexes of the same heparin polymer with 21 different Ampholine molecules (see Fig. 12.11). This was the 'catch 22': in those days, when binding of small molecules to macromolecules was suspected, all theoreticians had worked out a bimodal distribution, i.e. the bound vs. the unbound species! After our report, it was clear that multimodal distribution could be the real trend in IEF, so theoreticians like Cann et al. [165] hurried to change their models. It then became apparent that, in a specular fashion, also polycations would produce such an artefactual binding pattern. Except for these artefacts generated by peculiar structures, there was no practical evidence that carrier ampholytes would elicit the same multimodal distribution by interacting with proteins, so the general users widely accepted the technique.

While that might be generally true, recent evidence suggests that IEF could still be prone to problems. A case in point is shown in Fig. 12.12, which represents the

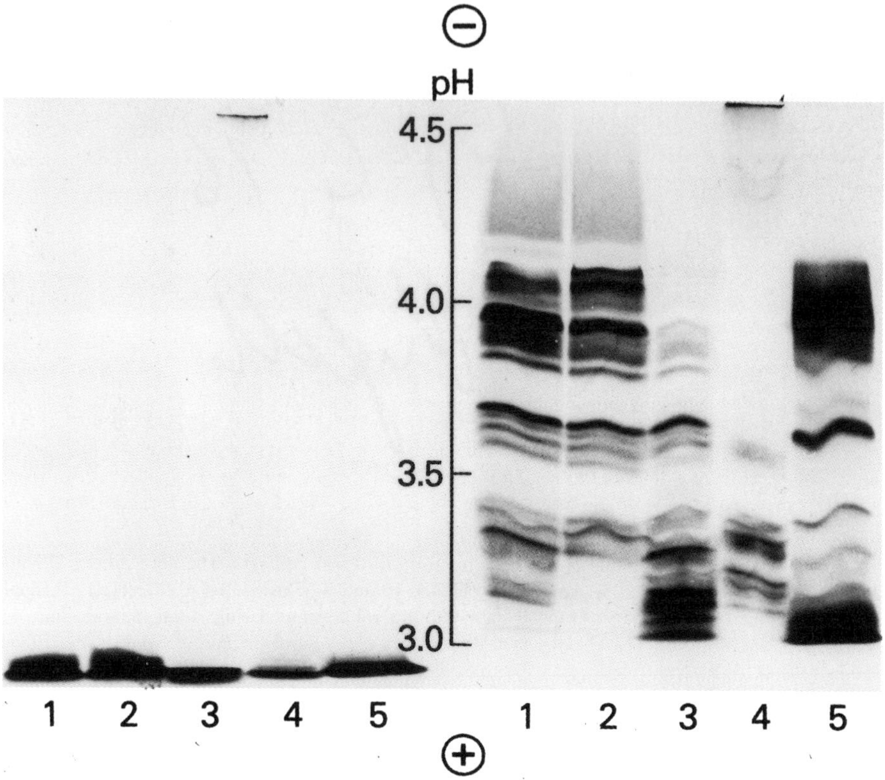

Fig. 12.11. IEF of polyanions in 5% *T* polyacrylamide gels, 2% Ampholine pH 2.5–5 in the absence (right side) and presence (left side) of 8 M urea. 1 = heparin A; 2 = heparin B; 3 = carboxyl-reduced heparin B; 4 = polygalacturonic acid; 5 = polyglutamic acid. Staining with toluidine blue. (From Ref. [164].)

banding pattern of diaspirin cross-linked haemoglobin (DCLHb), a haemoglobin-based oxygen carrier exhibiting near physiological oxygen binding capability and devoid of nephrotoxic side effects. DCLHb was previously found, by gel permeation, RP-HPLC and mass spectrometry, to consist of ca. 94% cross-linked product (reacted on the Lys 99 of two α-chains), accompanied by ca. 6% cross-linked Hb which also had reacted on the Lys 132 and/or Lys-144 of the β-chains and a small amount of intermolecularly cross-linked dimers. However, conventional IEF in carrier ampholyte buffers gave an unexpected spectrum of four major, almost equally represented, p*I* species in the pH range 6.82 to 7.01, a band of mid-intensity with a p*I* of 7.11 and two minor components with p*I* of 6.73 and 6.77. This extraordinary polydispersity was re-evaluated by other surface charge probes, such as immobilised pH gradients (IPG) and capillary zone electrophoresis (CZE) of native and denatured globin chains. IPGs of DCLHb gave the expected spectrum of bands, consisting of a main component (92%) with p*I* 7.337 and 3 additional minor bands, with lower p*I*s, representing ca. 8% of the total. These data were in agreement with CZE profiles of native DCLHb, which resolved, in addition to the main DCLHb peak, 3 to 4 minor components representing ca. 10% of the total. Also

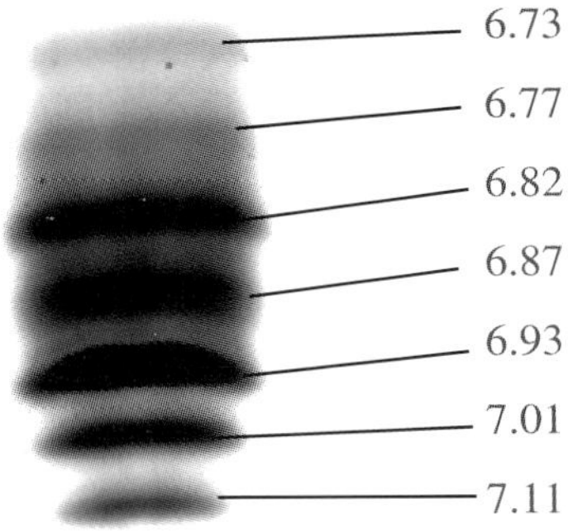

Fig. 12.12. Conventional IEF of diaspirin cross-linked haemoglobin (DCLHb) in pH 6–8 carrier ampholyte (CA) buffers. Gel: 5% *T*, 4% *C* polyacrylamide, containing 2% pH 6–8 admixed with 0.2% pH 3–10 CAs. The anolyte was 50 mM free acetic acid and the catholyte 50 mM Tris. After 30 min pre-focussing at 500 V, the sample (20 μl, corresponding to a total of 50 μg DCLHb) was loaded cathodically in surface basins. Focussing was continued for 90 min and then the gel was stained with Coomassie Brilliant Blue in the presence of Cu^{2+}. (From Ref. [166], with permission.)

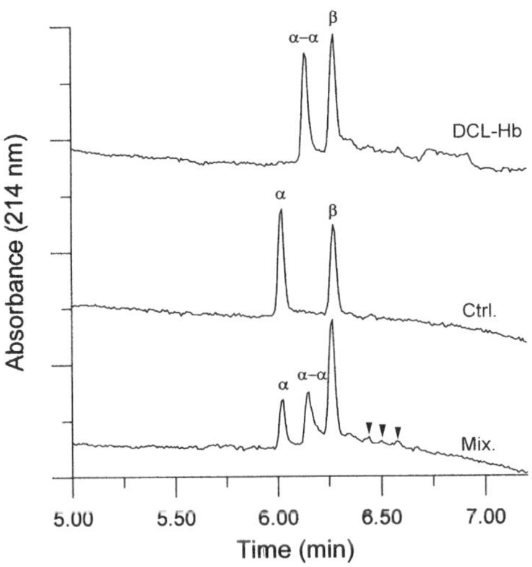

Fig. 12.13. CZE of haem-free globin chains. Thirty cm long, 100 μm I.D. uncoated capillaries were used. Background electrolyte: 50 mM IDA, 6 M urea, 0.5% HEC, pH 3.2. Run settings: 18 kV (360 V/cm), 22 μA, room temperature. Detection was at 214 nm. Top trace: DCL-Hb; middle profile: control, stroma-free haemoglobin (SFHb); bottom electropherogram: 1 : 1 mixture of DCL-Hb and SFHb (the arrowheads in this last tracing indicate the peaks of minor, 'decorated' Hbs). (From Ref. [166], with permission.)

CZE of denatured, haem-free globin chains gave the expected pattern with only traces of minor, extra-reacted species. The latter technique, in addition to resolving α- and β-globin chains in a 1 : 1 ratio in control Hb (Fig. 12.13, middle profile), resolved a free β- and the α–α-dimer in DCLHb (Fig. 12.13, upper tracing). In a 1 : 1 mixture of control and DCLHb, 3 peaks were observed: eluting in the order α-, α–α- and β-globin chains

(Fig. 12.13, bottom tracing). The identity of the major DCLHb and of the minor species was also ascertained by mass spectrometry [166]. So, it would appear that CA-IEF could be prone to odd results, although why this should happen with Hb, a very soluble and highly hydrophilic protein, remains a mystery.

Another case we have found disturbing is represented by the crystallisation of a basic fatty acid binding protein (FABP). FABP from chicken liver was recently purified to homogeneity in multicompartment electrolysers with isoelectric membranes. Large amounts of a p*I* 9.7 isoform were collected into a compartment delimited by p*I* 8.8 and 11.0 membranes. This isoform produced crystals with higher resolution than those obtained by purification via preparative IEF in soluble carrier ampholytes. In addition, a novel orthorhombic form with a different packing density was obtained. It is hypothesised that, when using conventional IEF, traces of carrier ampholytes could adhere to the protein, particularly in the hydrophobic ligand-binding pocket, thus blurring the X-ray signal [168]. Multicompartment electrolysers do not present this drawback, since they are based on insoluble buffering species. For these, and for other reasons discussed below, it is not surprising that the natural evolution of CA-IEF would be the Immobiline technology.

12.3. IMMOBILISED pH GRADIENTS

12.3.1. General considerations

As illustrated below, IPGs represent perhaps the ultimate development in all focussing techniques, a big revolution in the field, in fact. Due to the possibility of engineering the pH gradient at whim, from the narrowest (which, for practical purposes, has been set at 0.1 pH units over a 10 cm distance) to the widest possible one (a pH 2.5–12 gradient), IPGs permit the highest possible resolving power, on the one hand, and the widest possible collection of spots (in 2-D maps) on the other hand. The chemistry is precise and amply developed; so are all algorithms for implementing any possible width and shape of the pH gradient. Due to its unique performance, IPGs represent now the best possible first dimension for 2-D maps and are increasingly adopted for this purpose.

12.3.1.1. The problems of conventional IEF

Table 12.13 lists some of the major problems associated with conventional IEF using amphoteric buffers. Some of them are quite severe; for example (point 1) low ionic strength often induces near-isoelectric precipitation and smearing of proteins, even in analytical runs at low protein loads. The problem of uneven conductivity is magnified in poor ampholyte mixtures, like the Poly Sep 47 (a mixture of 47 amphoteric and non-amphoteric buffers, claimed to be superior to CAs) [169]. Due to their poor composition, huge conductivity gaps form along the migration path, against which proteins of different p*I*s are stacked. The results are simply disastrous [170]. Cathodic drift (point 6) is also a major unsolved problem of CA-IEF, resulting in extensive loss of proteins at the gel cathodic extremity upon prolonged runs. For all these reasons,

TABLE 12.13

PROBLEMS WITH CARRIER AMPHOLYTE FOCUSSING

1	Medium of very low and unknown ionic strength
2	Uneven buffering capacity
3	Uneven conductivity
4	Unknown chemical environment
5	Not amenable to pH gradient engineering
6	Cathodic drift (pH gradient instability)

in 1982 Bjellqvist et al. [59] launched the technique of immobilised pH gradients (IPGs).

12.3.1.2. The Immobiline matrix

IPGs are based on the principle that the pH gradient, which exists prior to the IEF run itself, is copolymerised, and thus insolubilised, within the fibres of a polyacrylamide matrix (see Fig. 12.14 for a pictorial representation). This is achieved by using, as buffers, a set of six non-amphoteric, weak acids and bases, having the following general chemical composition: CH_2=CH–CO–NH–R, where R denotes either two different weak carboxyl groups, with pKs 3.6 and 4.6 or four tertiary amino groups,

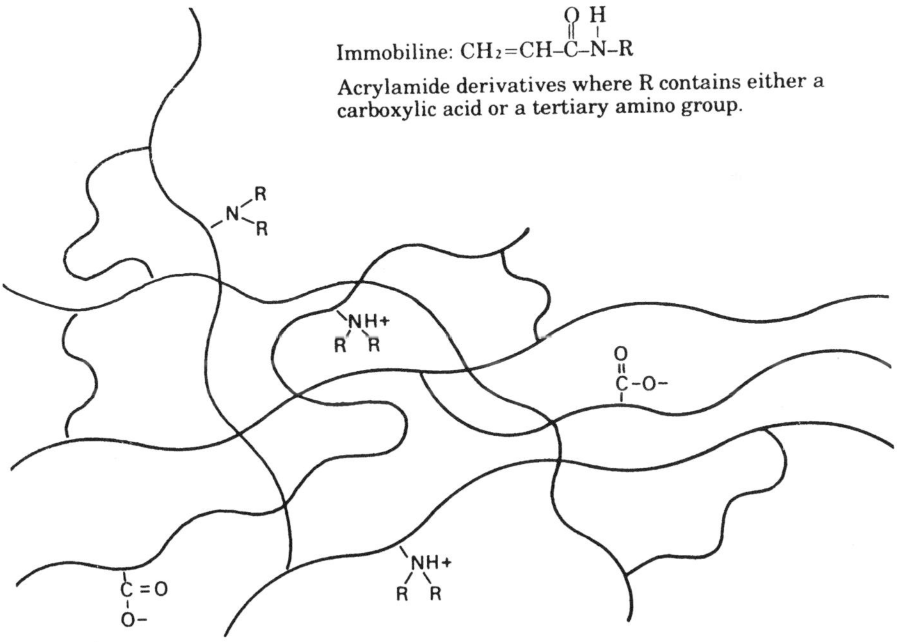

Fig. 12.14. Isoelectric focussing in immobilised pH gradients (IPG). A hypothetical gel structure is depicted, where the strings represent the neutral acrylamide residues, the cross-over points the Bis cross-linking and the positive and negative charges the grafted Immobiline molecules (by courtesy of LKB Produkter AB).

TABLE 12.14

ACIDIC ACRYLAMIDO BUFFERS

p*K*	Formula	Name	M_r
1.2	$CH_2{=}CH{-}CO{-}NH{-}C(CH_3)_2{-}CH_2{-}SO_3H$	2-acrylamido-2-methylpropane sulphonic acid	207
3.1	$CH_2{=}CH{-}CO{-}NH{-}CH(OH){-}COOH$	2-acrylamido-glycolic acid	145
3.6	$CH_2{=}CH{-}CO{-}NH{-}CH_2{-}COOH$	*N*-acryloyl-glycine	129
4.6	$CH_2{=}CH{-}CO{-}NH{-}(CH_2)_3{-}COOH$	4-acrylamido-butyric acid	157

with p*K*s 6.2, 7.0, 8.5 and 9.3 (available under the trade name Immobiline from Pharmacia-Upjohn and now also from Bio-Rad). Their synthesis has been described by Chiari et al. [173,174]. A more extensive set, comprising 10 chemicals (a p*K* 3.1 acidic buffer, a p*K* 10.3 basic buffer and two strong titrants, a p*K* 1 acid and a p*K* > 12 quaternary base) is available as 'p*I* select' from Fluka AG, Buchs, Switzerland (see Tables 12.14 and 12.15 for their formulas) [175]. All of the above chemicals have been reported by our group: e.g. the synthesis of the p*K* 3.1 buffer was utilised by Righetti et al. [172] for separation of isoforms of very acidic proteins (pepsin); the p*K* 10.3 species was first adopted by Sinha and Righetti [176] for creating alkaline gradients for separation of elastase isoforms. Over the years, we have reported the synthesis of a number of other buffering ions, produced with the aim of closing some gaps in between the available Immobilines, especially in the pH 7.0 to 8.5 interval. These are: a p*K* 6.6, 2-thiomorpholinoethylacrylamide and a p*K* 7.4, 3-thiomorpholinopropylacrylamide [177]; a p*K* 6.85, 1-acryloyl-4-methylpiperazine [179]; an alternative p*K* 7.0, 2-(4-imidazolyl)ethylamine-2-acrylamide [180] and a p*K* 8.05, *N,N*-bis(2-hydroxyethyl)*N'*-acryloyl-1,3-diaminopropane [178]. These compounds are listed in Table 12.16. Additional species have recently been described by Bellini and Manchester [181].

During gel polymerisation, these buffering species are efficiently incorporated into the gel (84–86% conversion efficiency at 50°C for 1 h) [182]. Immobiline-based pH gradients can be cast in the same way as conventional polyacrylamide gradient gels, using a density gradient to stabilise the Immobiline concentration gradient, with the aid of a standard, two-vessel gradient mixer. As shown in their formulas, these buffers are no longer amphoteric, as in conventional IEF, but are bifunctional. At one end of the molecule is located the buffering (or titrant) group, and at the other end is an acrylic double bond, which disappears during immobilisation of the buffer on the gel matrix. The three carboxyl Immobilines have rather small temperature coefficients (d*pK*/d*T*) in the 10–25°C range, due to their small standard heat of ionisation (~1 kcal/mol) and thus exhibit negligible p*K* variations in this temperature interval. On the other hand,

TABLE 12.15
BASIC ACRYLAMIDO BUFFERS

p*K*	Formula	Name	M_r
6.2	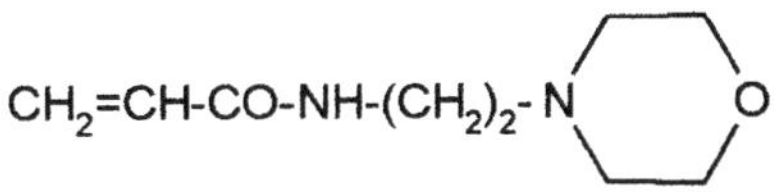	2-morpholinoethylacrylamide	184
7.0	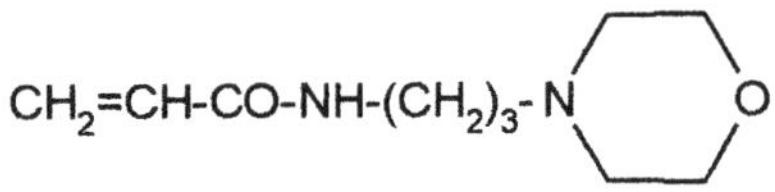	3-morpholinopropylacrylamide	199
8.5	$CH_2=CH-CO-NH-(CH_2)_2-N(CH_3)_2$	*N,N*-dimethyl aminoethyl acrylamide	142
9.3	$CH_2=CH-CO-NH-(CH_2)_3-N(CH_3)_2$	*N,N*-dimethyl aminopropyl acrylamide	156
10.3	$CH_2=CH-CO-NH-(CH_2)_3-N(C_2H_5)_2$	*N,N*-diethyl aminopropyl acrylamide	184
>12	$CH_2=CH-CO-NH-(CH_2)_2-N^+(C_2H_5)_3$	*N,N,N*-triethyl aminoethyl acrylamide	198

TABLE 12.16
NEW BASIC ACRYLAMIDO BUFFERS

p*K*[a]	Formula	Name	M_r	Ref.
6.6	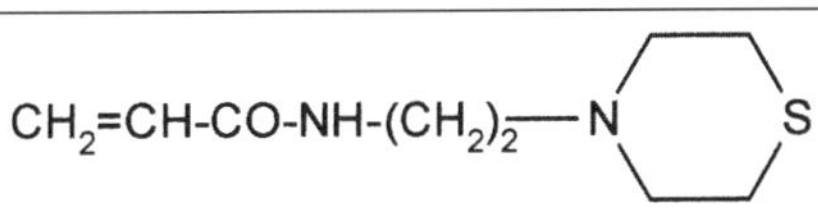	2-Tiomorpholinoethylacrylamide	200	[177]
6.85	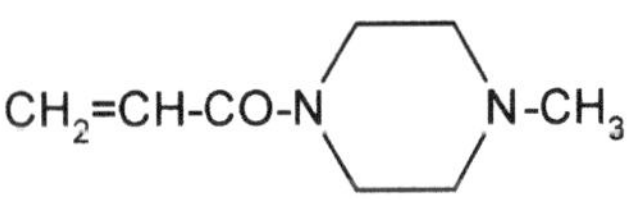	1-Acryloyl-4-methylpiperazine	154	[179]
7.0	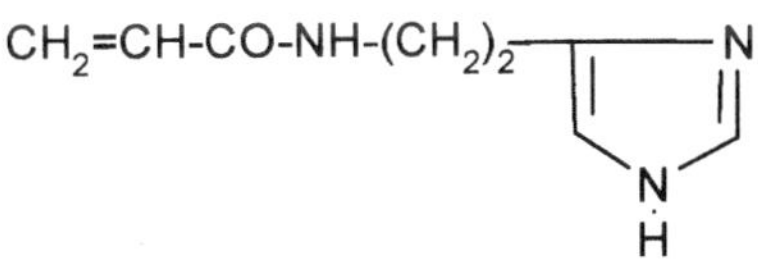	2-(4-Imidazolyl)ethylamine-2-acrylamide	165	[180]
7.4	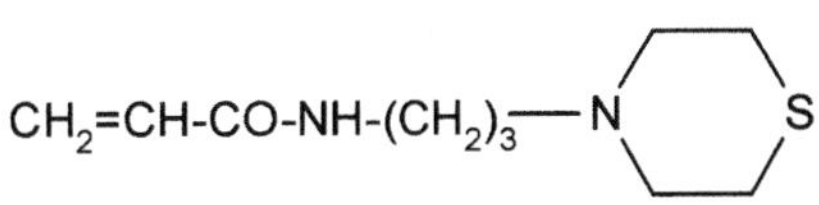	3-Thiomorpholinopropylacrylamide	214	[177]
8.05	$CH_2=CH-CO-NH-(CH_2)_3-N(CH_2CH_2OH)_2$	*N,N*-Bis(2-hydroxyethyl)-*N′*-acryloyl-1,3-diaminopropane	200	[178]

[a] All p*K* values measured at 25ºC.

the five basic Immobilines exhibit rather large ΔpKs (as much as $\Delta pK = 0.44$ for the pK 8.5 species) due to their larger heats of ionisation (6–12 kcal/mol). Therefore, for reproducible runs and pH gradient calculations, all the experimental parameters have been fixed at 10°C.

Temperature is not the only variable that affects Immobiline pKs (and therefore the actual pH gradient generated). Additives in the gel that change the water structure (chaotropic agents, e.g. urea) or lower its dielectric constant, and the ionic strength of the solution, alter their pK values. The largest changes, in fact, are due to the presence of urea: acidic Immobilines increase their pK in 8 M urea by as much as 0.9 pH units, while the basic Immobilines increase their pK by only 0.45 pH unit [183]. Detergents in the gel (2%) do not alter the Immobiline pK, suggesting that they are not incorporated into the surfactant micelle. For generating extended pH gradients, we use two additional chemicals which are strong titrants having pKs well outside the desired pH range. One is QAE (quaternary amino ethyl) acrylamide (p$K > 12$) and the other is AMPS (2-acrylamido-2-methyl propane sulphonic acid, pK 1.0) (see Tables 12.14 and 12.15).

As shown in Fig. 12.14, the proteins are placed on a gel with a preformed, immobilised pH gradient (represented by carboxyl and tertiary amino groups grafted to the polyacrylamide chains). When the field is applied only the sample molecules (and any ungrafted ions) migrate in the electric field. Upon termination of electrophoresis, the proteins are separated into stationary, isoelectric zones. Due to the possibility of designing stable pH gradients at will, separations have been reported in only 0.1 pH unit-wide gradients over the entire separation axis leading to an extremely high resolving power ($\Delta pI = 0.001$ pH unit, see also Section 12.2.1.3).

12.3.1.3. Narrow and ultra narrow pH gradients

We define the gradients from 0.1 to 1 pH unit as narrow (towards the 1 pH unit limit) and ultra-narrow (close to the 0.1 pH unit limit) gradients. Within these limits we work on a tandem principle — that is we choose a buffering Immobiline, either a base or an acid, with its pK within the pH interval we want to generate, and a non-buffering Immobiline, then an acid or a base, respectively, with its pK at least 2 pH units removed from either the minimum or maximum of our pH range. The titrant will provide equivalents of acid or base to titrate the buffering group but will not itself buffer in the desired pH interval. For these calculations, we used to resort to modified Henderson–Hasselbalch equations and to rather complex nomograms found in the LKB application note No. 321. In a later note (No. 324) are listed 58 gradients each 1 pH unit wide, starting with the pH 3.8–4.8 interval and ending with the pH 9.5–10.5 range, separated by 0.1 pH unit increments. In Table 12.17 are recipes giving the Immobiline volume which must be added to give 15 ml of mixture in the acidic (mixing) chamber to obtain pH_{min} and the corresponding volume for the basic (reservoir) chamber of the gradient mixer needed to generate pH_{max} of the desired pH interval. For 1 pH unit gradients between the limits pH 4.6–6.1 and pH 7.2–8.4 there are wide gaps in the pKs of neighbouring Immobilines and so three Immobilines need to be used to generate the desired pH_{min} and pH_{max} values (Table 12.17). As an example, take the pH 4.6–5.6 interval. There are no available Immobilines with pKs within this pH region, so the

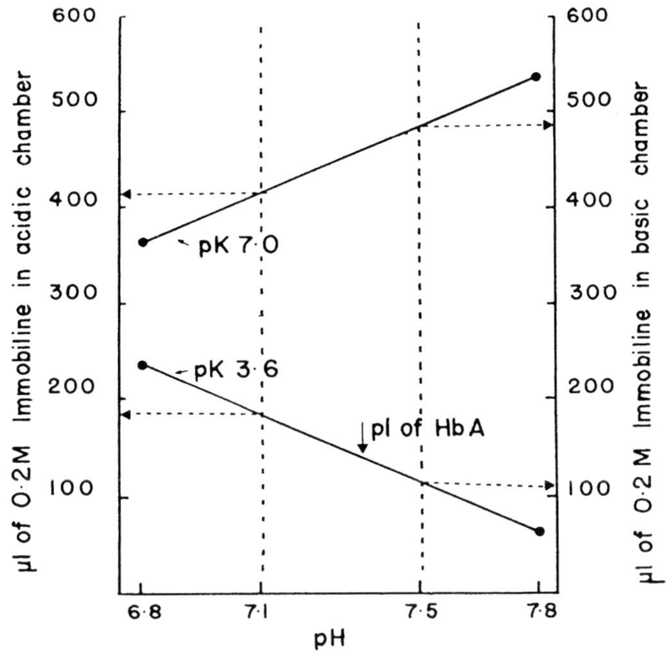

Fig. 12.15. Graphic representation of the preparation of narrow (up to 1 pH unit) IPG gradients on the 'tandem' principle. The limiting molarities of p*K* 7.0 (buffering species) and p*K* 3.6 (titrant) Immobilines needed to generate a pH 6.8–7.8 interval, as obtained directly from Table 12.19, are plotted on a graph. These points are joined by straight lines, and the new molarities needed to generate any narrower pH gradient within the stated pH intervals are obtained by simple linear interpolation (broken vertical and horizontal lines). In this example a narrow pH 7.1–7.5 gradient is graphically derived.

nearest species, p*K*s 4.6 and 6.2, will act as both partial buffers and partial titrants. A third Immobiline is needed in each vessel, a true titrant that will bring the pH to the desired value. As titrant for the acidic solution (pH_{min}) we use p*K* 3.6 Immobiline and for pH_{max} we use p*K* 9.3 Immobiline (Table 12.17).

If a narrower pH gradient is needed, it can be derived from any of the 58 one-pH intervals given in Table 12.17 by a simple linear interpolation of intermediate Immobiline molarities. Suppose that from a pH 6.8–7.8 range, which is excellent for most haemoglobin (Hb) analyses, we want to obtain a pH gradient of 7.1 to 7.5, which will resolve neutral mutants that co-focus with Hb A. Fig. 12.15 shows the graphic method. The limiting molarities of the two Immobilines in the 1 pH unit interval are joined by a straight line (because the gradient is linear), and then the new pH interval is defined according to experimental needs (in our case pH 7.1–7.5). Two new lines are drawn from the two new limits of the pH interval, parallel to the ordinates (broken vertical lines). Where they intersect the two sloping lines defining the two Immobiline molarities, four new lines (dashed) are drawn parallel to the abscissa and four new molarities of the Immobilines defining the new pH interval are read directly on the ordinates. This process can be repeated for any desired pH interval down to ranges as narrow as 0.1 pH units.

TABLE 12.17

1 pH UNIT GRADIENTS: VOLUMES OF IMMOBILINE FOR 15 mL OF EACH STARTING SOLUTION

Control pH at 20°C	Volume (μl) 0.2 M Immobiline p*K* acidic dense solution							pH range	Mid-point	Control pH at 20°C	Volume (μl) 0.2 M Immobiline p*K* basic light solution						
	3.6	4.4	4.6	6.2	7.0	8.5	9.3				3.6	4.4	4.6	6.2	7.0	8.5	9.3
3.84 ± 0.03	–	750	–	–	–	–	159	3.8–4.8	4.3	4.95 ± 0.06	–	750	–	–	–	–	591
3.94 ± 0.03	–	710	–	–	–	–	180	3.9–4.9	4.4	5.04 ± 0.07	–	810	–	–	–	–	667
4.03 ± 0.03	–	–	755	–	–	–	157	4.0–5.0	4.5	5.14 ± 0.06	–	–	745	–	–	–	584
4.13 ± 0.03	–	–	713	–	–	–	177	4.1–5.1	4.6	5.23 ± 0.07	–	–	803	–	–	–	659
4.22 ± 0.03	–	–	689	–	–	–	203	4.2–5.2	4.7	5.33 ± 0.08	–	–	884	–	–	–	753
4.32 ± 0.03	–	–	682	–	–	–	235	4.3–5.3	4.8	5.42 ± 0.10	–	–	992	–	–	–	871
4.42 ± 0.03	–	–	691	–	–	–	275	4.4–5.4	4.9	5.52 ± 0.12	–	–	1133	–	–	–	1021
4.51 ± 0.03	–	–	716	–	–	–	325	4.5–5.5	5.0	5.61 ± 0.14	–	–	1314	–	–	–	1208
4.64 ± 0.03	562	–	600	863	–	–	–	4.6–5.6	5.1	5.69 ± 0.04	–	–	863	863	–	–	105
4.75 ± 0.03	458	–	675	863	–	–	–	4.7–5.7	5.2	5.79 ± 0.04	–	–	863	863	–	–	150
4.86 ± 0.03	352	–	750	863	–	–	–	4.8–5.8	5.3	5.90 ± 0.04	–	–	863	863	–	–	202
4.96 ± 0.03	218	–	863	863	–	–	–	4.9–5.9	5.4	5.99 ± 0.03	–	–	863	863	–	–	248
5.07 ± 0.03	158	–	863	863	–	–	–	5.0–6.0	5.5	6.09 ± 0.04	–	–	863	863	–	–	338
5.17 ± 0.04	113	–	863	683	–	–	–	5.1–6.1	5.6	6.20 ± 0.04	–	–	863	713	–	–	443
5.24 ± 0.18	1251	–	–	1355	–	–	–	5.2–6.2	5.7	6.34 ± 0.04	337	–	–	724	–	–	–
5.33 ± 0.12	1055	–	–	1165	–	–	–	5.3–6.3	5.8	6.43 ± 0.03	284	–	–	694	–	–	–
5.43 ± 0.12	899	–	–	1017	–	–	–	5.4–6.4	5.9	6.53 ± 0.03	242	–	–	682	–	–	–
5.52 ± 0.09	775	–	–	903	–	–	–	5.5–6.5	6.0	6.63 ± 0.03	209	–	–	685	–	–	–
5.62 ± 0.07	676	–	–	817	–	–	–	5.6–6.6	6.1	6.73 ± 0.03	182	–	–	707	–	–	–
5.71 ± 0.06	598	–	–	755	–	–	–	5.7–6.7	6.2	6.82 ± 0.03	161	–	–	745	–	–	–
5.81 ± 0.06	536	–	–	713	–	–	–	5.8–6.8	6.3	6.92 ± 0.03	144	–	–	803	–	–	–
5.91 ± 0.05	486	–	–	689	–	–	–	5.9–6.9	6.4	7.02 ± 0.03	131	–	–	884	–	–	–
6.01 ± 0.05	447	–	–	682	–	–	–	6.0–7.0	6.5	7.12 ± 0.03	120	–	–	992	–	–	–
6.10 ± 0.04	416	–	–	691	–	–	–	6.1–7.1	6.6	7.22 ± 0.03	112	–	–	1133	–	–	–
6.11 ± 0.11	972	–	–	–	1086	–	–	6.2–7.2	6.7	7.21 ± 0.03	262	–	–	–	686	–	–
6.21 ± 0.09	833	–	–	–	956	–	–	6.3–7.3	6.8	7.31 ± 0.03	224	–	–	–	682	–	–
6.30 ± 0.08	722	–	–	–	857	–	–	6.4–7.4	6.9	7.41 ± 0.03	195	–	–	–	694	–	–
6.40 ± 0.07	635	–	–	–	783	–	–	6.5–7.5	7.0	7.50 ± 0.03	171	–	–	–	724	–	–
6.49 ± 0.06	565	–	–	–	732	–	–	6.6–7.6	7.1	7.60 ± 0.03	152	–	–	–	771	–	–

TABLE 12.17 (*continued*)

Control pH at 20°C	Volume (μl) 0.2 M Immobiline p*K* acidic dense solution							pH range	Mid-point	Control pH at 20°C	Volume (μl) 0.2 M Immobiline p*K* basic light solution						
	3.6	4.4	4.6	6.2	7.0	8.5	9.3				3.6	4.4	4.6	6.2	7.0	8.5	9.3
6.59 ± 0.05	509	–	–	–	699	–	–	6.7–7.7	7.2	7.70 ± 0.03	137	–	–	–	840	–	–
6.69 ± 0.05	465	–	–	–	683	–	–	6.8–7.8	7.3	7.80 ± 0.03	125	–	–	–	934	–	–
6.78 ± 0.04	430	–	–	–	684	–	–	6.9–7.9	7.4	7.80 ± 0.03	116	–	–	–	1058	–	–
6.88 ± 0.04	403	–	–	–	701	–	–	7.0–8.0	7.5	8.00 ± 0.03	108	–	–	–	1217	–	–
6.98 ± 0.04	381	–	–	–	736	–	–	7.1–8.1	7.6	8.09 ± 0.03	103	–	–	–	1422	–	–
7.21 ± 0.06	1028	–	–	–	750	750	–	7.2–8.2	7.7	8.36 ± 0.05	548	–	–	–	750	750	–
7.31 ± 0.06	983	–	–	–	750	750	–	7.3–8.3	7.8	8.46 ± 0.05	503	–	–	–	750	750	–
7.41 ± 0.05	938	–	–	–	750	750	–	7.4–8.4	7.9	8.56 ± 0.05	458	–	–	–	750	750	–
7.66 ± 0.15	1230	–	–	–	–	1334	–	7.5–8.5	8.0	8.76 ± 0.04	331	–	–	–	–	720	–
7.75 ± 0.12	1037	–	–	–	–	1149	–	7.6–8.6	8.1	8.85 ± 0.03	279	–	–	–	–	692	–
7.85 ± 0.10	885	–	–	–	–	1004	–	7.7–8.7	8.2	8.95 ± 0.03	238	–	–	–	–	682	–
7.94 ± 0.08	764	–	–	–	–	893	–	7.8–8.8	8.3	9.05 ± 0.06	206	–	–	–	–	687	–
8.04 ± 0.07	667	–	–	–	–	810	–	7.9–8.9	8.4	9.14 ± 0.06	180	–	–	–	–	710	–
8.13 ± 0.06	591	–	–	–	–	750	–	8.0–9.0	8.5	9.24 ± 0.06	159	–	–	–	–	750	–
8.23 ± 0.06	530	–	–	–	–	710	–	8.1–9.1	8.6	9.34 ± 0.06	143	–	–	–	–	810	–
8.33 ± 0.05	482	–	–	–	–	687	–	8.2–9.2	8.7	9.44 ± 0.06	130	–	–	–	–	893	–
8.43 ± 0.04	443	–	–	–	–	682	–	8.3–9.3	8.8	9.54 ± 0.06	119	–	–	–	–	1004	–
8.52 ± 0.04	413	–	–	–	–	692	–	8.4–9.4	8.9	9.64 ± 0.06	111	–	–	–	–	1149	–
8.62 ± 0.04	389	–	–	–	–	720	–	8.5–9.5	9.0	9.74 ± 0.06	105	–	–	–	–	1334	–
8.40 ± 0.14	1208	–	–	–	–	–	1314	8.6–9.6	9.1	9.50 ± 0.06	325	–	–	–	–	–	716
8.49 ± 0.12	1021	–	–	–	–	–	1133	8.7–9.7	9.2	9.59 ± 0.06	275	–	–	–	–	–	691
8.59 ± 0.10	871	–	–	–	–	–	992	8.8–9.8	9.3	9.69 ± 0.06	235	–	–	–	–	–	682
8.68 ± 0.08	753	–	–	–	–	–	884	8.9–9.9	9.4	9.79 ± 0.06	203	–	–	–	–	–	689
8.78 ± 0.07	659	–	–	–	–	–	803	9.0–10.0	9.5	9.88 ± 0.06	177	–	–	–	–	–	713
8.87 ± 0.06	584	–	–	–	–	–	745	9.1–10.1	9.6	9.98 ± 0.06	157	–	–	–	–	–	755
8.97 ± 0.05	525	–	–	–	–	–	707	9.2–10.2	9.7	10.08 ± 0.06	141	–	–	–	–	–	817
9.07 ± 0.04	478	–	–	–	–	–	686	9.3–10.3	9.8	10.18 ± 0.06	129	–	–	–	–	–	903
9.16 ± 0.07	440	–	–	–	–	–	682	9.4–10.4	9.9	10.28 ± 0.06	119	–	–	–	–	–	1017
9.26 ± 0.07	410	–	–	–	–	–	694	9.5–10.5	10.0	10.38 ± 0.06	111	–	–	–	–	–	1165

12.3.1.4. Extended pH gradients: general rules for their generation and optimisation

Linear pH gradients are obtained by arranging for an even buffering power throughout. The latter could be ensured only by ideal buffers spaced apart by $\Delta pK = 1$. In practice, there are only 8 buffering Immobilines with some wider gaps in ΔpKs; therefore other approaches must be used to solve this problem. Two methods are possible. In one approach (constant buffer concentration), the concentration of each buffer is kept constant throughout the span of the pH gradient and 'holes' of buffering power are filled by increasing the amounts of the buffering species bordering the largest ΔpKs. In the other approach (varying buffer concentration) the variation in concentration of different buffers along the width of the desired pH gradient results in a shift in each buffer's apparent pK, together with the ΔpK values evening out. The second approach is by far preferred, since it gives much higher flexibility in the computational approach. In a series of papers [170,184–194]) we have described a computer approach able to calculate and optimise any such pH interval, up to the most extended one (which can cover a span of pH 2.5–11). Tables for these recipes can be found in the book by Righetti [2] and in many of the above references. We prefer not to give here such recipes, since anyone can easily calculate any desired pH interval with the user-friendly computer program (written on a Windows platform) of Giaffreda et al. [195]. However, we will give here general guidelines for the use of such program and optimisation of various recipes.

(a) When calculating recipes up to 4 pH units, in the pH 4–9 interval, there is no need to use strong titrants. As most acidic and basic titrants the pKs 3.1 and 10.3 Immobilines can be used, respectively.

(b) When optimising recipes > 4 pH units (or close to the pH 3 or pH 11 extremes) strong titrants are preferably used, otherwise it will be quite difficult to obtain linear pH gradients, since weak titrants will act, per force, as buffering ions as well.

(c) When calculating recipes of 4 pH units, it is best to insert in the recipe all the 8 weak buffering Immobilines. The computer program will automatically exclude the ones not needed for optimisation.

(d) The program of Giaffreda et al. [195] can calculate not only linear, but also concave or convex exponential gradients (including sigmoidal ones). In order to limit consumption of Immobilines (at high concentration in the gel they could give rise to ominous reswelling and interact with the macromolecule separand via ion-exchange mechanisms), limit the total Immobiline molarity (e.g. to only 15–20 mM) and the average buffering power (β) (these two bits of information are specifically asked when preparing any recipe). In particular, please note that recipes with an average β value of only 1–2 mequiv. l^{-1} pH^{-1} are quite adequate in IPGs. The separand macroions, even at concentration > 10 mg/ml, rarely have β values > 1 microequiv. l^{-1} pH^{-1}.

(e) When working at acidic and alkaline pH extremes, however, please note that the average β power of the recipe should be progressively higher, so as to counteract the β value of bulk water. Additionally, at such pH extremes, the matrix acquires a net positive or negative charge and this gives rise to strong electroosmotic flow (EOF). In order to quench EOF, the washed and dried matrix should be reswollen against a gradient of viscous polymers (e.g. liquid linear polyacrylamide, hydroxyethyl cellulose) [196].

It might be asked how good such wide gradients could be made. Fig. 12.16 gives an example of a linear pH gradient (as calculated with the program of Giaffreda et al.

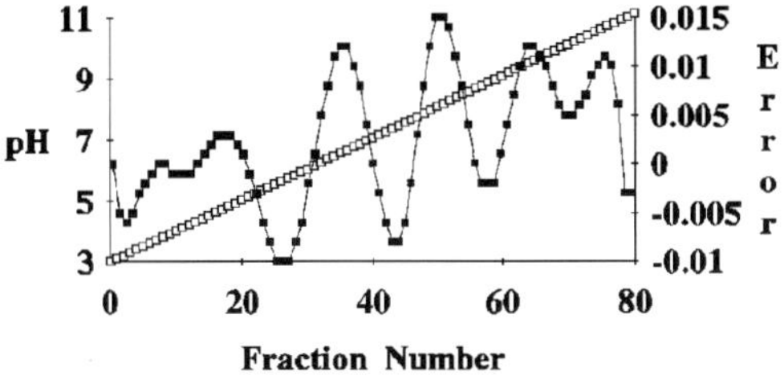

Fig. 12.16. Optimisation of a linear IPG interval spanning the pH 3–11 range. Notice the minute deviation from linearity. Open squares = pH gradient; solid squares = deviation. (From Ref. [195], with permission.)

[195]) covering the pH 3–11 interval: it can be appreciated that the gradient is indeed highly linear, the maximum deviation from the target function being at most ±0.01 pH units, but often considerably smaller than that.

12.3.1.5. Non-linear, extended pH gradients

Although originally most IPG formulations for extended pH intervals had been given only in terms of rigorously linear pH gradients, this might not be the optimal solution in some cases. The pH slope might need to be altered in pH regions that are overcrowded with proteins. This is particularly important in the general case involving the separation of proteins in a complex mixture, such as cell lysates and is imperative thus when performing two-dimensional (2-D) maps. We have computed the statistical distribution of the p*I*s of water-soluble proteins and plotted them in the histogram of Fig. 12.17. From the histogram, given the relative abundance of different species, it is clear that an optimally resolving pH gradient should have a gentler slope in the acidic portion and a steeper profile in the alkaline region. Such a course has been calculated by assigning

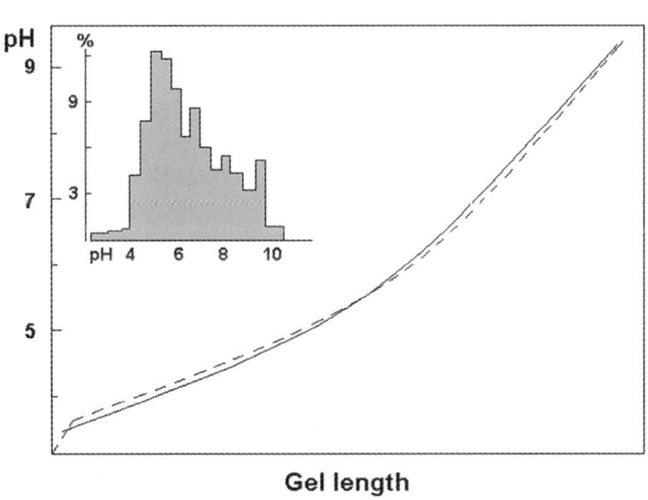

Fig. 12.17. Non-linear pH 4–10 gradient: 'ideal' (solid line) and actual (dashed line) courses. The shape of the 'ideal' profile was computed from data on statistical distribution of protein p*I*s. The relevant histogram is redrawn in the figure inset. (With permission from Ref. [185].)

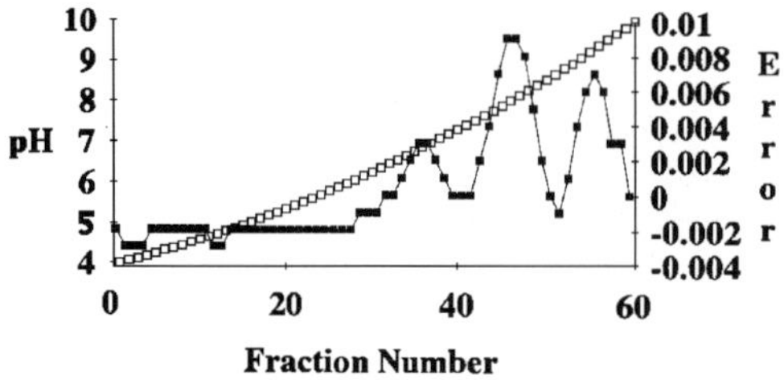

Fig. 12.18. Calculation of a non-linear, concave IPG interval spanning the pH 4–10 range. Notice the minute deviation from the target shape. Open squares = pH gradient; solid squares = deviation. (From Ref. [195], with permission.)

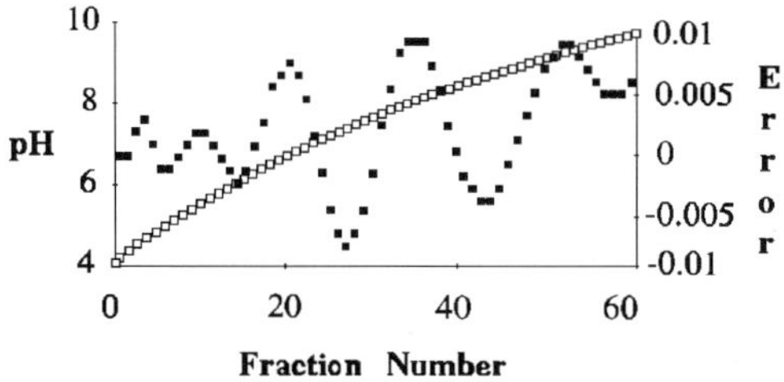

Fig. 12.19. Calculation of a non-linear, convex IPG interval spanning the pH 4–10 range. Notice the minute deviation from the target function. Open squares = pH gradient; solid squares = deviation. (From Ref. [195], with permission.)

to each 0.5 pH unit interval in the pH 3.5–10 region a slope inversely proportional to the relative abundance of proteins in that interval. This generated the ideal curve (dotted line) in Fig. 12.17. Here too it might be asked how good is the gradient compared with the target function. Although that gradient had been optimised with somewhat more primitive algorithms (it was in 1985) we have recalculated it with the program of Giaffreda et al. [195]: as shown in Fig. 12.18, this non-linear, concave gradient, can be made with extreme accuracy, the error, at least in the acidic region of the pH interval, being close to zero. Just as our program can optimise non-linear, concave gradients, it can also generate convex gradients, as shown in Fig. 12.19; here too the maximum error being barely ±0.01. What is also important, in the present case, is the establishment of a new principle in IPG technology, namely that the pH gradient and the density gradient stabilising it need not be co-linear, because the pH can be adjusted by localised flattening for increased resolution while leaving the density gradient unaltered. Although non-linear gradients, of just about any shape, are much in vogue in chromatography, it should be emphasised that the gradients displayed here are somewhat unique, in that they are created an optimised simply with the aid of a two-vessel gradient mixer, by manipulating the composition (in Immobilines) of the two limiting solutions. On the contrary, in chromatography, this is achieved by drawing solutions of different composition from several vessels and/or altering the pumping speed of said solutions to be then mixed in a mixing chamber. Conversely, our set-up, for casting Immobiline gels of any pH shape, is gadget-free! Though we have considered only examples of extended pH gradients, narrower pH intervals can be treated in the same fashion.

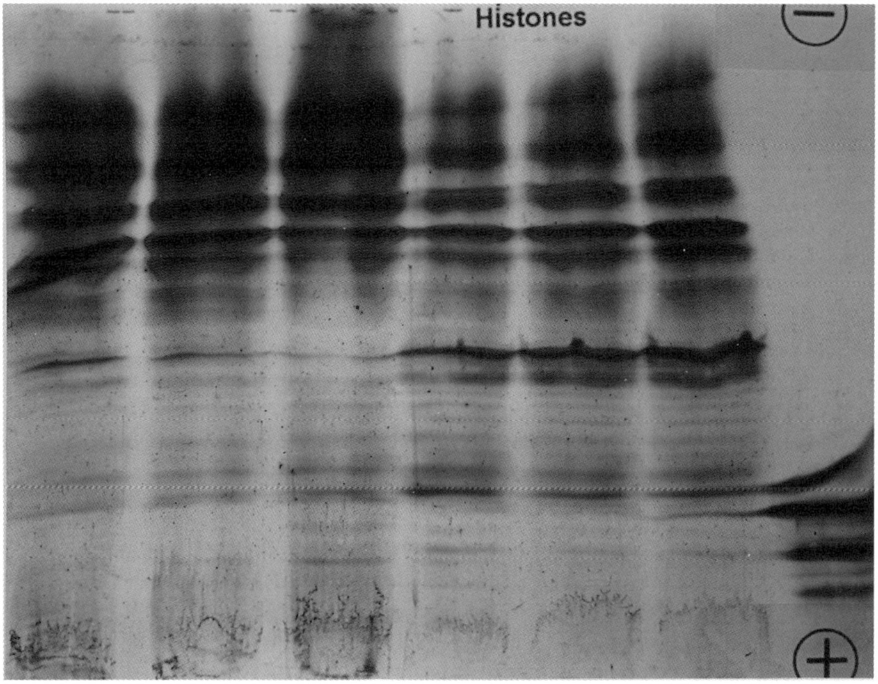

Fig. 12.20. Focussing of histones in a pH 10–12 interval. Gel: 6% *T*, 4% *C* polyacrylamide matrix, containing an IPG 10–12 gradient, reswollen in 7 M urea, 1.5% Nonidet P-40 and 0.5% Ampholine pH 9–11. The gel was run at 10ºC under a layer of light paraffin oil at 500 V for the first hour, followed by increasing voltage gradients, after sample penetration, up to 1300 V for a total of 4 h. The samples (2 mg/ml; 50 μl seeded) were loaded in plastic wells at the anodic gel surface. Staining with Coomassie Brilliant Blue R-250 in Cu^{2+}. Samples (from left to right): tracks 1–3: VIII-S histones; 4–6: II-AS histones; 7: cytochrome C (the main upper band has a p*I* of 10.6). (From Ref. [198], with permission.)

12.3.1.6. Extremely alkaline pH gradients

We have recently optimised a recipe for producing an extremely alkaline immobilised pH gradient, covering non-linearly the pH 10–12 interval, for separation of very alkaline proteins, such as subtilisins and histones [197,198]. Successful separations were obtained in 6% *T*, 4% *C* polyacrylamide matrices, reswollen in 8 M urea, 1.5% Tween 20, 1.5% Nonidet P-40 and 0.5% Ampholine pH 9–11. Additionally, in order to quench the very high conductivity of the gel region on the cathodic side, the reswelling solution contained a 0–10% (anode to cathode) sorbitol gradient (or an equivalent 0–1% HEC gradient). Best focussing was obtained by running the gel at 17ºC, instead of the customary 10ºC temperature. In the case of histones, all their major components had p*I* values between pH 11 and 12 and only minor components (possibly acetylated and phosphorylated forms) focussed below pH 11. By summing up all bands in Arg- and Lys-rich fractions, 8–10 major components and at least 12 minor zones were clearly resolved (Fig. 12.20). This same recipe could be used as first dimension run for a 2-D separation of histones [199].

Protocol 9. IPG flow sheet

1. Assemble the gel mould (Protocol 10) and mark the polarity of the pH gradient on the back of the supporting plate.
2. Mix the required amounts of Immobilines. Fill to one half of the final volume with distilled water.
3. Check the pH of the solution and adjust as required.
4. Add the correct volume of 30% *T* acrylamide–bisacrylamide monomer (Table 12.5), glycerol (0.2–0.3 ml/ml of the 'dense' solution only) and TEMED (Table 12.18) and bring to final volume with distilled water.
5. For ranges removed from neutrality, titrate to about pH 7.5 using Tris–base for acidic solution and acetic acid for alkaline solutions.
6. Transfer the denser solution to the mixing chamber and the lighter solution to the other reservoir of the gradient mixer. Centre the mixer on a magnetic stirrer and check for the absence of air bubbles in the connecting duct.
7. Add ammonium persulphate to the solutions as specified in Table 12.18.
8. Allow the gradient to pour into the mould from the gradient mixer.
9. Allow the gel to polymerise for 1 h at 50°C.
10. Disassemble the mould and weigh the gel.
11. Wash the gel for 1 h for three times (20 min each) with 200 ml of distilled water with gentle shaking.
12. Reduce the gel back to its original weight using a non-heating fan.
13. Transfer the gel to the electrophoresis chamber (at 10°C) and apply the electrodic strips (as described in Protocol 4).
14. Load the samples and start the run.
15. After the electrophoresis, stain the gel to detect the separated proteins.

12.3.2. IPG methodology

The overall procedure is outlined in Protocol 9. Note that the basic equipment required is the same as for conventional CA-IEF gels. Thus the reader should consult Sections 12.2.2.1 and 12.2.2.2. In addition, as we essentially use the same polyacrylamide matrix, the reader is referred to Section 12.2.3 for a description of its general properties.

12.3.2.1. Casting an Immobiline gel

When preparing for an IPG experiment, two pieces of information are required: the total liquid volume needed to fill the gel cassette, and the required pH interval. Once the first is known, this volume is divided into two halves: one half is titrated to one extreme of the pH interval, the other to the opposite extreme. As the analytical cassette usually has a thickness of 0.5 mm and, for the standard 12 $\times$ 25 cm size (see Fig. 12.5) contains 15 ml of liquid to be gelled, in principle two solutions, each of 7.5 ml, should be prepared. However, because the volume of some Immobilines to be added to 7.5 ml might sometimes be rather small (i.e. <50 μl), we prefer to prepare a double volume, which will be enough for casting two gel slabs. The Immobiline solutions (mostly the basic ones) tend to leave droplets on the plastic disposable tips of micropipettes. For

Protocol 10. Assembling the mould

Equipment
- Gel mould (Pharmacia)
- Dymo tape
- Repel-silane (ready to be used from Sigma or as described in Protocol 1)
- Gel Bond PAG film (Pharmacia)

Method

1. Wash the glass plate bearing the U-frame with detergent and rinse with distilled water.
2. Dry with paper tissue.
3. To mould sample application slots in the gel, apply suitably-sized pieces of Dymo tape to the glass plate with the U-frame; a 5 × 3 mm slot can be used for sample volumes between 5 and 20 μl (this step is only necessary when preparing a new mould or re-arranging an old one; see Fig. 12.4A). To prevent the gel from sticking to the glass plates with U-frame and slot former, coat them with Repel-Silane according to Protocol 1. Make sure that no dust or fragments of gel from previous experiments remain on the surface of the gasket, since this can cause the mould to leak.
4. Use a drop of water on the Gel Bond PAG film to determine the hydrophilic side. Apply a few drops of water to the plain glass plate and carefully lay the sheet of Gel Bond PAG film on top with the hydrophobic side down (see Fig. 12.4B). Avoid touching the surface of the film with fingers. Allow the film to protrude 1 mm over one of the long side of the plate, as a support for the tubing from the gradient mixer when filling the cassette with gel solution (but only if using a cover plate without any V-indentations). Roll the film flat to remove air bubbles and to ensure good contact with the glass plate.
5. Clamp the glass plates together with the Gel Bond PAG film and slot former on the inside, by means of clamps placed all along the U-frame, opposite to the protruding film. To avoid leakage, the clamps must be positioned so that the maximum possible pressure is applied (see Fig. 12.4C).

---

accurate dispensing, therefore, we suggest rinsing the tips once or twice with distilled water after each measurement. The polymerisation cassette is filled with the aid of a two-vessel gradient mixer and thus the liquid elements which fill the vertically standing cassette have to be stabilised against remixing by a density gradient. In Table 12.17 the two solutions are called 'acidic dense' and 'basic light' solutions. This choice is, however, a purely conventional one, and can be reversed, provided one marks the bottom of the mould as the cathodic side. In order to understand the sequence of steps needed, let us refer to Protocol 11 (as a general example of an IPG protocol) and to Fig. 12.18 for the final gel assembly.

(i) *Preparation of the gel mould.* Fig. 12.21 gives the final assembly for cassette and gradient mixer. The cassette is assembled as described in Protocol 10. Note that inserting the capillary tubing conveying the solution from the mixer into the cassette is greatly facilitated when using a cover plate bearing 3 V-shaped indentations. As for the gradient mixer, it should be noted that one chamber contains a magnetic stirrer, while in the reservoir is inserted a plastic cylinder having the same volume, held by a trapezoidal rod. The latter, in reality, is a 'compensating cone' needed to raise the liquid level to

Protocol 11. Polymerisation of a linear pH gradient gel

Equipment and reagents

- Gradient maker (the Pharmacia-Hoefer catalogues offer a large choice of gradient-mixers, in this case very suitable, the model having vessels of 15 or 30 ml volume)
- Magnetic stirrer and stirring bar
- Gel mould with V-indentations (see text and Fig. 12.4A)
- Oven at 50°C
- Basic light and acidic dense gel mixtures (see Section 12.3.2.1); 7.5 ml of each
- Non-heating fan

Method

1. Check that the valve in the gradient mixer and the clamp on the outlet tubing are both closed.
2. Transfer 7.5 ml of the basic, light solution, to the reservoir chamber.
3. Slowly open the valve just enough to fill the connecting channel with the solution, and quickly close it again. Then transfer 7.5 ml of the acidic dense solution to the mixing chamber.
4. Place the prepared mould upright on a levelled surface. The optimum flow rate is obtained when the outlet of the gradient mixer is 5 cm above the top of the mould. Open the clamp of the outlet tubing, fill the tubing halfway with the dense solution, and close the clamp again.
5. Switch on the stirrer, and set to a speed of about 500 rpm.
6. Add the catalysts to each chamber as specified in Table 12.18.
7. Insert the free end of the tubing between the glass plates of the mould at the central V-indentation (Fig. 12.21).
8. Open the clamp on the outlet tubing, then immediately open the valve between the dense and light solutions so that the gradient solution starts to flow down into the mould by gravity. Make sure that the levels of liquid in the two chambers fall at the same rate. The mould will be filled within 5 min. To assist the mould to fill uniformly across its width, the tubing from the mixer may be substituted with a 2- or 3-way outlet assembled from small glass or plastic connectors (e.g. spare parts of chromatographic equipment) and butterfly needles.
9. When the gradient mixer is empty, carefully remove the tubing from the mould. After leaving the cassette to rest for 5 min, place it on a levelled surface in an oven at 50°C. Polymerisation is allowed to continue for 1 h. Meanwhile, wash and dry the mixer and tubing.
10. When polymerisation is complete, remove the clamps and carefully take the mould apart. Start by removing the glass plate from the supporting foil. Then hold the remaining part so that the glass surface is on top and the supporting foil underneath. Gently peel the gel away from the slot former, taking special care not to tear the gel around the slots.
11. Weigh the gel and then place it in 300 ml of distilled water for 1 h to wash out any remaining ammonium persulphate, TEMED and unreacted monomers and Immobilines. Change the water three times (changes are every 20 min).
12. After washing the gel, carefully remove any excess water from the surface with a moist paper tissue. To remove the water absorbed by the gel during the washing step, leave it at room temperature until the weight has returned to within 5% of the original weight. To shorten the drying time, use a non-heating fan placed at about 50 cm from the gel to increase the rate of evaporation. Check the weight of the gel after 5 min and, from this, estimate the total drying time. The drying step is essential, as a gel containing too much water will 'sweat' during the electrofocussing run and droplets of water will form on the surface. However, if the gel dries too much, the value of $\%T$ will be increased, resulting in longer focussing times and a greater sieving effect.

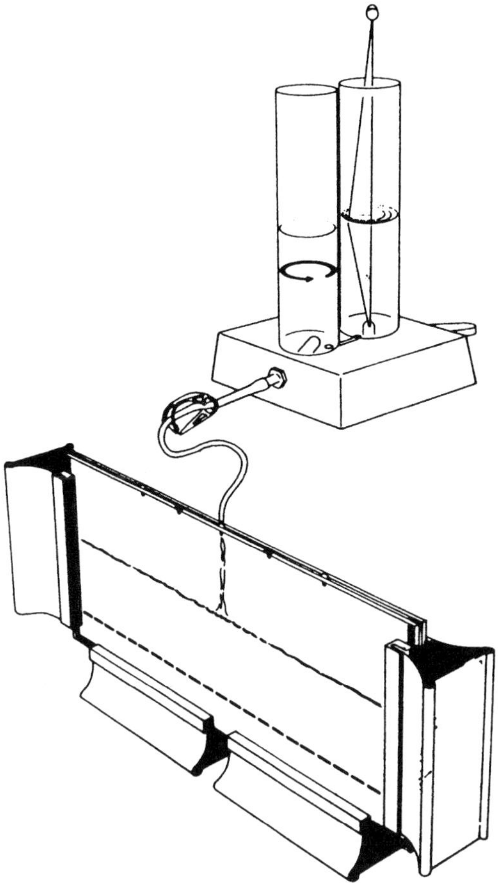

Fig. 12.21. Set-up for casting an IPG gel. A linear pH gradient is generated by mixing equal volumes of a dense and light solution, titrated to the extremes of the desired pH interval. Note the 'compensating' rod in the reservoir, used as a stirrer after addition of catalysts and for hydrostatically equilibrating the two solutions. Insertion of the capillary conveying the solution from the mixer to the cassette is greatly facilitated by using modern cover plates, bearing 3 V-shaped indentations. (By courtesy of LKB Produkter AB.)

such an extent that the two solutions (in the mixing chamber and in the reservoir) will be hydrostatically equilibrated. In addition, this plastic rod can also be utilised for manually stirring the reservoir after addition of TEMED and persulphate.

(ii) *Polymerisation of a linear pH gradient.* It is preferable to use 'soft' gels, i.e. with a low $\%T$. Originally, all recipes were given for 5% T matrices, but today we prefer 3.5–4% T gels. These 'soft' gels can be easily dried without cracking and allow better entry of larger proteins. In addition, the local ionic strength along the polymer coil is increased, and this permits sharper protein bands due to increased solubility at the pI. A linear pH gradient is generated by mixing equal volumes of the two starting solutions in a gradient mixer. It is a must, for any gel formulation removed from neutrality (pH 6.5–7.5) to titrate the two solutions to neutral pH, so as to ensure reproducible polymerisation

TABLE 12.18

WORKING CONCENTRATIONS FOR THE CATALYSTS

Gel type	TEMED (μl/ml)		40% w/v Ammonium persulphate (μl/ml)
	acidic pH	basic pH	
Lower limit	0.5	0.3	0.6
Standard, %T = 5 [a]	0.5	0.3	0.8
Standard, %T = 3	0.7	0.5	1.0
For 5–10% alcohol	0.7	0.5	1.0
Higher limit [b]	0.9	0.6	1.4

[a] From ref. [59].

[b] From LKB Application Note no. 321.

conditions and avoid hydrolysis of the five alkaline buffering Immobilines. If the pH interval used is acidic, add Tris, if it is basic, add formic acid. We recommend that a minimum of 15 ml of each solution (enough for two gels) is prepared and that the volumes of Immobiline needed are measured with a well-calibrated microsyringe to ensure high accuracy. Prepare the acidic, dense solution and the basic, light solution for the pH gradient as described in Protocol 9 (stock acrylamide solutions are given in Table 12.2; for the catalysts, refer to Table 12.18). If the same gradient is to be prepared repeatedly, the buffering and non-buffering Immobiline and water mixtures can be prepared as stock solutions and stored according to the recommendations for Immobiline. Prepared gel solutions must not be stored. However, gels with a pH less than 8 can be stored in a humidity chamber for up to one week after polymerisation. An example of a preparation of a linear pH gradient is given in Protocol 11.

12.3.2.2. Reswelling dry Immobiline gels

Precast, dried Immobiline gels, encompassing a number of ranges, are now available from Amersham Pharmacia Biotech and from Bio-Rad. They all contain 4% T and they span the following pH narrow ranges: pH 3.5–4.5; pH 4.0–5.0 (e.g. for α_1-antitrypsin analysis); pH 4.5–5.5 (e.g. for Gc screening); pH 5.0–6.0 (e.g. for transferrin analysis); pH 5.5–6.7 (e.g. for phosphoglucomutase screening). In addition, there are a number of wide pH ranges: pH 4–7L; pH 6–9; pH 6–11; pH 3–10L and pH 3–10NL (L = linear; NL = non-linear). Some of them are available in 7, 11, 13 and 18 cm in length, whereas all the narrow ranges are cast only as long (18 cm) strips. All of them are 3 mm wide and, when reswollen, 0.5 mm thick (gel layer). Pre-cast, dried IPG gels in alkaline narrow ranges should be handled with care, because at high pHs the hydrolysis of both the gel matrix and the Immobiline chemicals bound to it is much more pronounced.

It has been found that the diffusion of water through Immobiline gels does not follow a simple Fick's law of passive transport from high (the water phase) to zero (the dried gel phase) concentration regions, but it is an active phenomenon: even under isoionic conditions, acidic ranges cause swelling 4–5 times faster than alkaline ones [200]. Given these findings, it is preferable to reswell dried Immobiline gels in a cassette similar to the one for casting the IPG gel. Fig. 12.22 shows the reswelling system: the dried gel is inserted in the cassette, which is clamped and allowed to stand on the short side. The

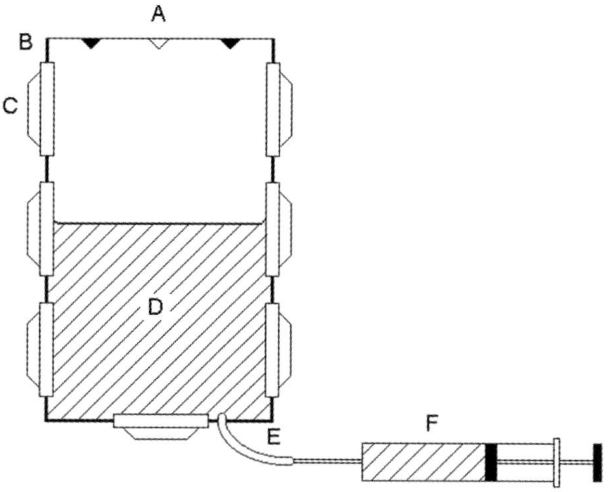

Fig. 12.22. Reswelling cassette for dry IPG gels. The dried IPG gel (on its plastic backing) is inserted in the cassette, which is then gently filled with any desired reswelling solution via a bottom hole with the help of tubing and a syringe. A = gel cassette; B = gasket; C = clamps; D = gel solution; E = tubing; F = syringe.

reswelling solution is gently injected into the chamber via a small hole in the lower right side using a cannula and a syringe, until the cassette is completely filled. As the system is volume-controlled, it can be left to reswell overnight, if needed. Gel drying and reswelling is the preferred procedure when an IPG gel containing additives is needed. In this case it is always best to cast an 'empty' gel, wash it, dry it and then reconstitute it in presence of the desired additive (e.g. urea, alkyl ureas, detergents, carrier ampholytes and mixtures thereof).

12.3.2.3. Electrophoresis

A list of the electrode solutions in common use can be found in Table 12.19. A common electrophoresis protocol consists of an initial voltage setting of 500 V, for 1–2 h, followed by an overnight run at 2–2500 V. Ultranarrow gradients are further subjected to a couple of hours at 5000 V, or better at about 1000 V/cm across the region containing the bands of interest.

TABLE 12.19

IPG ELECTRODE SOLUTIONS

Substance	Application	Concentration
Glutamic acid	anolyte	10 mM
Lysine	catholyte	10 mM
Carrier ampholytes [a,b]	both electrolytes	0.3–1%
Distilled water	both electrolytes	–

[a] Of the same or of a narrower range than the IPG.
[b] For mixed-bed gels or for samples with high salt concentration.

Protocol 12. Silver staining procedure for IPG gels

Equipment and reagents

• Ethanol, acetic acid, glutaraldehyde, silver nitrate, NH_4OH, sodium hydroxide, sodium thiosulphate pentahydrate, formaldehyde 37%, citric acid.

Method

The procedure requires to expose the gels to the series of steps listed below:

1. fixation, 40% ethanol, 10% acetic acid 4 × 5 min
2. rinse, 20% ethanol, 5 min
3. rinse, water, 5 min
4. sensitisation, 12.5% glutaraldehyde, 15 min
5. rinse, water, 2 × 10 min
6. rinse, 20% ethanol, 2 × 10 min
7. silver, $AgNO_3$ 2 g/l, 5% NH_4OH 10 ml, NaOH 50 ml/l (in 20% ethanol), 15 min
8. rinse, 20% ethanol 2 × 5 min
9. rinse, 150 mg/l sodium thiosulphate pentahydrate in 20% ethanol 30 s
10. development, formaldehyde 1 ml/l, citric acid 0.1 g/l in 20% ethanol 2–5 min
11. stop, 5% acetic acid, 20% ethanol 30 min
12. store before drying, 20% ethanol >1 h

12.3.2.4. Staining and pH measurements

IPGs tend to bind strongly to dyes, so the gels are better stained for a relatively short time (30–60 min) with a stain of medium intensity, e.g. the second method listed in Table 12.11. For silver staining, a novel recipe optimised for IPG gels has been published [201] (see Protocol 12). A novel fluorescent staining protocol, exploiting the dye SYPRO Ruby, has been reported in Section 12.2.5.6 and is valid for both conventional IEF and IPGs. It appears to have the highest possible sensitivity so far reported for any staining procedure.

Accurate pH measurements are virtually impossible by equilibration between a gel slice and excess water, and not very reliable with a contact electrode, although, in mixed Immobiline–Ampholine gels this is feasible, due to elution of the soluble CA buffers into the supernatant [189,202,203]. One can preferably either refer to the banding pattern of a set of marker proteins, or elute CAs from a mixed bed gel, when applicable (see Section 12.3.2.6; for the correction factors to be applied for temperature of the run different from 10°C and for the effect of urea, see Section 12.3.1.2).

12.3.2.5. Storage of the Immobiline chemicals

There are two major problems with the Immobiline chemicals, especially with the alkaline ones: hydrolysis and spontaneous auto-polymerisation. Hydrolysis is quite a nuisance because then only acrylic acid is incorporated into the IPG matrix, with a strong acidification of the calculated pH gradient. Hydrolysis is an auto-catalysed process for the basic Immobilines, since it is pH-dependent. For the pK 8.5 and 9.3 species, such a cleavage reaction on the amido bond can occur even in the frozen state, at a rate of about 10% per year [204]. Auto-polymerisation is also quite deleterious for the IPG technique. Again, this reaction occurs particularly with alkaline Immobilines, and is purely auto-catalytic, as it is greatly accelerated by deprotonated amino groups. Oligomers and *n*-mers are formed which stay in solution and can even be incorporated into the IPG gel. These products of auto-polymerisation, when added to proteins in solution, are able to bridge them via two unlike binding surfaces. A lattice is formed and the proteins (especially larger ones, like ferritin, α_2-macroglobulin, thyroglobulin) are precipitated out of solution. This precipitation effect is quite strong and begins even at the level of short oligomers (>decamer) [205–207]. A curious anecdote is due here: these problems of auto-polymerisation and oligomers formation, although discovered only in 1987, originated in 1986, since a few scientists lamented, at that time, loss of large serum proteins from 2-D maps [208]. It took us one year of hard work to find out what caused the problem and suggest remedies. When we finally discovered the oligomers and published our first paper on Electrophoresis, we called it the 'Chernobyl effect'. It should be remembered that the radioactive cloud, meandering on the skies of Europe in April 1986, also navigated over the skies of Stockholm (and Bromma, a suburb where LKB labs. were located and all Immobiline bottles stored). There is nothing like gamma rays to induce polymerisation of even the most stubborn double bonds. However, since we had no direct proof of it, the editor of Electrophoresis (the Berlin wall was still in its place, two German countries still existed and the journal was printed in Germany), afraid of creating a diplomatic accident, had any reference to Chernobyl deleted. It is a fact, however, that this problem rarely surfaced after those disastrous years (and probably after all the bottles full of oligomers had been either destroyed by the producer or consumed by customers), and was rarely reported after that (the last sight was from Esteve-Romero et al. [209]). One clear fact has emerged: the species most prone to auto-polymerisation is the p*K* 7.0 Immobiline and, in present-day Immobilines, we could detect, by mass spectrometry, only traces (<0.5%) of barely dimers (Hamdan and Righetti, unpublished observations, 1999).

These problems with the basic Immobilines could potentially remove one of the major advantages of the IPG technique, namely its high reproducibility from run to run. As a remedy to these drawbacks, it has been shown that, when dissolved in anhydrous *n*-propanol (containing a maximum of 60 ppm water), these species are stabilised against both hydrolysis and auto-polymerisation for a virtually unlimited period of time (less than 1% degradation per year even when stored at +4°C). Thus, present-day alkaline Immobiline bottles are now supplied as 0.2 M solutions in *n*-propanol. The acidic Immobilines, being much more stable, are available as water solutions laced with 10 ppm of an inhibitor [211].

12.3.2.6. Mixed-bed, CA-IPG gels

In CA-IPG gels the primary, immobilised pH gradient, is admixed with a secondary, soluble carrier ampholyte-driven pH gradient. It sounds strange that, given the problems connected with the CA buffers (discontinuities along the electrophoretic path, pH gradient decay, etc.), which the IPG technique was supposed to solve, one should resurrect this past methodology. In fact, when working with membrane proteins ([212]) and with microvillar hydrolases, partly embedded in biological membranes [69], we found that the addition of CAs to the sample and IPG gel would increase protein solubility, possibly by forming mixed-micelles with the detergent used for membrane solubilisation or by directly complexing with the protein itself. It is a fact that, in the absence of CAs, these same proteins essentially fail to enter the gel and mostly precipitate or give elongated smears around the application site (in general cathodic sample loading). More recently, it was found that, on a relative hydrophobicity scale, the five basic Immobilines (p*K*s 6.2, 7.0, 8.5, 9.3 and 10.3) are decidedly more hydrophobic than their acidic counterparts (p*K*s 3.1, 3.6 and 4.6). Upon incorporation in the gel matrix, the phenomenon becomes cooperative and could lead to the formation of hydrophobic patches on the surface of such a hydrophilic gel as polyacrylamide. As the strength of a hydrophobic interaction is directly proportional to the product of the cavity area times its surface tension, it is clear that experimental conditions which lead to a decrement of molecular contact area axiomatically weaken such interactions. Thus CAs might quench the direct hydrophobic protein–IPG matrix interaction, effectively detaching the protein from the surrounding polymer coils and allowing good focussing into sharp bands. For this to happen, the CA shielding species should already be impregnated in the Immobiline gel and present in the sample solution as well. In other words, CAs can only prevent the phenomenon and cannot cure it a posteriori. It has been additionally found that addition of CAs to the sample protects it from strongly acidic and alkaline boundaries originating from the presence of salts in the sample zone (especially 'strong' salts, such as NaCl, phosphates, etc.) [171].

A note of caution should be mentioned concerning the indiscriminate use of the CA-IPG technique: at high CA levels (>1%) and high voltages (>100 V/cm) these gels start exuding water with dissolved carrier ampholytes, with severe risks of short-circuits, sparks and burning on the gel surface. The phenomenon is minimised by chaotropes (e.g. 8 M urea) by polyols (e.g. 30% sucrose) and by lowering the CA molarity in the gel [213]. As an answer to the basic question of when and how much CAs to add, we suggest the following guidelines: (a) if your sample focusses well as such, ignore the mixed-bed technique (which presumably will be mostly needed with hydrophobic proteins and in alkaline pH ranges); (b) add only the minimum amount of CAs (in general around 1%) needed for avoiding sample precipitation in the pocket and for producing sharply focussed bands.

12.3.3. Troubleshooting

One could cover pages with a description of all the troubles and possible remedies in any methodology. However, we will just list here only the major ones encountered with the IPG technique and we refer the readers to Table 12.20 for all the possible causes and remedies suggested. We highlight the following points.

(a) When the gel is gluey and there is poor incorporation of Immobilines, the biggest offenders are generally the catalysts (e.g. too old persulphate, crystals wet due to adsorbed humidity; wrong amounts of catalysts added to the gel mix); check in addition the polymerisation temperature and the pH of the gelling solutions.

(b) Bear in mind the last point in Table 12.20: if you have done everything right, and still you do not see any focussed protein, you might have simply positioned the platinum

TABLE 12.20

TROUBLESHOOTING GUIDE FOR IPGs

Symptom	Cause	Remedy
Drifting of pH during measurement of basic starting solution	Inaccuracy of glass pH electrodes (alkaline error)	Consult information supplied by electrode manufacturer
Leaking mould	Dust or gel fragment on the gasket	Carefully clean the gel plate and gasket
The gel consistency is not firm, gel does not hold its shape after removal from the mould	Inefficient polymerisation	Prepare fresh ammonium persulphate and check that the recommended polymerisation conditions are being used
Plateau visible in the anodic and/or cathodic section of the gel during electrofocussing, no focussing proteins seen in that part of the gel	High concentration of salts in the system	Check that the correct amounts of ammonium persulphate and TEMED are used
Overheating of gel near sample application when beginning electrofocussing	High salt content in the sample	Reduce salt concentration by dialysis or gel filtration
Non-linear pH gradient	Back-flow in the gradient mixer	Find and mark the optimal position for the gradient mixer on the stirrer
Refractive line at pH 6.2 in the gel after focussing	Unincorporated polymers	Wash the gel in 2 l of distilled water; change the water once and wash overnight
Curved protein zones in that portion of the gel which was at the top of the mould during polymerisation	Too rapid polymerisation	Decrease the rate of polymerisation by putting the mould at −20°C for 10 min before filling it with the gel solution, or place the solutions at 4°C for 15 min before casting the gel
Uneven protein distribution across a zone	Slot or sample application not perpendicular to running direction	Place the slot or sample application pieces perfectly perpendicular to the running direction
Diffuse zones with unstained spots, or drops of water on the gel surface during electrofocussing	Incomplete drying of the gel after the washing step	Dry the gel until it is within 5% of its original weight
No zones detected	Gel is focussed with the wrong polarity	Mark the polarity on the gel when removing it from the mould

wires on the gel with the wrong polarity. Unlike conventional IEF gel, in IPGs the anode has to be positioned at the acidic (or less alkaline) gel extremity, while the cathode has to be placed at the alkaline (or less acidic) gel end.

12.3.4. Some analytical results with IPGs

We will limit this section to some examples of separations in ultranarrow pH intervals, where the tremendous resolving power (ΔpI) of IPGs can be fully appreciated.

The (ΔpI) is the difference, in surface charge, in pI units, between two barely resolved protein species. Rilbe has defined (ΔpI) as:

$$\Delta(pI) = 3.07\sqrt{\frac{D[d(pH)/dx]}{E[-du/d(pH)]}}$$

where D and $du/d(pH)$ are the diffusion coefficient and titration curve of proteins, E is the voltage gradient applied and $d(pH)/dx$ is the slope of the pH gradient over the separation distance. Experimental conditions that minimise ΔpI will maximise the resolving power. Ideally, this can be achieved by simultaneously increasing E and decreasing $d(pH)/dx$, an operation for which IPGs seem well suited. As stated previously (see Section 12.2.1.3), with conventional IEF it is very difficult to engineer pH gradients that are narrower than 1 pH unit. One can push the ΔpI, in IPGs, to the limit of 0.001 pH unit, the corresponding limit in CA-IEF being only 0.01 pH unit. We began to investigate the possibility of resolving neutral mutants, which carry a point mutation involving amino acids with non-ionisable side chains and are, in fact, described as 'electrophoretically silent' because they cannot be distinguished by conventional electrophoretic techniques. The results were quite exciting. As shown in Fig. 12.23, Hb F Sardinia (which carries an Ile → Thr substitution in γ-75) is not quite resolved from the wild-type Hb F in a 1-pH unit span CA-IEF (top panel). In a shallow IPG range spanning only 0.25 pH units Hb F Sardinia and the wild type are now well resolved (central panel). There is, however, a more subtle mutation that could not be resolved in the present case. As shown in both panels, the lower band is actually an envelope of two components, called Aγ and Gγ, carrying a Gly → Ala mutation in γ-136. These two tetramers, normal components during fetal life, are found in approximately an 80 : 20 ratio. If the pH gradient is further decreased to 0.1 pH unit (over a standard 10-cm migration length), even these two tetramers can be separated (bottom panel) with a resolution close to the practical limit of $\Delta pI = 0.001$ [214].

12.4. CAPILLARY ISOELECTRIC FOCUSSING (cIEF)

Although cIEF still is beleaguered by some problems (notably how to cohabit with the ever present hazard of electroosmotic flow), it will be seen that the technique is gaining momentum and is becoming quite popular, especially in the analysis of r-DNA products and of heterogeneity due to differential glycosylation patterns and to 'protein ageing', i.e. Asn and Gln deamidation in vitro. A spin-off of cIEF is zone electrophoresis in zwitterionic, isoelectric buffers, a technique that exploits all the basic concepts of IEF and offers unrivalled resolution due to fast analysis in high-voltage gradients.

12.4.1. General considerations

In addition to the reviews suggested at the beginning of this chapter [3–5], we recommend, as further readings, the following Refs.: [215–225].

Capillary electrophoresis offers some unique advantages over conventional gel slab

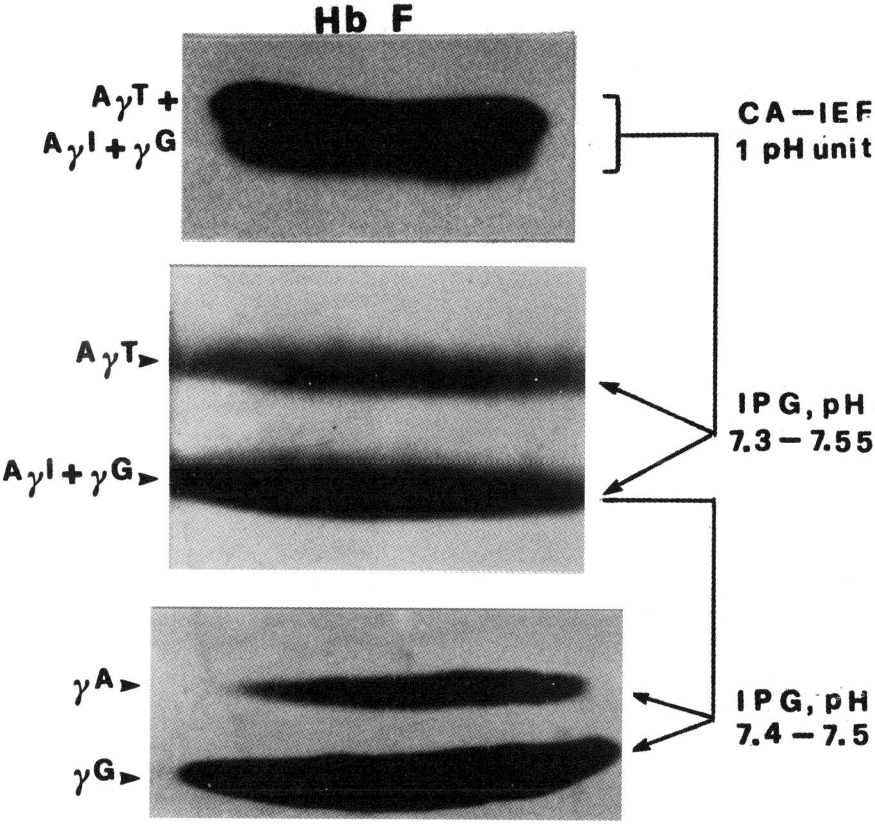

Fig. 12.23. Focussing of umbilical cord lysates from an individual heterozygous for fetal haemoglobin (HbF) Sardinia (for simplicity, only the HbF bands are shown, and not the two other major components of cord blood, i.e. HbA and HbF_{ac}). Upper panel: focussing performed in a 1-pH unit span in CA-IEF. Note that broadening of the HbF occurs, but not the splitting into well-defined zones. Central panel: same sample as above, but focussed over an IPG range spanning 0.25 units. Bottom panel: same as above, but in an IPG gel spanning 0.1 pH unit. The resolved $A\gamma/G\gamma$ bands are in a 20 : 80 ratio, as theoretically predicted from gene expression. Their identity was established by eluting the two zones and fingerprinting the γ chains. (With permission from Ref. [214].)

techniques: the amount of sample required is truly minute (a few μl at the injection port, but only a few nl in the moving zone); the analysis time is in general very short (often just a few minutes) due to the very high voltages applicable; analyte detection is on-line and is coupled to a fully instrumental approach (with automatic storage of electropherograms on a magnetic support).

A principal difference between IEF in a gel and in a capillary is that, in the latter, mobilisation of the focussed proteins past the detector has to be carried out if an on-line imaging detection system is not being used. Mainly three techniques are used: chemical and hydrodynamic flow mobilisation (in coated capillaries) and mobilisation utilising the electroosmotic flow (EOF, in uncoated or partially coated capillaries). We do not particularly encourage the last approach [226], since the transit times of the focussed zones change severely from run to run, thus we will only describe the cIEF approach in well-coated capillaries, where EOF is completely suppressed.

12.4.2. cIEF methodology

Table 12.21 gives a typical methodology for cIEF in a coated capillary. Since a tremendous amount of procedures for silanol deactivation has been reported [227] and good coating practice is very difficult to achieve in a general biochemical lab, we recommend buying pre-coated capillaries, e.g. from Beckman and from Bio-Rad.

12.4.2.1. General guidelines for cIEF

The following general guidelines are additionally suggested.

(a) All solutions should be degassed.

(b) The ionic strength of the sample may influence dramatically the length of the focussing step and also completely ruin the separation; therefore, sample desalting prior to focussing or a low buffer concentration (ideally made of weak buffering ion and weak count-ion) is preferable. Easy sample desalting can be achieved via centrifugation through Centricon membranes (Amicon).

(c) The hydrolytic stability of the coating is poor at alkaline pH; therefore mobilisation with NaOH may destroy such coatings after a few runs.

(d) Ideally, non-buffering ions should be excluded in all compartments for cIEF. This means that in the electrodic reservoirs one should use weak acids (at the anode) and weak bases (at the cathode) instead of phosphoric acid and NAOH, as adopted today by most cIEF users. This includes the use of zwitterions (e.g. Asp, pI = 2.77, or Glu pI = 3.25 at the anode and Lys pI = 9.74 or Arg pI = 10.76 at the cathode).

TABLE 12.21

cIEF SEPARATION OF PROTEINS IN COATED CAPILLARIES

1	Ampholytes: commercially available ampholyte solutions (e.g. Pharmalyte, Biolyte, Servalyte, etc.) in the desired pH range.
2	Sample: 1–2 mg/ml protein solution, mixed with 3–4% carrier ampholytes (final concentration) in the desired pH range. The sample solution should be desalted or equilibrated in a weak buffer–counterion system.
3	Capillary: 25–30 cm long, 50 to 75 μm I.D., coated, filled with the sample-CA solution.
4	Anolyte: 10 mM phosphoric acid (or any other suitable weak acid, such as acetic, formic acids) or low-pI zwitterions.
5	Catholyte: 20 mM NaOH (or any other suitable weak base, such as Tris, ethanolamine) or high-pI zwitterions.
6	Focussing: 8–10 kV, constant voltage, for 5–10 min.
7	Mobiliser: 50–80 mM NaCl, or 20 mM NaOH (cathodic) or 20 mM NaOH (anodic).
8	Mobilisation: 5–6 kV, constant voltage.
9	Detection: 280 nm, near to the mobiliser (or any appropriate visible wavelength for coloured proteins).
10	Washing: after each run, with 1% neutral detergent (such as Nonidet P-40, Triton X-100) followed by distilled water.

(e) When eluting the focussed bands past the detector, we have found that resolution is maintained better by a combination of salt elution (e.g. adding 20 mM NaCl or Na phosphate to the appropriate compartment) and a siphoning effect, obtained by having a higher liquid level in one compartment and a lower level in the other. The volumes to use will depend on the apparatus. For the BioFocus 2000 apparatus from Bio-Rad, the volumes used are 650 μl and 450 μl, respectively.

12.4.2.2. Increasing the resolution by altering the slope of the pH gradient

Methods have not yet been devised for casting IPGs in a capillary format and so it is difficult in cIEF to achieve the resolution typical of IPGs, namely $\Delta pI = 0.001$. Nevertheless, one efficient way for incrementing the resolving power of CA-IEF is to add 'spacers' to a regular 2-pH unit's interval. Cossu et al. [158] reported in fact a modified gel-slab 'spacer' IEF technique for screening for β-thalassaemia by measuring the relative proportions of 3 major Hbs present in cord blood of newborns: adult (A), fetal (F) and acetylated fetal (F_{ac}). We have applied this know-how to cIEF of cord blood. As shown in Fig. 12.24A, a standard IEF run in a pH 6–8 gradient (the best for focussing any Hb variant), while allowing ample separation of F_{ac}, could hardly separate the fetal from the adult Hbs (F/A). A simple addition of 100 mM β-Ala (Fig. 12.24B) brought about excellent resolution among the 3 species, thus allowing proper quantitation and correct diagnosis of thalassaemic conditions [228].

Another case in point is the separation of Hb from its glycated form, Hb A_{1c}. Determination of the percentage of Hb A_{1c} in adult blood is important for evaluation of some pathological alterations of the glycosidic metabolic pathways. In particular, the percentage of Hb A_{1c} is routinely used for assessing the degree of diabetes, by providing an integrated measurement of blood glucose according to the red cell life span. Hb A and A_{1c} have minute pI differences, of the order of 0.01 and less. In conventional IEF in gel slabs, Cossu et al. [229] again solved this problem by resorting to the same combination of 0.2 M β-alanine and 0.2 M 6-amino caproic acid, which provided base-line resolution between Hb A and Hb A_{1c}. Conti et al. [230] have thus re-applied this pH gradient manipulation procedure to cIEF with very good results: as shown in Fig. 12.25B, base-line separation was easily obtained between A and A_{1c}, thus allowing for correct evaluation of the ratio of these two peaks (whereas, as shown in Fig. 12.25A, the peak of A_{1c} was not quite distinguishable in a control run). An artificial mixture of 65% A and 35% A_{1c}, as purified by IPGs, gave the expected pattern with full base-line resolution (Fig. 12.25B).

Another interesting application of cIEF, in combination with CZE, is the study of protein–protein interaction. This is usually done by immobilising in the capillary one of the reactants and letting the interacting species to move through the immobilised zone by electrophoretic transport. But there is another, efficient and simpler procedure: one of the two reactants could be immobilised 'as a temporal' event, in a pH gradient. The ligand could then be swept through the stationary zone and the stoichiometric complex, provided its pI value is outsides the bounds of the pH gradient created in the capillary, emerge at the detector window and thus be quantified. We have applied this concept to the study of haptoglobin (Hp)–haemoglobin (Hb) complex formation [223]. A known amount of Hb is focussed in a capillary in a pH 6–8 range (pI of Hb = 7.0)

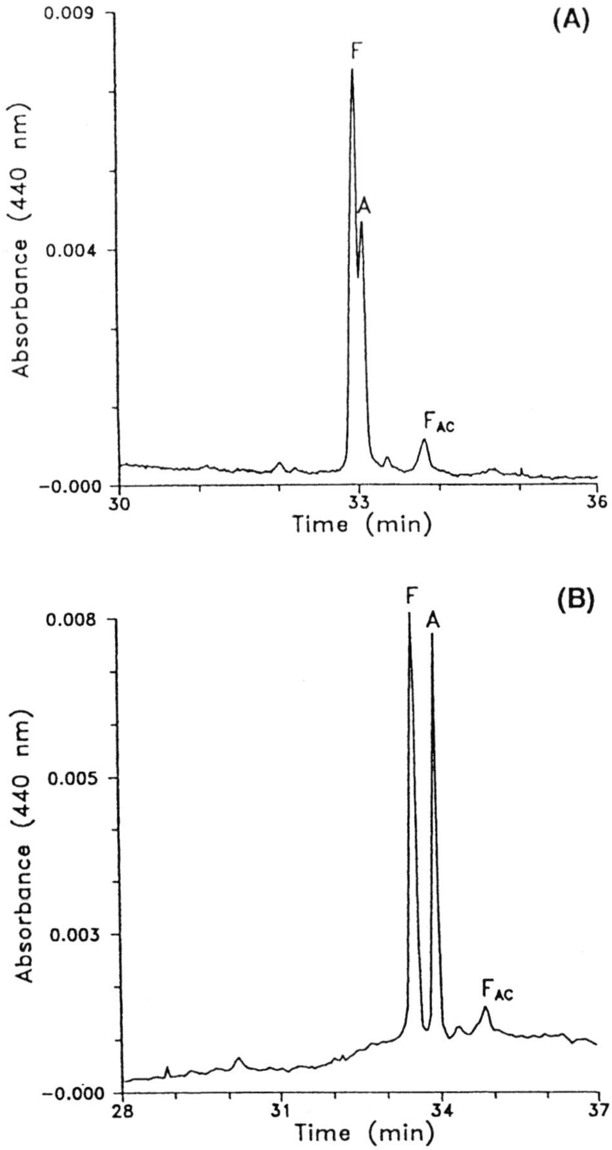

Fig. 12.24. Separation of Hb F, A and F_{ac} by capillary IEF. Background electrolyte: 5% Ampholine, pH 6–8, added with 0.5% TEMED (panel A) and additionally with 3% short-chain polyacrylamide and 100 mM β-Ala (panel B). Anolyte: 20 mM H_3PO_4; catholyte: 40 mM NaOH. Sample loading: by pressure, for 60 s. Focussing run: 20 kV constant at 7 μA (initial) to 1 μA (final current), 20°C in a Bio-Rad Bio Focus 2000 unit. Capillary: coated with poly(*N*-acryloyl amino ethoxy ethanol), 25 μm I.D., 23.6/19.1 total/effective length. Mobilisation conditions: with 200 mM NaCl added to anolyte, 22 kV. Detection at 415 nm. (From Ref. [228], with permission.)

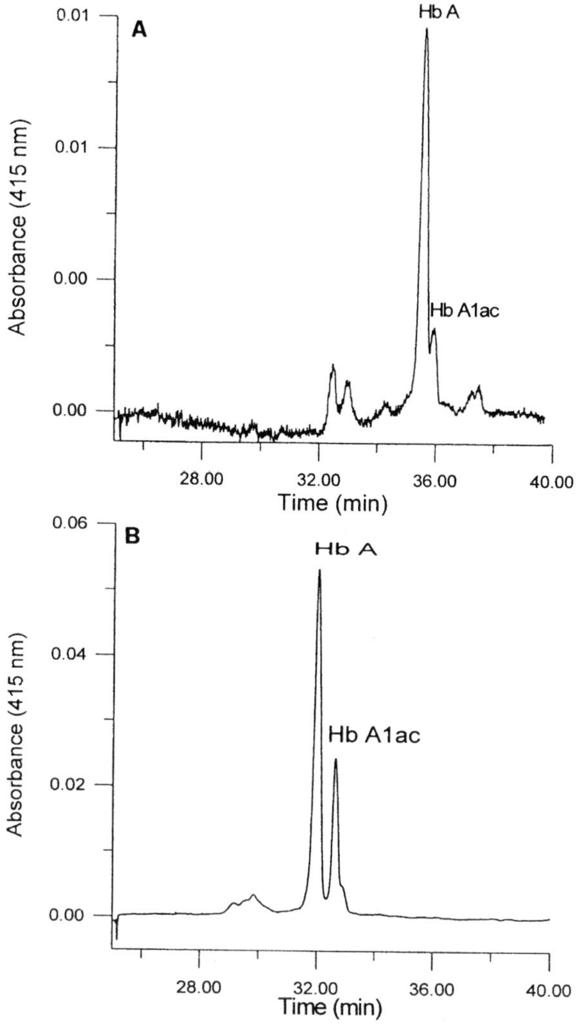

Fig. 12.25. Separation of Hb A from A_{1c} by capillary IEF in the absence (A) and in presence (B) of 3% short-chain polyacrylamide and an equimolar mixture of 'separators', 0.33 M β-Ala and 0.33 M 6-ACA. The latter is a separation of an artificial mixture of 65% Hb A and 35% Hb A_{1c} purified by small-scale, preparative IPGS, in the same 'separator' cocktail. Background electrolyte: 5% Ampholine, pH 6–8, added with 0.5% TEMED. Anolyte: 20 mM H_3PO_4; catholyte: 40 mM NaOH. Sample loading: by pressure, for 60 s. Focussing run: 20 kV constant at 7 μA (initial) to 1 μA (final current), 20°C in a Bio-Rad Bio Focus 2000 unit. Capillary: coated, 25 μm I.D., 23.6/19.1 total/effective length. Mobilisation conditions: with 200 mM NaCl added to anolyte, 22 kV. Detection at 415 nm. (From Ref. [230], with permission.)

and thus kept temporarily 'immobilised' in the electrophoretic chamber. Subsequently, increasing amounts of ligand (Hp) are loaded cathodically and allowed to sweep past the focussed Hb zone. As the complex formed has a pI value well outside the bounds of such a pH gradient (the 1 : 1 molar Hb/Hp complex has a pI of 5.5; the 1 to 1/2

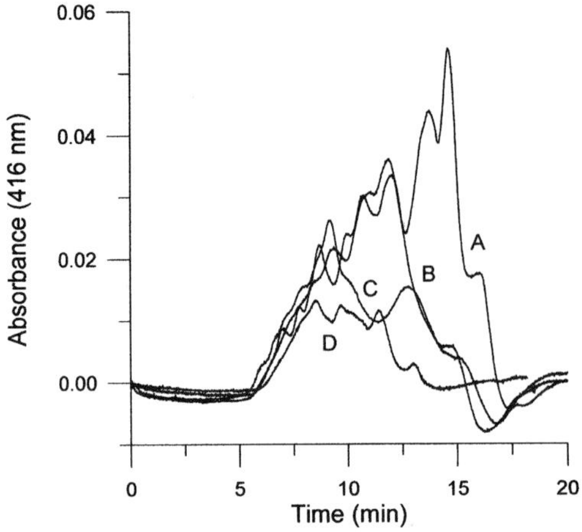

Fig. 12.26. CZE pattern of the Hp/Hb complex. IEF was performed in 75 μm I.D. and 24/19.4 cm total/effective length capillaries, coated with poly(AAP). Isoelectric Lys (50 mM, pH 9.7) and acetic acid (50 mM, pH 3.5) were used as catholyte and anolyte, respectively. The carrier ampholyte consisted of 5% Ampholine pH 6–8, added with 0.5% pH 3–10 range. Focussing was performed for 600 s at 10 kV (13 μA initial, 3 μA at steady state). After focussing free Hb, free haptoglobin was added to the cathodic capillary end and swept through the focussed ligand zone (at 14 kV). Lowest curve (D): addition of 1/4 molar amount of Hp; uppermost curve (A): addition of a 1 : 1 molar amount of Hp; (B) addition of 1/3 molar amount of Hp; (C) addition of 1/2 molar amount of Hp to the Hb zone. Detection at 416 nm. (From Ref. [223], with permission.)

molar Hp/Hb complex has a p*I* of 5.0) it escapes immobilisation and moves past the detector window, where it is monitored and quantified (Fig. 12.26). Since the detector is set at 416 nm, where only Hb absorbs, and since the molar extinction coefficient of Hb is well known, it is quite easy to calculate the molar amount of Hb bound to the complex. As an additional check, the amount of unreacted Hb can now be mobilised by disrupting the pH gradient and allowing this residual-free Hb to also reach the detector and be quantified.

12.4.2.3. On the problem of protein solubility at their pI

One of the most severe shortcomings of all IEF techniques (whether in gel slabs or capillaries, in soluble buffers or IPGs) is protein precipitation at the p*I* value. This problem is aggravated by increasing sample concentrations (overloading is often necessary in order to reveal minor components) and by decreasing the ionic strength (*I*) of the background electrolyte. In this last case, it has been calculated that a 1% carrier ampholyte solution, once focussed, would exhibit a remarkably low *I* value [76], of the order of 0.5 mequiv. l^{-1}. As demonstrated by Grönwall [231], the solubility of an isoionic protein, plotted against pH near the isoionic point, is a parabola, with a fairly narrow minimum at relatively high *I*, but with progressively wider minima, on the pH axis, at decreasing *I* values. This means that, in unfavourable conditions, protein precipitation will not simply occur at a precise point of the pH scale (the p*I*), but it will occur in the form of smears covering as much as 0.5 pH units.

In the past we used, with some success, glycerol, ethylene and propylene glycols [232], when purifying substantial amounts of proteins by preparative IPGs. However, we recently found a number of proteins completely insensitive to these solubilisers. Of course, one could always use the classical mixture of 8 M urea and 2% detergents, as routinely adopted for solubilising entire cell lysates in two-dimensional maps. Yet, we were looking for non-denaturing solubilisers, so that proteins could be recovered in a native form or enzyme assays could be performed in the capillary. Recently, we have obtained particularly encouraging results with the use of zwitterions in cIEF, especially at high concentrations (ca. 1 M) [233]. When attempting to focus the flavoprotein LASPO (L-aspartate oxidase) only massive precipitates were obtained. The only mixture that could restore full solubility was a combination of non-detergent sulphobetaines (NDSB), 0.5 M of the M_r 195 and 0.5 M of the M_r 256 compounds. Interestingly, a very similar pattern was obtained by using, as additive, 1 M bicine, i.e. one of the Good's buffers. Another difficult case was the analysis of thermamylase, an endoamylase that catalyses the hydrolysis of α-D-(1,4) glycosidic linkages in starch components, which produced only smears and precipitates upon IEF. Excellent resolution and focussing patterns were finally obtained in mixtures of neutral additives (typically 20% sucrose, but also sorbitol and, to a lesser extent, sorbose) and zwitterions, in particular 0.1 M taurine [210].

It is noteworthy that the use of zwitterions had been advocated long ago by Alper et al. [234]. Although this use had fallen into oblivion, recent reports by Vuillard et al. [235–237] suggest that this was indeed an avenue worth exploring, as their results with this novel class of zwitterions synthesised by them (called non-detergent sulphobetaines, NDSB) have been encouraging not only in focussing of mildly hydrophobic membrane proteins, but also in improving protein crystal growth. What did not seem to have been explored so far was the idea of mixing different solubilisers, a winning strategy in the present case. For example, mixtures of 30–40% sugars and 0.2 M taurine were highly effective in maintaining sample solubility at the p*I* value even under relatively high protein loads [210,238]. A nice explanation for this solubilising power could come from the work of Timasheff and Arakawa [239] on stabilisation of protein structure by solvents. They explored the physical basis of the stabilisation of native protein structures in aqueous solution by the addition of co-solvents at high (ca. 1 M) concentration. According to them, class I stabilisers (such as sucrose and zwitterions, e.g. amino acids, taurine) act by increasing the surface tension of water and by being preferentially excluded from the hydration shell of the protein. In fact, all these chemicals show a negative value of the binding parameters, signifying that there is an excess of water in the domain of the protein, i.e. that the macromolecule is preferentially hydrated. It is of interest to note that all these phenomena occur at high concentrations of the co-solvents, typically above 1 M, as found in the present report. As a result, the protein is in a state of 'superhydration', which might prevent binding to Immobilines in the gel matrix and might markedly improve solubility at the p*I* value in cIEF. It goes without saying that these additives fully maintain enzyme activity throughout the IEF process.

It should finally be noted that, often, protein precipitation and denaturation could be induced by the presence of high salt levels in the sample, as typically occurring in biological fluids (e.g. urine, cerebrospinal fluid, CSF). Thus, desalting prior to the IEF

References pp. 208–215

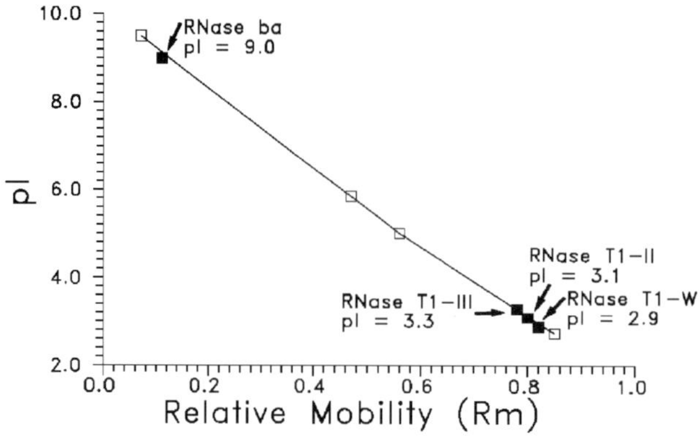

Fig. 12.27. Calibration graph for p*I* determination using a set of marker proteins. The markers (open squares) are: ribonuclease a (p*I* 9.45); carbonic anhydrase (p*I* 5.90); β-lactoglobulin B (p*I* 5.1) and unsulphated cholecystikinin flanking peptide (p*I* 2.75). The four solid squares represent four unknown proteins, whose p*I*s have been determined by linear interpolation in the calibration graph. (From Ref. [241], with permission.)

step is often a must. Manabe et al. [240] reported a method for micro-dialysis of CSF, able to process al little as a 20–30 μl volume, coupled to a conductivity device. With this pre-cIEF step, these authors were able to successfully fractionate CSF and resolve as many as 70 peaks.

12.4.2.4. Assessment of pH gradients and pI values in cIEF

In conventional IEF in gel slabs, protein p*I* markers are commonly offered by a number of suppliers (e.g. Amersham Pharmacia Biotech, Bio-Rad). In its simplest approach, unknown p*I* values can be assessed also in cIEF by plotting the p*I* values of a set of markers, co-focussed with the proteins under investigation, vs. their relative mobility upon elution. In the vacuum method proposed by Chen and Wiktorowicz [241] this plot is linear and thus a high precision (±0.1 pH unit) is obtained (Fig. 12.27) According to these authors, p*I* values as low as 2.9 (for RNase T1-wild type) and 2.75 (for unsulphated cholecystikinin flanking peptide) could be determined. In another, more intriguing approach, in the focussing of transferrin, Kilàr [242] has proposed a novel method for p*I* assessments: monitoring the current in the mobilisation step. If one records simultaneously the peaks of the mobilised stack of proteins and the rising current due to passage of the salt wave in the capillary, one can correlate a given p*I* value (which should be known from the literature a priori) with a given current associated with the transit of a given peak at the detector port. The system can thus be standardised and used for constructing a calibration graph to be adopted in further work, without resorting to 'internal standards'. One such graph correlating current with p*I* values is shown in Fig. 12.28: this appears to be a precise method, since the error is given as only about ±0.03 pH units.

The use of protein markers has problems, though, since not only they are difficult to be obtained as single p*I* components (they are often a family of closely related species), but they are also subjected to ageing, due to hydrolysis of side chain amides

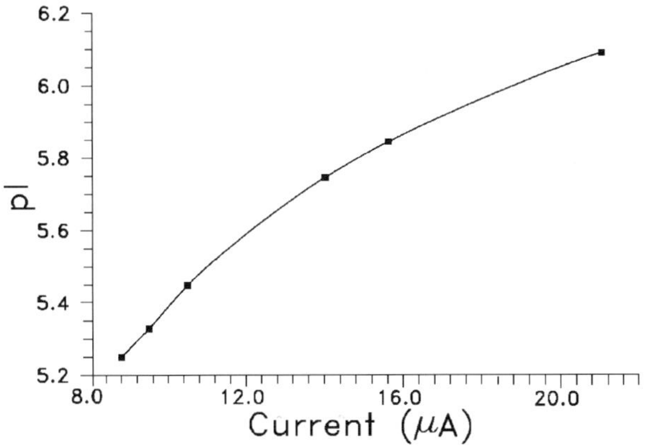

Fig. 12.28. Calibration graph for p*I* determination using the current, during the mobilisation step, as a parameter in cIEF. The six experimental points represent six forms of transferrin, containing different amounts of sialic acid and of iron. (From Ref. [242], with permission.)

in Asn and Gln residues. As an alternative to such proteinaceous material, a number of other markers have been proposed, such as amphoteric dyes [243], phenanthroline iron complexes [244] and substituted aminomethyl phenols (the latter specifically for cIEF; [245]). They do not seem to be quite appropriate for cIEF, though, due to the risk of adsorption of such compounds to the silica wall, with drastic changes of EOF. A recent report by Shimura et al. [246] offers an interesting solution to all these problems. These authors have synthesised a set of sixteen peptides (trimers to examers), which cover the grounds quite evenly from p*I* 3.38 up to p*I* 10.17. Each peptide contains one Trp residue (for on-line detection at 280 nm) and other amino acid with ionisable side chains, responsible for a good titration curve around their p*I* values. The sharp focussing, stability, high purity and high solubility of these synthetic p*I* markers should facilitate the profiling of a pH gradient in cIEF and thus the determination of protein p*I* values.

12.5. SEPARATION OF PEPTIDES AND PROTEINS BY CZE IN ISOELECTRIC BUFFERS

This is an interesting development of cIEF, whereby zone electrophoresis can be performed in isoelectric, very-low-conductivity buffers, allowing the highest possible voltage gradients and thus much improved resolution of peptides and proteins due to reduced diffusion in short analysis times. Although originally described in alkaline pH values (notably in Lys, pI = 9.74) and at neutral pH (notably His buffers, pH = pI = 7.6, albeit in this last case mostly for DNA and oligonucleotides [247–249]), we have found that acidic zwitterions offer an extra bonus: they allow the use of uncoated capillaries, due to protonation of silanols at the prevailing pH of the background electrolyte.

12.5.1. General properties of amphoteric, isoelectric buffers

Although not strictly related to IEF per se, the use of isoelectric buffers stems from the IEF know-how and is having a unique impact in CZE, thus we feel it is appropriate to end this review with a glimpse at this field. The physico-chemical properties of such buffers (especially in regard to their buffering power) have been already discussed in Section 12.2.2. Moreover, Stoyanov et al. [250], additionally, introduced a new parameter for evaluating the performance of amphoteric buffers: the β/λ ratio, i.e. the ratio between the molar buffering power and their conductivity. Ideal buffers are those with the highest possible β/λ ratio (for non-zero I values, of course!), since they allow delivering very high voltage gradients with minimal joule effects. In the field of proteins, and other small M_r compounds, Hjertèn et al. [251] have explored a number of different amphoteric compounds and given proper guidelines for their use. These authors reported separations at voltage gradients as high as 2000 V/cm. Their results with protein separations have been modelled also by Blanco et al. [252] who obtained simulated protein separations in close agreement with the experimental ones of Hjertèn et al. [251]. More recently, Righetti and Nembri [253] have generated peptide maps in isoelectric aspartic acid and shown that such maps could be developed in only 8–10 min, as opposed to 70–80 min in the standard pH 2.0 phosphate buffer, with much superior resolution. Isoelectric Asp, at 50 mM concentration, produces a pH in solution almost coincided with its pI value (pI = 2.77 at 25ºC). At this pH value, some of the large peptides (tryptic digests of casein) analysed were strongly adsorbed by the uncoated capillary wall. Generation of peptide maps in isoelectric Asp would be great if one could use plain, uncoated capillaries, since the technique would be extremely simple (no sample derivatisation, due to reading at 214 nm, where the adsorption of the peptide bond is quite strong, use of unmodified capillaries, very short analysis times, very high resolution). After many attempts, a buffer mixture was optimised, comprising: 50 mM Asp, pH 2.77, 0.5% hydroxyethyl cellulose (HEC, for dynamic coating of the silanols) and 5% 2,2,2-trifluoroethanol (TFE, for modulating peptide mobility). We have now applied this system to the routine analysis of tryptic digests of α and β globin chains, so as to identify point mutations producing amino acid substitutions [254]. Yet, this system was able to resolve only 11 out of 13 fragments present in the β-chain digest. In attempts at ameliorating the system, we have finally adopted the following buffer mixture: 30 mM Asp, pH 2.97, 0.5% HEC, 10% TFE and 50 mM NDSB-195. This buffer mixture performed extremely well, fully resolving 13 out of 13 peptides, in a total time window of 15–16 min in a 75 μm I.D. capillary at 600 V/cm (see Fig. 12.29). The inset shows, on an enlarged area, the spectrum of the first 8 peptides eluted in the 3 to 6 min time window. But one could do better than that: by adopting a 50 μm I.D. capillary and at 900 V/cm (and doubling the injection time from 15 to 30 s, which corresponds to a sample plug of 17 nl) one can accomplish the analysis in only 9 min, with excellent resolution. One can appreciate the remarkable speed, resolution and sensitivity offered by such a simple technique. One last, very important remark: it is not true that, by adopting a single isoelectric buffer, one has to necessarily work at a fixed pH value (which would greatly limit the usefulness of the technique). One basic rule should be remembered: the pH produced in solution by an ampholyte has two limits: on the one side (the

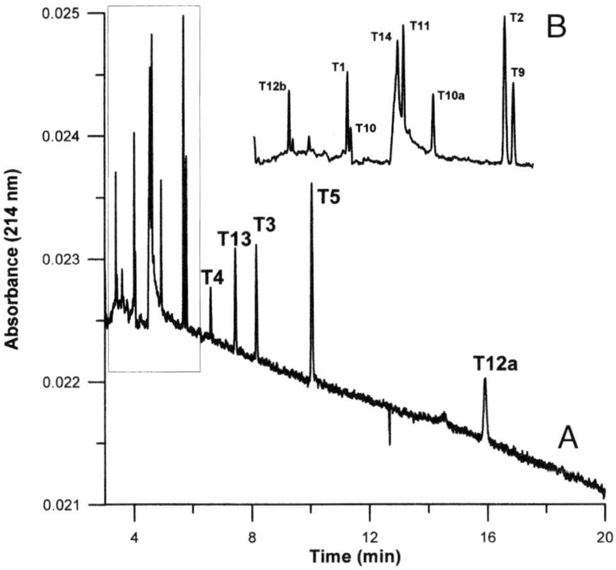

Fig. 12.29. Separation of 13 peptides of tryptic digest of human α-globin chains in isoelectric buffers. Background electrolyte: 30 mM Asp, pH 2.97, 0.5% HEC, 10% TFE and 50 mM NDSB-195. Capillary: 30 cm long (23 cm effective length), 75 μm I.D. Run at 600 V/cm in a Bio-Rad Bio Focus 2000 unit. Sample injection: 30 s, corresponding to a plug of 17 nl. (A) Entire electropherogram with the 13 eluted peaks. (B) Expanded-scale representation of the boxed area in (A). Note that the third eluting peak from left (4 min transit time), almost invisible in panel A, is in fact completely resolved from its neighbour (T_1/T_{10} bands). (From Ref. [254], with permission.)

upper bounds for an acidic, the lower bounds for an alkaline amphotere) is located the pI of water (i.e. pH 7.0), on the opposite side, the true pI of the ampholyte [255]. These two limits can be reached by modulating the concentration of the amphotere in solution. At extreme dilutions (practically useless, of course, since the β power would approach zero!) the pI of the ampholyte will be that of water, i.e. pH 7.0. At the correct concentration (which is not a universal value, it depends on ΔpK, i.e. on how 'good' or 'bad' is a carrier ampholyte) the pH in solution will approach the pI of the amphotere. We have exploited this subtle rule for implementing the separation shown in Fig. 12.29: note that here, by lowering the concentration of Asp from 50 mM (pH = 2.77) to 30 mM (pH = 2.97) we have in fact moved the pH of the background electrolyte by as much as 0.2 pH units. With this simple modification, we could move along the pH/mobility curves of the 13 peptides and find a pH window were no nodal (or cross-over) points existed among all the curves, thus ensuring separation of all 13 peptides. Soon after the report on Asp, we described another amphoteric buffer: iminodiacetic acid (IDA), whose physico-chemical parameters were found, by theoretically modelling and experimental verification, to be: pI 2.23 (at 100 mM concentration), pK_1 1.73 and a pK_2 = 2.73 (no attempts were made at measuring the pK of the primary amino group, since such a low pI value would be compatible with any pK value of the basic group, down to as low as pK 5.5) [256]. IDA was found to be compatible with most hydro-organic solvents, including trifluoroethanol (TFE), up to at least 40% (v/v), typically used for modulating peptide mobility. In naked capillaries, a buffer comprising 50 mM IDA, 10% TFE and 0.5% hydroxyethyl cellulose (HEC) allows generation of peptide maps with high resolution, reduced transit times and no interaction of even large peptides with the wall. However, the best background electrolyte was found to be a solution of 50 mM IDA in 0.5% HEC and 6–8 M urea, one of the best solubilisers of proteins and peptides known. IDA thus appears to be another valid isoelectric buffer system, operating in a different pH window (pH 2.33 in 50 mM IDA) as compared to the other amphotere previously adopted (50 mM Asp, pH 2.77) for the same kind of analysis [256]. Soon after IDA, a novel amphoteric, isoelectric, acidic buffer was reported for separation of oligo- and poly-peptides by capillary zone electrophoresis: cysteic acid (Cys-A). Cys-A, at 200 mM concentration, exhibited a pI of 1.80; given a $\Delta pK = 0.6$, the pK of the carboxyl was assessed as 2.1 and the pK of the sulphate group as 1.50. At 100 mM concentration, this buffer provided an extraordinary buffering power: 140×10^{-3} equivalents l^{-1} pH^{-1}. In presence of 30% (v/v) hexafluoro-2-propanol (HFP), this buffer did not change its apparent pI value, but drastically reduced its conductivity. In Cys-A/HFP buffer, small peptides exhibited a mobility closely following the Offord equation, i.e. proportional to the ratio $M^{2/3}/Z$). When added with 4–5 M urea, there was an inversion in the mobility of some peptides, suggesting strong pK changes as an effect of urea addition. It was found that the minimum mass increment, for proper peptide separation, was $\Delta M_r = \sim 1\%$. In case of simultaneous M_r and pK changes, the minimum ΔM_r is reduced to only 0.6%, provided that a concomitant minimum $\Delta pK = 0.08$ took place. When separating large peptides (human globin chains) in 100 mM Cys-A, 30% HFP and 7 M urea, the β-chain was found to co-elute with the α-chain, suggesting a subtle interplay between the helix-forming (HFP) and helix-breaking (urea) agents. When HFP was omitted, the original globin separation could be restored [257]. In the latest paper of this series, the properties of four acidic, isoelectric buffers were summarised: cysteic acid (Cys-A, pI 1.85), iminodiacetic acid (IDA, pI 2.23), aspartic acid (Asp, pI 2.77) and glutamic acid (Glu, pI 3.22). These four buffers allow to explore an acidic portion of the titration curves of macroions, covering about 1.6 pH units (from pH 1.85 to ca. 3.45), thus permitting resolution of compounds having coincident titration curves at a given pH value. Given the rather acidic pI values of these buffers, their long-term stability has been investigated, by monitoring pH and conductivity changes upon increasing storage times. When dissolved in plain water, all four buffers appear to give constant pH and conductivity readings up to 15 days; after that, the conductivity keeps steadily increasing in a similar fashion. The same parameters, when the same buffers are dissolved in 6 M urea, appear to be stable for only one week, with the conductivity progressively augmenting after this period. By mass spectrometry, Cys-A shows minute amounts (ca. 1%) of a degradation product after ageing for 3 weeks; in the same time period, Glu is extensively degraded (20%). No degradation species could be detected in IDA and Asp solutions. It was additionally shown that the acidic buffers are not quite stationary in the electric field, but can be transported at progressively higher rates (according to the pI value) from the cathodic to the anodic vessel. This is due to the fact that, at their respective pIs, a fraction of the amphotere has to be negatively charged in order to provide counterions to the excess of protons due to bulk water dissociation [167].

12.5.2. Examples of some separations of proteins in isoelectric buffers

The same CZE technique in isoelectric buffers was successfully applied to the screening of wheat cultivars, via analysis of a fraction of gluten proteins present in the endosperm, the gliadins. Previously published procedures recommended a 100 mM phosphate buffer, added with 0.05% hydroxypropyl methyl cellulose and 20% acetonitrile, in uncoated capillaries. Due to the very high conductivity of such a buffer (4.7 mmhos at 25°C) high speed separations (10–12 min analysis time at 800 V/cm) could only be elicited in 20 μm I.D. capillaries, at the expense of sensitivity. Capelli et al. [258] have optimised the following background electrolyte: 40 mM aspartic acid (pH = pI = 2.77) in presence of 7 M urea and 0.5% short-chain hydroxyethyl cellulose (27,000 Da in Mn) (apparent pH in 7 M urea: 3.9). As an alternative recipe, the same isoelectric buffer can be added with a mixed organic solvent composed of 4 M urea and 20% acetonitrile (apparent pH: 3.66). Due to the much lower conductivity (0.7 mmhos), separations could be carried out at 1000 V/cm in only 10 min, but in larger-bore capillaries (50 μm I.D.), ensuring a five-times higher sensitivity. As shown in Fig. 12.30, the gliadin patterns thus obtained are species-specific and allow easy identification of all cultivars tested of durum and bread wheat. No adsorption of proteins

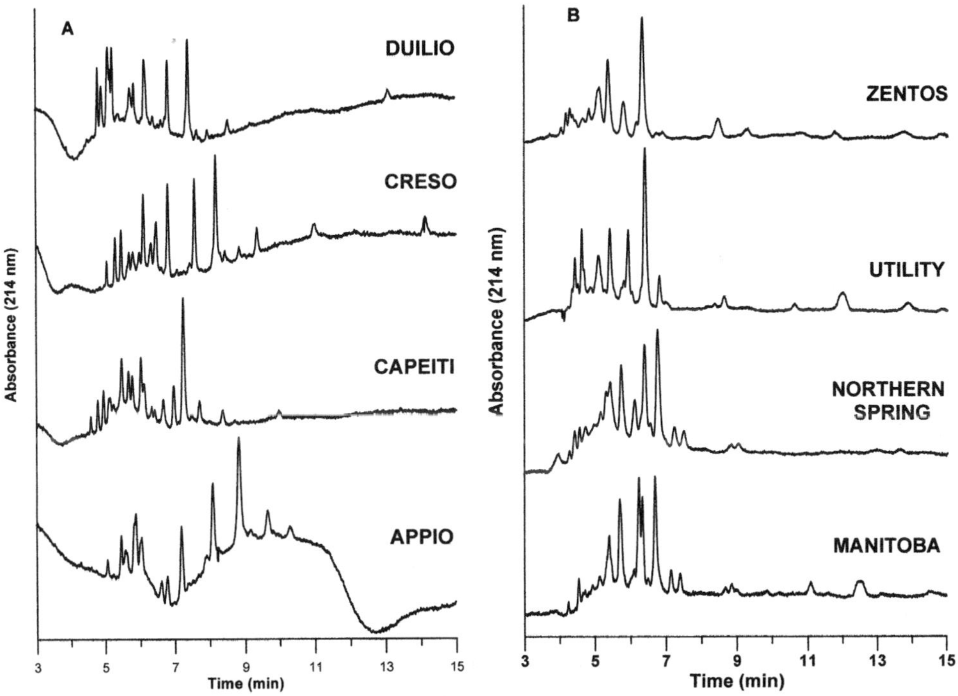

Fig. 12.30. Representative CZE run of gliadins, obtained by sequential extraction of flour, from four different cultivars of durum (panel A) and of soft (panel B) wheats. Runs performed in 40 mM Asp buffer, added with 7 M urea and 0.5% HEC (apparent pH: 3.9). Conditions: 50 μm I.D., 30 cm long capillary, run at 1000 V/cm at room temperature. Detection at 214 nm. (From Ref. [258], with permission.)

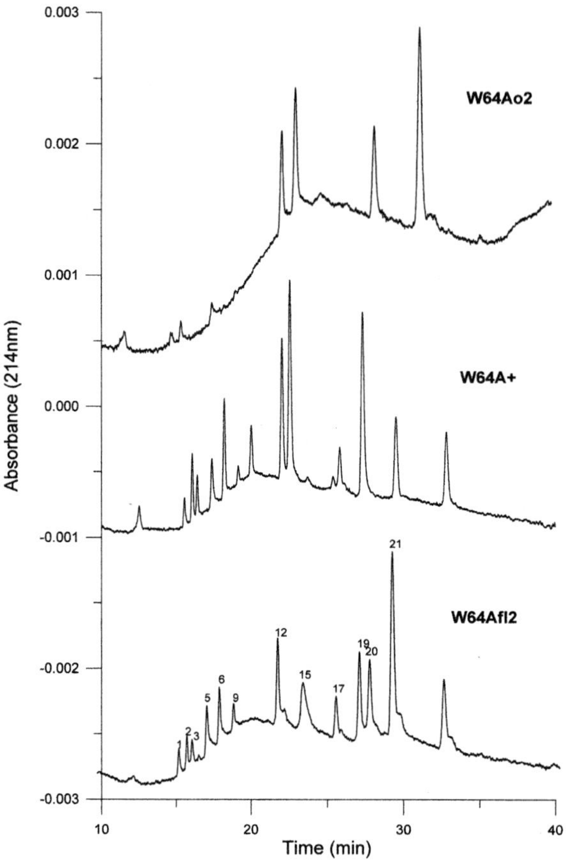

Fig. 12.31. Representative CZE runs of zeins, obtained by 70% ethanol, 2% β-mercaptoethanol extraction of ground endosperm, from three different lines of maize. Run performed in 40 mM Asp buffer, 6 M urea and 0.5% HEC (apparent pH: 3.7). Conditions: 50 μm I.D., 30 cm long capillary, run at 800 V/cm and 30°C. Detection at 214 nm. From top to bottom: W64Ao2; W64+; W64Afl2. Peak numbering as obtained by multivariate statistical analysis. (From Ref. [259], with permission.)

to the silica wall seemed to occur and high reproducibility in peak areas and transit times was obtained.

Another interesting application in food analysis has been the screening of different maize lines via CZE profiling of zeins. Zeins (the prolamins or seed storage proteins in maize) have been traditionally used to characterise and identify different genotypes. Zeins have been fractionated by capillary zone electrophoresis in acidic, amphoteric buffers, which represent a medium of moderate conductivity and thus compatible with relatively higher voltage gradients. The running buffer consisted of 40 mM isoelectric aspartic acid, in presence of 6 M urea and 0.5% hydroxyethyl cellulose (apparent pH: 3.7; p*I* in the absence of urea: 2.77). A total of 31 different zein peaks were mapped out of a total of 21 different maize lines. Each line exhibited typically seven to twelve peaks, with some lines showing up to 20 zein bands (Fig. 12.31). Due to slightly changing elution times, caused by a lack of reproducibility of the electroendoosmotic flow in uncoated silica surfaces, correct peak assignment and alignment among different runs was obtained by multivariate statistical analysis. The present method compares well, both in resolution and total number of peaks, with current protocols presently adopted for screening of maize inbreds, which consist of isoelectric focussing in agarose gels ([259,260]).

An improved method for determination of cow's milk in non-bovine cheese could additionally be devised: electrophoresis of whey proteins in acidic, isoelectric buffers. Two background electrolytes (BGE) were tested: (i) 50 mM iminodiacetic acid (pH = pI = 2.30 at 25°C), 0.5% hydroxyethylcellulose, 0.1% Tween 20 and 6 M urea (apparent pH 3.1), $E = 300$ V/cm, for the separation of α-lactalbumins (α-LA); (ii) a BGE with the same composition, but supplemented with 10% Tween 20, $E = 450$ V/cm, for the fractionation of β-lactoglobulin (β-LG). Surfactants have a discriminating effect on the retention behaviour of the bovine α-LA and β-LG proteins, owing to the different strength of the protein–surfactant association complexes, and are needed for separating these two proteins from small peaks in the electropherograms generated by degradation of casein during cheese ripening. Novel equations are given for deriving the ratio of the area (or height) of bovine α-LA, or β-LG, to the area (or height) of ovine or caprine α-LA or β-LG (such ratios being typically used to determine the percentage of cow's milk in dairy products), since previous equations had marked drawbacks, such as non-linearity of the plots with increasing slopes at large cow's milk percentages, and too broad confidence limits at high cow's milk contents, where the peak area (or height) ratio tends asymptotically to infinite. A representative separation is shown in Fig. 12.32. With the novel procedures here reported, contents of cow's milk as low as 1% can be quantified in goat's and ewe's cheeses. The present protocols give lower detection limits, are cheaper and more rapid than any other methodology reported in the literature, and can be easily applied to the routine quality control of binary and ternary cheeses [262,263].

Another interesting application has been the separation of globin chains in umbilical cord blood and for screening for point mutations in α and β human globin chains. A solution of 50 mM iminodiacetic acid (pI 2.23) containing 7 M urea and 0.5% hydroxyethylcellulose (apparent pH 3.2) is used as background electrolyte for fast separation of haem-free, denatured globin (α and β) chains. Here too, due to the low conductivity of such buffers, high voltage gradients (600 V/cm) can be applied, thus reducing the separation time to only a few minutes (also thanks to the very high positive charge of globin chains). Fig. 12.33 gives the separation of α, β and γ globin chains from umbilical cord blood: it is seen that the order of elution follows the order of pI values, the α chains, with a pI of ca. 10, being the first ones to be eluted. In presence of neutral to neutral amino acid substitutions, it was additionally shown that the inclusion of 3% surfactant (Tween 20) in the sample and background electrolyte induced the separation of the wild-type and mutant chains, probably by a mechanism of hydrophobic interaction of the more hydrophobic mutant with the detergent micelle, via a mechanism similar to 'micellar electrokinetic chromatography'. At this low operative pH, however, charged mutants, involving substitutions of acidic amino acids (Glu and Asp) could not be detected, since these residues are extensively protonated. Curiously, however, they were still separated in presence of detergent, due to the large variation in hydrophobicity

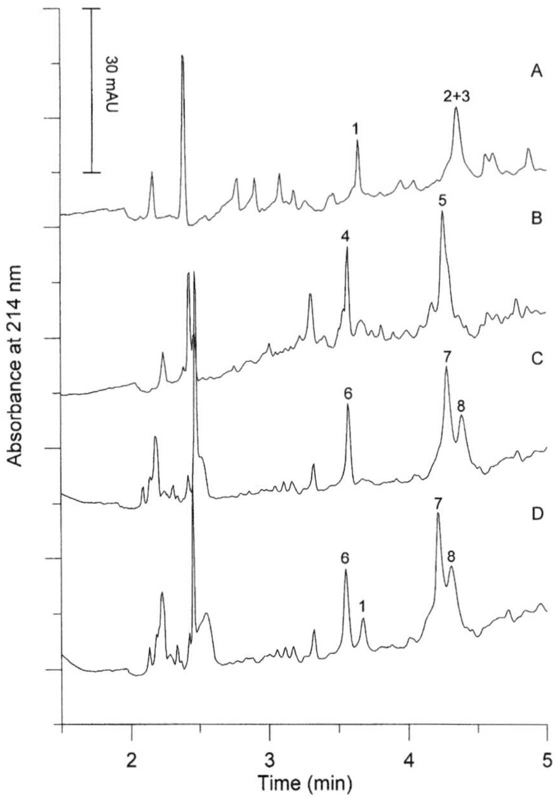

Fig. 12.32. Electropherograms of whey from (A) cow (B) goat and (C) ewe ripened cheeses. Electropherogram (D): ewe's extract spiked with 200 mg/l of bovine α-LA. Capillary: 33.5 cm (effective length 25 cm) $\times$ 50 μm I.D.; hydrodynamic injection: 50 mbar $\times$ 3 s; running buffer: 50 mM IDA, 6 M urea and 0.5% HEC; electric field: 700 V/cm (resulting in a current of 27 μA); detection at 214 nm. Peak identification: 1 = bovine α-LA; 2 = bovine β-LG A; 3 = bovine β-LG B; 4 = caprine α-LA; 5 = caprine β-LG; 6 = ovine α-LA; 7 = ovine β-LG A; 8 = ovine β-LG B. (From Ref. [262], with permission.)

involved in such mutations (Fig. 12.34). Of the 19 mutants analysed, all but one were resolved: Hb St Nazaire (β103 Phe $\rightarrow$ Ile). This is due to the fact that the ΔG (in kcal/mol) in the substitution Phe $\rightarrow$ Ile is zero, thus no separation can possibly take place between two chains exhibiting the same hydrophobicity parameter [261,264].

12.5.3. Troubleshooting for CZE in isoelectric buffers

Although the technique seems to be working quite well, one should be aware of the following problems.

(a) As stated above, these rather acidic buffers might not be very stable in solution. Some of them are amino acids (Asp, Glu) and could be good pabulum for bacterial growth. Thus, unless these solutions are made sterile and kept that way, they should not be used for more than a week and made fresh after that.

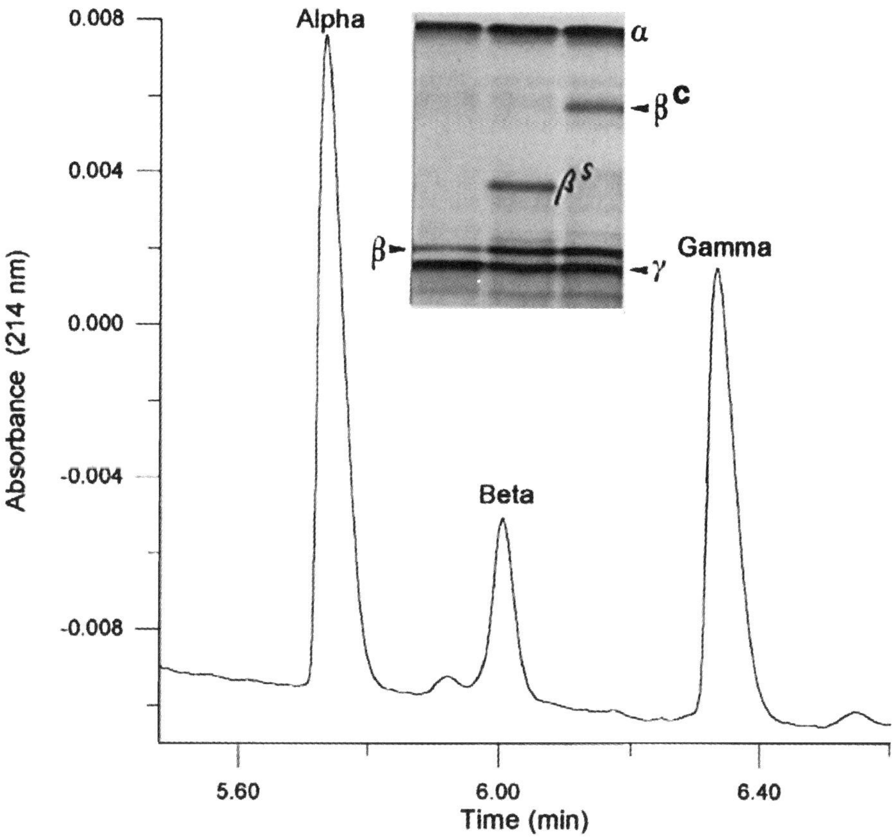

Fig. 12.33. CZE of haem-free, globin chains from normal newborns at delivery (cord blood). Background electrolyte: 50 mM IDA, added with 7 M urea and 0.5% hydroxyethyl cellulose (pH 3.2). Uncoated capillary of 30 cm length, 50 μm I.D. Sample injection: 2 psi for 1 s. Running conditions: 600 V (22 μA). Detection at 214 nm. Insert: IEF in IPGs of different globin chains. The gel contained an immobilised pH 6.5–10 gradient. Staining with Coomassie Brilliant Blue R-250. Sample tracks (from right to left): a mixture of fetal, adult and Hb C chains; a mixture of fetal, adult and Hb S chains; a mixture of fetal and adult chains. (From Ref. [261], with permission.)

(b) Although zwitterionic buffers, used at pH = p*I*, should in theory be stationary, this is less and less valid the more acidic is their p*I* value, since a fraction of these species must be negatively charged to act as counterions to the excess of protons in solution. This would result in net migration of such 'isoelectric buffers' from cathode to anode. In addition, because the buffer chambers in CZE are rather small (often less than 1 ml in volume), depletion of the buffering ions could ensue after only a few runs. Thus, the electrodic reservoirs might have to be replenished after only 3–4 runs. An easy check for that is to measure the pH and conductivity of both vessels and change the solutions when these values vary by a given amount (e.g. 15% of the original value).

Although in rather acidic zwitterions (Cys-A, IDA) addition of 0.5% HEC seems to be quite effective in preventing binding of polypeptide chains to the silica wall, this will

50. M.J. Dunn (Guest Ed.), From Genome to Proteome: Proceedings of the Third Siena Two-Dimensional Electrophoresis Meeting, Electrophoresis, 20 (1999) 643–1122.
51. F. Lottspeich (Guest Ed.), Paper Symposium: Electrophoresis and Amino Acid Sequencing, Electrophoresis, 17 (1996) 811–966.
52. B. Tümmler (Guest Ed.), Microbial Genomes: Biology and Technology, Electrophoresis 19 (1998) 467–624.
53. K.L. Williams (Guest Ed.), Strategies in Proteome Research, Electrophoresis 19 (1988) 1853–2050.
54. A. Humphery-Smith (Guest Ed.), Paper Symposium: Microbial Proteomes, Electrophoresis, 18 (1997) 1207–1497.
55. R.D. Appel, M.J. Dunn and D.F. Hochstrasser (Guest Eds.), Paper Symposium: Biomedicine and Bioinformatics, Electrophoresis, 18 (1997) 2703–2842.
56. R.D. Appel, M.J. Dunn and D.F. Hochstrasser (Guest Eds.), Paper Symposium: Biomedicine and Bioinformatics. Electrophoresis, 20 (1999) 3481–3686.
57. P. Cash (Guest Ed.), Paper Symposium: Microbial Proteomes, Electrophoresis, 20 (1999) 2149–2310.
58. J.E. Celis (Guest Ed.), Paper Symposium: Two Dimensional Gel Protein Databases, Electrophoresis, 16 (1995) 2175–2264.
59. B. Bjellqvist, K. Ek, P.G. Righetti, E. Gianazza, A. Görg, W. Postel and R. Westermeier, J. Biochem. Biophys. Methods, 6 (1982) 317–339.
60. B.J. Radola, Electrophoresis, 1 (1980) 43–56.
61. P.G. Righetti and M. Bello, Electrophoresis, 13 (1992) 275–279.
62. H. Rilbe, Ann. N.Y. Acad. Sci., 209 (1973) 11–22.
63. E. Gianazza, F. Chillemi, C. Gelfi and P.G. Righetti, J. Biochem. Biophys. Methods, 1 (1979) 237–251.
64. E. Gianazza, F. Chillemi and P.G. Righetti, J. Biochem. Biophys. Methods, 3 (1980) 135–141.
65. E. Gianazza, F. Chillemi, M. Duranti and P.G. Righetti, J. Biochem. Biophys. Methods, 8 (1984) 339–351.
66. E. Gianazza and P.G. Righetti, J. Chromatogr., 193 (1980) 1–8.
67. R. Krishnamoorthy, A. Bianchi-Bosisio, D. Labie and P.G. Righetti, FEBS Lett., 94 (1978) 319–323.
68. P.G. Righetti and C. Tonani, in F. Dondi and G. Guiochon (Eds.), Theoretical Advancement in Chromatography and Related Separation Techniques, NATO ASI Series C: Mathematical and Physical Sciences, Kluwer, Dordrecht, Vol. 383, 1992, pp. 581–605.
69. P.K. Sinha and P.G. Righetti, J. Biochem. Biophys. Methods, 12 (1986) 289–297.
70. O. Vesterberg, Ann. N.Y. Acad. Sci., 209 (1973) 23–33.
71. L. Söderberg, D. Buckley, G. Hagström and J. Bergström, Prot. Biol. Fluids, 27 (1980) 687–691.
72. N. Grubhofer and C. Borja, in B.J. Radola and D. Graesslin (Eds.), Electrofocusing and Isotachophoresis, de Gruyter, Berlin, 1974, pp. 111–120.
73. P. Pogacar and R. Jarecki, in R. Allen and H. Maurer (Eds.), Electrophoresis and Isoelectric Focusing in Polyacrylamide Gels, de Gruyter, Berlin, 1977, pp. 153–158.
74. A. Bianchi-Bosisio, R.S. Snyder and P.G. Righetti, J. Chromatogr., 209 (1981) 265–272.
75. E. Galante, T. Caravaggio and P.G. Righetti, in P.G. Righetti (Ed.), Progress in Isoelectric Focusing and Isotachophoresis, Elsevier, Amsterdam, 1975, pp. 3–12.
76. P.G. Righetti, J. Chromatogr., 190 (1980) 275–282.
77. H. Rilbe, pH and Buffer Theory. A New Approach, Wiley, Chichester, 1996, 192 pp.
78. A. Bianchi-Bosisio, C. Loehrlein, R.S. Snyder and P.G. Righetti, J. Chromatogr., 189 (1980) 317–330.
79. M. Chiari, C. Micheletti, M. Nesi, M. Fazio and P.G. Righetti, Electrophoresis, 15 (1994) 177–186.
80. E. Simò-Alfonso, C. Gelfi, R. Sebastiano, A. Citterio and P.G. Righetti, Electrophoresis, 17 (1996) 723–731.
81. E. Simò-Alfonso, C. Gelfi, R. Sebastiano, A. Citterio and P.G. Righetti, Electrophoresis, 17 (1996) 732–737.
82. P.G. Righetti, B.C. Brost and R.S. Snyder, J. Biochem. Biophys. Methods, 4 (1981) 347–363.
83. C. Gelfi and P.G. Righetti, Electrophoresis, 2 (1981) 213–219.
84. M. Chiari, P.G. Righetti, A. Negri, F. Ceciliani and S. Ronchi, Electrophoresis, 13 (1992) 882–884.
85. E. Bordini, M. Hamdan and P.G. Righetti, Rapid Commun. Mass Spectrom., 13 (1999) 1818–1827.

86. E. Bordini, M. Hamdan and P.G. Righetti, Rapid Commun. Mass Spectrom., 13 (1999) 2209–2215.
87. A. Bossi and P.G. Righetti, Electrophoresi, 16 (1995) 1930–1934.
88. T. Låås and I. Olsson, Anal. Biochem., 114 (1981) 167–172.
89. K. Altland and M. Kaempfer, Electrophoresis, 1 (1980) 57–62.
90. M.L. Caspers, Y. Posey and R.K. Brown, Anal. Biochem., 79 (1977) 166–180.
91. P.G. Righetti, G. Tudor and E. Gianazza, J. Biochem. Biophys. Methods, 6 (1982) 219–227.
92. J.G. Yao and R. Bishop, J. Chromatogr., 234 (1982) 459–462.
93. M.D. Zhu, R. Rodriguez and T. Wehr, J. Chromatogr., 559 (1991) 479–485.
94. P.G. Righetti and F. Chillemi, J. Chromatogr., 157 (1978) 243–251.
95. P.G. Righetti and J.W. Drysdale, J. Chromatogr., 98 (1974) 271–321.
96. V. Neuhoff, R. Stamm and H. Eibl, Electrophoresis, 6 (1985) 427–437.
97. C.R. Merril, D. Goldman, S.A. Sedman and M.H. Ebert, Science, 211 (1981) 1438–1440.
98. B.J. Radola, Biochim. Biophys. Acta, 295 (1973) 412–428.
99. R.F. Riley and M.K. Coleman, J. Lab. Clin. Med., 72 (1968) 714–720.
100. O. Vesterberg, Biochim. Biophys. Acta, 257 (1972) 11–22.
101. P.J. Wirth and A. Romano, J. Chromatogr. A, 698 (1995) 123–143.
102. C.R. Merril and K.M. Washart, in B.D. Hames (Ed.), Gel Electrophoresis of Proteins: a Practical Approach, Oxford University Press, Oxford, 1998, pp. 53–91.
103. C.R. Merril and K.M. Washart, in B.D. Hames (Ed.), Gel Electrophoresis of Proteins: a Practical Approach, Oxford University Press, Oxford, 1998, pp. 319–343.
104. A.A. Waheed and P.D. Gupta, Anal. Biochem., 233 (1996) 249–252.
105. J.K. Choi, S.H. Yoon, H.Y. Hong, D.K. Choi and G.S. Yoo, Anal. Biochem., 236 (1996) 82–84.
106. J.M. Myers, A. Veis, B. Sabsay and A.P. Wheeler, Anal. Biochem., 240 (1996) 300–302.
107. T.H. Steinberg, L.J. Jones, R.P. Haugland and V.L. Singer, Anal. Biochem., 239 (1996) 223–237.
108. T.H. Steinberg, R.P. Haugland and V.L. Singer, Anal. Biochem., 239 (1996) 238–245.
109. F. Javier Alba, A. Bermudez, S. Bartolome and J.R. Daban, BioTechniques, 21 (1996) 625–626.
110. T. Rabilloud, Anal. Chem., 72 (2000) 48A–55A.
111. H.F. Bosshard and A. Datyner, Anal. Biochem., 82 (1977) 327–333.
112. V.E. Urwin and P. Jackson, Anal. Biochem., 209 (1993) 57–62.
113. M. Unlii, E.M. Morgan and J.S. Minden, Electrophoresis, 18 (1997) 2071–2077.
114. W.M. Bonner and R.A. Laskey, Eur. J. Biochem., 46 (1974) 83–88.
115. R.F. Johnston, S.C. Pickett and D.L. Barker, Electrophoresis, 11 (1990) 355–360.
116. G. Charpak, W. Dominik and N. Zaganidis, Proc. Natl. Acad. Sci. USA, 86 (1989) 1741–1745.
117. V.E. Urwin and P. Jackson, Anal. Biochem., 195 (1991) 30–37.
118. P. Jackson, V.E. Urwin and C.D. Mackay, Electrophoresis, 9 (1988) 330–339.
119. S. Fazekas de St Groth, R.G. Webster and A. Datyner, Biochim. Biophys. Acta, 71 (1963) 377–391.
120. H. Chen, H. Cheng and M. Bjerknes, Anal. Biochem., 212 (1993) 295–296.
121. P.M. Horowitz and S. Bowman, Anal. Biochem., 165 (1987) 430–434.
122. F.J. Alba, A. Bermudez, S. Bartolome and J.R. Daban, BioTechniques, 21 (1996) 625–626.
123. M.L. Ortiz, M. Calero, C. Fernandez-Patron, L. Castellanos and E. Mendez, FEBS Lett., 296 (1992) 300–304.
124. C. Fernandez-Patron, L. Castellanos-Serra, E. Hardy, M. Guerra, E. Estevez, E. Mehl and R.W. Frank, Electrophoresis, 19 (1998) 2398–2406.
125. T. Rabilloud, L. Vuillard, C. Gilly and J.J. Lawrence, Cell. Mol. Biol., 40 (1994) 57–75.
126. J. Christiansen and G. Houen, Electrophoresis, 13 (1992) 179–183.
127. M.J. Lim, W.F. Patton, M.F. Lopez, K.H. Spofford, N. Shojaee and D. Shepro, Anal. Biochem., 245 (1997) 184–195.
128. J.M. Coull and D.J.C. Pappin, J. Prot. Chem., 9 (1990) 259–260.
129. V. Neuhoff, N. Arold, D. Taube and W. Ehrhardt, Electrophoresis, 9 (1988) 255–262.
130. S. Haebel, T. Albrecth, K. Sparbier, P. Walden, R. Korner and M. Steup, Electrophoresis, 19 (1998) 679–686.
131. C. Scheler, S. Lamer, Z. Pan, X. Li, J. Salnikov and P. Jungblut, Electrophoresis, 19 (1998) 918–927.
132. S. Patterson and R. Aebersold, Electrophoresis, 16 (1995) 1791–1814.
133. M. Hunkapillar, E. Lujan, F. Ostrander and L. Hood, Methods Enzymol., 91 (1983) 227–236.

134. T. Rabilloud, Electrophoresis, 11 (1990) 785–794.
135. A. Shevchenko, M. Wilm and M. Mann, Anal. Chem., 68 (1996) 850–858.
136. T. Rabilloud, S. Kieffer, V. Procaccio, M. Louwagie, P.L. Courchesne, S.D. Patterson, P. Martinez, J. Garin and J. Lunardi, Electrophoresis, 19 (1998) 1006–1014.
137. F. Gharahdaghi, C.R. Weinberg, D.A. Meagher, B.S. Imai and S.M. Mische, Electrophoresis, 20 (1999) 601–605.
138. T.H. Steinberg, E. Chernokalskaya, K. Berggren, M.F. Lopez, Z. Diwu, R.P. Haugland and W.F. Patton, Electrophoresis, 21 (2000) 8–496.
139. A. Bermudez, J. Daban, J. Garcia and E. Mendez, BioTechniques, 16 (1994) 621–624.
140. J.P. Hebert and B. Strobbel LKB Application Note #151, 1974.
141. W.J. Godolphin and R.A. Stinson, Clin. Chim. Acta, 56 (1974) 97–103.
142. R.A. Laskey and A.D. Mills, Eur. J. Biochem., 56 (1975) 335–341.
143. R.A. Laskey, Methods Enzymol., 65 (1980) 363–371.
144. H. Harris and D.A. Hopkinson, Handbook of Enzyme Electrophoresis in Human Genetics, Elsevier, Amsterdam, 1976.
145. R.F. Richtie and R. Smith, Clin. Chem., 22 (1976) 497–499.
146. P. Arnaud, G.B. Wilson, J. Koistinen and H.H. Fudenberg, J. Immunol. Methods, 16 (1977) 221–231.
147. H. Towbin, T. Staehelin and J. Gordon, Proc. Natl. Acad. Sci. USA, 76 (1979) 4350–4352.
148. P.G. Righetti and C. Macelloni, J. Biochem. Biophys. Methods, 6 (1982) 1–15.
149. E. Gianazza, J. Chromatogr. A, 705 (1995) 67–87.
150. J.R. Cann, Electrophoresis, 19 (1998) 1577–1585.
151. H.F. Bunn and J.W. Drysdale, Biochim. Biophys. Acta, 229 (1971) 51–57.
152. C.M. Park, Ann. N.Y. Acad. Sci., 209 (1973) 237–256.
153. M. Perrella, L. Cremonesi, L. Benazzi and L. Rossi-Bernardi, J. Biol. Chem., 256 (1981) 11098–11103.
154. J.B. Clegg, M.A. Naughton and D.J. Weatherall, J. Mol. Biol., 19 (1966) 91–100.
155. P.G. Righetti, E. Gianazza, A.M. Gianni, P. Comi, B. Giglioni, S. Ottolenghi, C. Secchi and L. Rossi-Bernardi, J. Biochem. Biophys. Methods, 1 (1979) 47–59.
156. G. Saglio, G. Ricco, U. Mazza, C. Camaschella, P.G. Pich, A.M. Gianni, E. Gianazza, P.G. Righetti, B. Giglioni, P. Comi, M. Gusmeroli and S. Ottolenghi, Proc. Natl. Acad. Sci. USA, 76 (1979) 3420–3424.
157. K. Valkonen, E. Gianazza and P.G. Righetti, Clin. Chim. Acta, 107 (1980) 223–229.
158. G. Cossu, M. Manca, P.M. Gavina, R. Bullitta, A. Bianchi-Bosisio, E. Gianazza and P.G. Righetti, Am. J. Haematol., 13 (1982) 149–157.
159. Z.L. Awdeh, Sci. Tools, 16 (1969) 42–43.
160. R.B. Merrifield, Adv. Enzymol., 32 (1969) 221–241.
161. P.G. Righetti and A. Bianchi-Bosisio, Electrophoresis, 2 (1981) 65–75.
162. E.A. Johnson and B. Mulloy, Carbohydr. Res., 51 (1976) 119–127.
163. P.G. Righetti and E. Gianazza, Biochim. Biophys. Acta, 532 (1978) 137–146.
164. P.G. Righetti, R. Brown and A.L. Stone, Biochim. Biophys. Acta, 542 (1978) 222–231.
165. J.R. Cann, in P.G. Righetti, C.J. Van Oss and J.W. Vanderhoff (Eds.), Electrokinetic Separation Methods, Elsevier, Amsterdam, 1979, pp. 369–388.
166. A. Bossi, M.J. Patel, E.J. Webb, M.A. Baldwin, R.J. Jacob, A.L. Burlingame and P.G. Righetti, Electrophoresis, 20 (1999) 2810–2817.
167. A. Bossi, E. Olivieri, L. Castelletti, C. Gelfi, M. Hamdan and P.G. Righetti, J. Chromatogr. A, 853 (1999) 71–82.
168. M. Perduca, H. Monaco, L. Goldoni, A. Bossi and P.G. Righetti, Electrophoresis, 21 (2000) 2316–2320.
169. C.B. Cuono and G.A. Chapo, Electrophoresis, 3 (1982) 65–70.
170. P.G. Righetti, M. Fazio and C. Tonani, J. Chromatogr., 440 (1988) 367–377.
171. P.G. Righetti, M. Chiari and C. Gelfi, Electrophoresis, 9 (1988) 65–73.
172. P.G. Righetti, M. Chiari, P.K. Sinha and E. Santaniello, J. Biochem. Biophys. Methods, 16 (1988) 185–192.
173. M. Chiari, E. Casale, E. Santaniello and P.G. Righetti, Theor. Appl. Electr., 1 (1989) 99–102.

174. M. Chiari, E. Casale, E. Santaniello and P.G. Righetti, Theor. Appl. Electr., 1 (1989) 103–107.
175. P.G. Righetti, C. Gelfi and M. Chiari, in B.L. Karger and W.S. Hancock (Eds.), Methods in Enzymology: High Resolution Separation and Analysis of Biological Macromolecules, Part A: Fundamentals, Vol. 270, Academic Press, San Diego, CA, 1996, pp. 235–255.
176. P.K. Sinha and P.G. Righetti, J. Biochem. Biophys. Methods, 15 (1987) 199–206.
177. M. Chiari, P.G. Righetti, P. Ferraboschi, T. Jain and R. Shorr, Electrophoresis, 11 (1990) 617–620.
178. M. Chiari, L. Pagani, P.G. Righetti, T. Jain, R. Shorr and T. Rabilloud, J. Biochem. Biophys. Methods, 21 (1990) 165–172.
179. M. Chiari, C. Ettori, A. Manzocchi and P.G. Righetti, J. Chromatogr., 548 (1991) 381–392.
180. M. Chiari, M. Giacomini, C. Micheletti and P.G. Righetti, J. Chromatogr., 558 (1991) 285–295.
181. M.P. Bellini and K.L. Manchester, Electrophoresis, 19 (1998) 1590–1595.
182. P.G. Righetti, K. Ek and B. Bjellqvist, J. Chromatogr., 291 (1984) 31–42.
183. E. Gianazza, G. Artoni and P.G. Righetti, Electrophoresis, 4 (1983) 321–324.
184. E. Gianazza, F. Celentano, G. Dossi, B. Bjellqvist and P.G. Righetti, Electrophoresis, 5 (1984) 88–97.
185. E. Gianazza, P. Giacon, B. Sahlin and P.G. Righetti, Electrophoresis, 6 (1985) 53–56.
186. E. Gianazza, S. Astrua-Testori and P.G. Righetti, Electrophoresis, 6 (1985) 113–117.
187. R.A. Mosher, M. Bier and P.G. Righetti, Electrophoresis, 7 (1986) 59–66.
188. P.G. Righetti, E. Gianazza and F. Celentano, J. Chromatogr., 356 (1986) 9–14.
189. P.G. Righetti, A. Morelli, C. Gelfi and R. Westermeier, J. Biochem. Biophys. Methods, 13 (1986) 151–159.
190. F. Celentano, E. Gianazza, G. Dossi and P.G. Righetti, Chemometr. Intell. Lab. Syst., 1 (1987) 349–358.
191. F.C. Celentano, C. Tonani, M. Fazio, E. Gianazza and P.G. Righetti, J. Biochem. Biophys. Methods, 16 (1988) 109–128.
192. E. Gianazza, F.C. Celentano, S. Magenes, C. Ettori and P.G. Righetti, Electrophoresis, 10 (1989) 806–808.
193. C. Tonani and P.G. Righetti, Electrophoresis, 12 (1991) 1011–1021.
194. P.G. Righetti and C. Tonani, Electrophoresis, 12 (1991) 1021–1027.
195. E. Giaffreda, C. Tonani and P.G. Righetti, J. Chromatogr., 630 (1993) 313–327.
196. A. Bianchi-Bosisio, P.G. Righetti, N.B. Egen and M. Bier, Electrophoresis, 7 (1986) 128–133.
197. A. Bossi, P.G. Righetti, G. Vecchio and S. Severinsen, Electrophoresis, 15 (1994) 1535–1540.
198. A. Bossi, C. Gelfi, A. Orsi and P.G. Righetti, J. Chromatogr., A 686 (1994) 121–128.
199. P.G. Righetti, A. Bossi, A. Görg, C. Obermaier and G. Boguth, J. Biochem. Biophys. Methods, 31 (1996) 81–91.
200. C. Gelfi and P.G. Righetti, Electrophoresis, 5 (1984) 257–262.
201. T. Rabilloud, V. Brodard, G. Peltre, P.G. Righetti and C. Ettori, Electrophoresis, 13 (1992) 264–266.
202. C. Gelfi, A. Morelli, E. Rovida and P.G. Righetti, J. Biochem. Biophys. Methods, 13 (1986) 113–124.
203. E. Rovida, C. Gelfi, A. Morelli and P.G. Righetti, J. Chromatogr., 353 (1986) 159–171.
204. S. Astrua-Testori, J.J. Pernelle, J.P. Wahrmann and P.G. Righetti, Electrophoresis, 7 (1986) 527–529.
205. T. Rabilloud, C. Gelfi, M.L. Bossi and P.G. Righetti, Electrophoresis, 8 (1987) 305–312.
206. P.G. Righetti, C. Gelfi, M.L. Bossi and E. Boschetti, Electrophoresis, 8 (1987) 62–70.
207. P.G. Righetti, C. Gelfi and M.L. Bossi, J. Chromatogr., 392 (1987) 123–132.
208. D. Hochstrasser, V. Augsburger, M. Funk, R. Appel, C. Pellegrini and A.F. Muller, in M.J. Dunn (Ed.), Electrophoresis '86, VCH, Weinheim, 1986, pp. 566–568.
209. J. Esteve-Romero, E. Simò-Alfonso, A. Bossi, F. Bresciani and P.G. Righetti, Electrophoresis, 17 (1996) 704–708.
210. J.S. Esteve-Romero, A. Bossi and P.G. Righetti, Electrophoresis, 17 (1996) 1242–1247.
211. B.M. Gåveby, P. Pettersson, J. Andrasko, L. Ineva-Flygare, U. Johannesson, A. Görg, W. Postel, A. Domscheit, P.L. Mauri, P. Pietta, E. Gianazza and P.G. Righetti, J. Biochem. Biophys. Methods, 16 (1988) 141–164.
212. M. Rimpilainen and P.G. Righetti, Electrophoresis, 6 (1985) 419–422.
213. S. Astrua-Testori and P.G. Righetti, J. Chromatogr., 387 (1987) 121–127.
214. G. Cossu and P.G. Righetti, J. Chromatogr., 398 (1987) 211–216.

215. J.R. Mazzeo and I.S. Krull, in A.N. Guzman (Ed.), Capillary Electrophoresis Technology, Dekker, New York, 1993, pp. 795–818.
216. P.G. Righetti and M. Chiari, in A.N. Guzman (Ed.), Capillary Electrophoresis Technology, Dekker, New York, 1993, pp. 89–116.
217. S. Hjertèn, in P.D. Grossman and J.C. Colburn (Eds.), Capillary Electrophoresis: Theory and Practice, Academic Press, San Diego, CA, 1992, pp. 191–214.
218. T.J. Pritchett, Electrophoresis, 17 (1996) 1195–1201.
219. F. Kilàr, in J.P. Landers (Ed.), Handbook of Capillary Electrophoresis, CRC Press, Boca Raton, FL, 1994, pp. 95–109.
220. T. Wehr, M.D. Zhu and R. Rodriguez-Diaz, in B.L. Karger and W.S. Hancock (Eds.), Methods in Enzymology: High Resolution Separation and Analysis of Biological Macromolecules, Part A: Fundamentals, Vol. 270, Academic Press, San Diego, CA, 1996, pp. 358–374.
221. R. Rodriguez-Diaz, M. Zhu and T. Wehr, J. Chromatogr. A, 772 (1997) 145–160.
222. R. Rodriguez-Diaz, M. Zhu and T. Wehr, Electrophoresis, 18 (1997) 2134–2144.
223. P.G. Righetti, M. Conti and C. Gelfi, J. Chromatogr. A, 767 (1997) 255–262.
224. P.G. Righetti, C. Gelfi and M. Conti, J. Chromatogr. B, 699 (1997) 91–104.
225. M. Taverna, N.T. Tran, T. Merry, E. Horvath and D. Ferrier, Electrophoresis, 19 (1998) 2527–2594.
226. W. Thormann, J. Caslavska, S. Molteni and J. Chmelik, J. Chromatogr., 589 (1992) 321–327.
227. M. Chiari, M. Nesi and P.G. Righetti, in P.G. Righetti (Ed.), Capillary Electrophoresis in Analytical Biotechnology, CRC Press, Boca Raton, FL, 1996, pp. 1–36.
228. M. Conti, C. Gelfi and P.G. Righetti, Electrophoresis, 16 (1995) 1485–1491.
229. G. Cossu, M. Manca, M.G. Pirastru, R. Bullitta, A. Bianchi-Bosisio and P.G. Righetti, J. Chromatogr., 307 (1984) 103–110.
230. M. Conti, C. Gelfi, A. Bianchi-Bosisio and P.G. Righetti, Electrophoresis, 17 (1996) 1590–1596.
231. A. Grönwall, C. R. Lab. Carlsberg, Ser. Chem., 24 (1942) 185–195.
232. C. Ettori, P.G. Righetti, C. Chiesa, F. Frigerio, G. Galli and G. Grandi, J. Biotechnol., 25 (1992) 307–318.
233. M. Conti, M. Galassi, A. Bossi and P.G. Righetti, J. Chromatogr. A, 757 (1997) 237–245.
234. C.A. Alper, M.J. Hobart and P.J. Lachmann, in J.P. Arbuthnott and J.A. Beeley (Eds.), Isoelectric Focusing, Butterworths, London, 1975, pp. 306–312.
235. L. Vuillard, T. Rabilloud, R. Leberman, C. Berthet-Colominas and S. Cusack, FEBS Lett., 353 (1994) 294–296.
236. L. Vuillard, C. Braun-Breton and T. Rabilloud, Biochem. J., 305 (1995) 337–343.
237. L. Vuillard, N. Marret and T. Rabilloud, Electrophoresis, 16 (1995) 295–297.
238. J. Herrero-Martinez, E. Simò-Alfonso, G. Ramis-Ramos, C. Gelfi and P.G. Righetti, J. Chromatogr. A, 878 (2000) 261–271.
239. S.N. Timasheff and T. Arakawa, in T.E. Creighton (Ed.), Protein Structure, a Practical Approach, Oxford, IRL Press, 1989, pp. 331–345.
240. T. Manabe, H. Miyamoto, K. Inoue, M. Nakatsu and M. Arai, Electrophoresis, 20 (1999) 3677–3683.
241. S.M. Chen and J.E. Wiktorowicz, Anal. Biochem., 206 (1992) 84–90.
242. F. Kilàr, J. Chromatogr., 545 (1991) 403–410.
243. A. Conway-Jacobs and L.A. Lewin, Anal. Biochem., 43 (1971) 394–400.
244. E.T. Nahkleh, S.A. Samra and Z.L. Awdeh, Anal. Biochem., 49 (1972) 218–224.
245. K. Slais and Z. Friedl, J. Chromatogr. A., 661 (1994) 249–256.
246. K. Shimura, Z. Wang, H. Matsumoto and K.I. Kasai, Electrophoresis, 21 (2000) 603–610.
247. C. Gelfi, M. Perego and P.G. Righetti, Electrophoresis, 17 (1996) 1470–1475.
248. C. Gelfi, M. Perego, P.G. Righetti, S. Cainarca, S. Firpo, M. Ferrari and L. Cremonesi, Clin. Chem., 44 (1998) 906–913.
249. S. Magnusdottir, C. Gelfi, M. Hamdan and P.G. Righetti, J. Chromatogr. A, 859 (1999) 87–98.
250. A.V. Stoyanov, C. Gelfi and P.G. Righetti, Electrophoresis, 18 (1997) 717–723.
251. S. Hjertèn, L. Valtcheva, K. Elenbring and J.L. Liao, Electrophoresis, 16 (1995) 584–594.
252. S. Blanco, J.M. Clifton, J.L. Loly and G. Peltre, Electrophoresis, 17 (1996) 1126–1133.
253. P.G. Righetti and F. Nembri, J. Chromatogr. A, 772 (1997) 203–211.
254. L. Capelli, A.V. Stoyanov, H. Wajcman and P.G. Righetti, J. Chromatogr. A, 791 (1997) 313–322.

255. A.V. Stoyanov and P.G. Righetti, J. Chromatogr. A, 790 (1997) 169–176.
256. A. Bossi and P.G. Righetti, Electrophoresis, 18 (1997) 2012–2018.
257. A. Bossi and P.G. Righetti, J. Chromatogr. A, 840 (1999) 117–129.
258. L. Capelli, F. Forlani, F. Perini, N. Guerrieri, P. Cerletti and P.G. Righetti, Electrophoresis, 19 (1998) 311–318.
259. E. Olivieri, A. Viotti, M. Lauria, E. Simò-Alfonso and P.G. Righetti, Electrophoresis, 20 (1999) 1595–1604.
260. P.G. Righetti, E. Olivieri and A. Viotti, Electrophoresis, 19 (1998) 1738–1741.
261. P.G. Righetti, A. Saccomani, A.V. Stoyanov and C. Gelfi, Electrophoresis, 19 (1998) 1733–1737.
262. J. Herrero-Martinez, E. Simò-Alfonso, G. Ramis-Ramos, C. Gelfi and P.G. Righetti, J. Chromatogr. A, 878 (2000) 261–271.
263. J. Herrero-Martinez, E. Simò-Alfonso, G. Ramis-Ramos, C. Gelfi and P.G. Righetti, Electrophoresis, 21 (2000) 633–640.
264. A. Saccomani, C. Gelfi, H. Wajcman and P.G. Righetti, J. Chromatogr. A, 832 (1999) 225–238.
265. B. Verzola, C. Gelfi and P.G. Righetti, J. Chromatogr. A, 868 (2000) 85–99.
266. B. Verzola, C. Gelfi and P.G. Righetti, J. Chromatogr. A, 874 (2000) 293–303.
267. C.S. Effenhauser, G.J.M. Bruin and A. Paulus, Electrophoresis, 18 (1997) 2203–2213.
268. P.G. Righetti, M. Fazio, C. Tonani, E. Gianazza and F.C. Celentano, J. Biochem. Biophys. Methods, 16 (1988) 129–140.

CHAPTER 13

Sodium Dodecyl Sulphate Polyacrylamide Gel Electrophoresis (SDS-PAGE)

CONTENTS

13.1. INTRODUCTION

SDS-PAGE is perhaps the most popular and direct method for assessing, in a fast and reproducible manner, the M_r of denatured polypeptide chains and the purity of a protein preparation. Accounts on SDS-PAGE can be found in e.g. [1–5]. In small or large (or very large) gel slab formats, it is the standard second dimension of 2-D maps, and that is why it will be treated in detail in this book. Typically, in disc electrophoresis, proteins migrate according to both surface charge and mass, so that discriminating the two contributions is not an easy task, although it can be done in a number of mathematical treatments, such as the Ferguson [6] plots. A possible approach to M_r measurements of proteins would be to cancel out differences in molecular charge by chemical means, so that migration would then occur solely according to size. SDS, an amphipatic molecule, is known to form complexes with both non-polar side chains and charged groups of amino acid residues in polypeptides of all possible sizes and shapes without rupturing polypeptide bonds. Surprisingly large amounts of SDS can be bound to proteins, assessed at about 1.4 g SDS per gram of protein by a number of authors [7–10]. This means that the number of SDS molecules is of the order of half the number of amino acid residues in the polypeptide chain. This amount of highly charged surfactant molecules is sufficient to overwhelm effectively the intrinsic charges on the polymer chain, so that the net charge per unit mass becomes approximately constant. Electrophoretic migration is then proportional to the effective molecular radius, i.e. to the molecular mass of the polypeptide chain [11]. Although it is an oversimplification to which a number of exceptions are to be found, the relationship does in fact hold true for a very large number of proteins [12,13] and the method has become one of the most widely used for measurement of protein molecular mass. The unique properties of SDS are due to its long, flexible alkyl tail, which is able to establish hydrophobic interactions with all combinations of amino acids, which leads to massive unfolding of proteins. The action of SDS is also due to its ionic head, which can break ionic interactions between proteins and drive an important electrostatic repulsion between SDS–protein complexes. This prevents reassociation of such complexes, even at the very high concentrations encountered in gel electrophoresis. Another important property of SDS lies in the fact that the ionic head is a strong electrolyte, so that it is fully ionised in the pH 2–12 interval, where most of the biochemical separations take place. According to Reynolds and Tanford [9] a concentration of SDS above 0.5 mM is sufficient for binding 1.4 g SDS per gram of protein in a primarily hydrophobic way, provided the disulphide bridges are reduced and the polymer chain is thus in an extended conformation. Under the influence of SDS, proteins assume the shape of rod-like particles, the length of which varies uniquely with the molecular mass of the protein moiety, occupying 0.074 nm (1.4 g SDS bound) per amino acid residue. Certain proteins, such as papain, pepsin and glucose oxidase [14] and two different classes of proteins, the glycoproteins [15] and the histones [16], show an 'anomalous' behaviour towards SDS. Either they bind a relatively low amount of detergent, or SDS cannot compensate for their intrinsic charges, such as in the case of histones. But these examples are not sufficient to cast severe doubts on the validity of M_r estimation by SDS [17,18], and in fact SDS-PAGE has become by far the most popular method for M_r assessments of denatured polypeptide chains.

The reasons for this rapid acceptance are not difficult to see. The apparatus required is readily available in most laboratories and is inexpensive. Once learnt, the procedure is straightforward and highly reproducible, and results can be obtained within a few hours using only a few micrograms of material. As with other PAGE methods, in many cases the samples need not to be totally pure. The degree of purity required depends largely upon the sample being studied and the ease with which the component of interest can be identified on the final gel pattern.

In summary, SDS-PAGE is used mainly for the following purposes: (a) estimation of protein size; (b) assessment of protein purity; (c) protein quantitation; (d) monitoring protein integrity; (e) comparison of the protein composition of different samples; (f) analysis of the number and size of polypeptide subunits; (g) when using, e.g. Western blotting; (h) as a second dimension of 2-D maps.

13.2. SDS-PROTEIN COMPLEXES: A REFINEMENT OF THE MODEL

Over the years, some models for the structure of complexes between proteins and SDS have been proposed:

(1) A 'rod-like particle model', proposed on the basis of hydrodynamic measurements. This was one of the earliest models suggested [9] and it hypothesised that, upon binding of SDS, the polypeptide would form 'rod-like' structures, about 3.6 nm in diameter and 0.074 nm/amino acid residue in length. In a more recent reinterpretation, the particle was described as a "short, rigid, rod-like segment, with intervening regions possessing some flexibility" [19].

(2) A 'necklace model', in contradiction with the 'rod-like particle' above. In this second model, "the polymer chain is flexible and micelle-like clusters of SDS are scattered along the chain" [20,21]. In an improved version, SDS binds to the protein in the form of spherical micelles and the polypeptide forms α-helices mostly in the hydrophobic region of the micelles.

(3) A flexible 'α-helix/random coil' structure, suggested on the basis of circular dichroism changes of proteins upon SDS binding [22].

(4) A flexible helix model, in which the polypeptide chain is helically coiled around an SDS micelle, attached by hydrogen bonds between sulphate group oxygens and peptide bond nitrogens [23].

Over the years, Lundahl's group has refined this last model [24–28] which has now become the 'protein-decorated, micelle model'. Based on small-angle neutron scattering data, this model proposes that adjacent, protein-decorated, spherical micelles are formed, rather than cylindrical structures, as previously suggested by the same group [23]. These authors have studied the formation of SDS–protein complexes at SDS concentrations in the proximity of its critical micellar concentration (1.8 mM SDS in 100 mM buffer). They propose the following series of events in complex formation. First the dodecyl chains of the detergent penetrate the surface of the protein and come into contact with the hydrophobic interior of the protein or of its domains. As a consequence, polypeptide segments from the interior of the protein become displaced toward the surface of the complex, since they are less hydrophobic than the dodecyl chain. Many

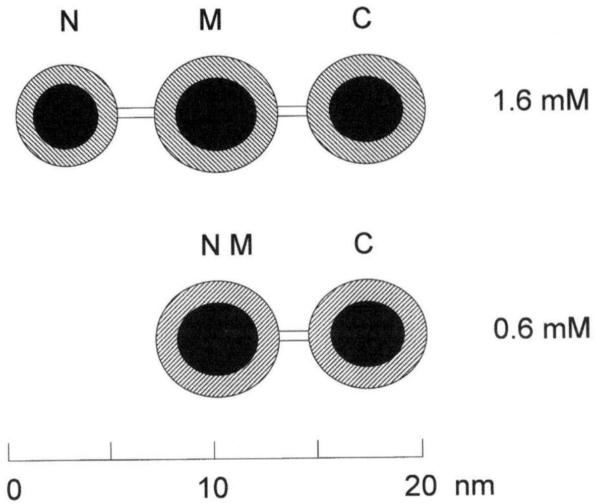

Fig. 13.1. Schematic scale models of the mutual disposition of the three-protein decorated micelles of the enzyme *N*-5′-phosphoribosylanthranilate isomerase/indole-3-glycerol-phosphate synthase in buffers with SDS concentrations approaching the CMC (1.6 mM SDS, top) and of the two protein-decorated micelles in buffers below the CMC (0.6 mM SDS, bottom). The independent micelles N (N-terminal), M (middle) and C (C-terminal) of the three-micelle complex are connected, at saturation, by flexible oligopeptide linkers about 5 or 6 amino acid residues long; with decreasing SDS concentrations to subsaturating levels, the small micelle N coalesces with M and micelle C approaches M. The hydrophobic micelle core is shown in black, whereas the outer hydrophilic shell, formed by hydrophilic stretches of the polypeptide chain and by the SDS sulphate groups is shaded. (From [28], with permission.)

SDS molecules become locked in the inserted position by ion-pair formation and hydrogen bonding. Additional SDS molecules then become included in the complex until a spherical SDS micelle is completed, around which the polypeptide is wound. Any length of polypeptide that, for steric reasons, cannot be accommodated in direct contact with the micelles, forms the core for growth of another protein-covered micelle. This process is repeated until the whole polypeptide is coiled around adjacent SDS micelles that are linked with short polypeptide segments. For a protein of average size (50,000 Da) it is believed that this complex could be composed of three protein-decorated SDS micelles, not necessarily of equal size. For example, in the case of the enzyme *N*-5′-phosphoribosylanthranilate isomerase/indole-3-glycerol-phosphate synthase (M_r 49,484 Da) the SDS–protein complex contained the dodecyl hydrocarbon moieties in three globular cores, of which the central one was the largest (see Fig. 13.1; notice though that, at SDS levels below the CMC, the micelles can be reduced to only two). Each core was surrounded by a hydrophilic shell, formed by the hydrophilic and amphiphilic stretches of the polypeptide chain, and by the sulphate groups of the detergent, whereas, presumably, most or all non-polar side groups of the polypeptide chain penetrate partly or completely into the hydrophobic micelle cores [26]. The model has received support from a recent paper by Westerhuis et al. [29], re-evaluating the migration behaviour of SDS–protein complexes. These authors have detected at least two independent electrophoretic migration mechanisms for SDS–protein micelles:

(i) for proteins in the 14–65 kDa range in a $<15\%\ T$ polyacrylamide matrix, linear Ferguson plots suggested that they migrated ideally and that their effective radii could be estimated in this manner; (ii) concave plots at higher gel concentrations, and for complexes derived for larger proteins, indicated that migration in these cases could be described by reptation theory. Migration of the large proteins at lower gel concentrations and small proteins at higher gel concentrations was not well described by either theory, representing intermediate behaviour not contemplated by these mechanisms. Such data support the model in which all but the smallest SDS–protein complexes adopt a necklace-like structure in which spherical micelles are distributed along the unfolded polypeptide chains, as proposed by Lundahl's group and also by Samsò et al. [30].

13.3. THEORETICAL BACKGROUND OF M_R MEASUREMENT BY SDS-PAGE

Methods for measurements of molecular size by electrophoresis fall into two main categories, namely those using a relationship between mobility and various gel concentrations [6] and those for which a single concentration is used. The latter can only be applied to families of molecules with the same charge/mass ratio, such as nucleic acids, or to molecules in which uniform charge densities have been produced by binding large amounts of a charge ligand, such as in the case of SDS.

Ferguson plots try to discriminate the contributions of charge and mass to the electrophoretic mobility of proteins, according to the following relationship:

$$\log R_f = \log Y_o - K_R T$$

where R_f is the relative protein mobility (corrected to the mobility of a fast migrating dye), Y_o the extrapolated free mobility on the y-axis, T the polyacrylamide gel concentration and K_R (coefficient of retardation) is the slope of the curve obtained in a plot of $\log R_f$ vs. $\%T$ and is proportional to the mass of the macromolecule subjected to electrophoresis in a series of gels at various $\%T$. When a series of Ferguson plots of R_f vs. $\%T$ were constructed for a number of SDS-treated proteins [31], it was found that the intercept at $T = 0$, which is the apparent free mobility (Y_o), was almost identical for all the species examined. This demonstrates that, for SDS-laden proteins, the effective charge to mass ratio is, to a first approximation, constant. In order to determine M_r from K_R values, all that is necessary is to show that there is a uniform dependence of M_r on K_R. Rodbard and Chrambach [32–34] have shown that K_R is dependent on the effective molecular radius. Since Reynolds and Tanford [10] and Fish et al. [8] have shown that the hydrodynamic properties of protein–SDS complexes are a unique function of polypeptide chain length, K_R must clearly be dependent on this also. In practice, plots of $\log K_R$ vs. $\log M_r$ are generally linear with slopes proportional to the relationship between M_r and the Stokes radii. It must be emphasised that $\log M_r$ vs. R_f plots (in gels of constant $\%T$) are linear only over a certain range of mass values, typically between 15,000 and 70,000 Da (see Fig. 13.2); however, if porosity gradients are used (e.g. gels ranging in concentration between 5 and 20% T) plots of $\log M_r$ vs. $\log \%T$ are linear

TABLE 13.2

PRELABELLING OF PROTEINS PRIOR TO SDS-PAGE

Method	λ^{a}_{exc}	λ^{b}_{emis}
Remazol dye		
1 Mix 0.2 ml protein (2–10 mg/ml) in 0.15 M NaCl with 0.05 ml 1 M Na phosphate (pH 7.2–9.2)	–	–
2 Add 0.05 ml Remazol BBR (10 mg/ml) in 10% SDS		
3 Heat at 100°C for 5 min or 56°C for 10 min; add 1% β-mercaptoethanol and repeat the heating cycle		
4 Apply 4 μl samples to gels		
Dansylation		
1 Mix protein solution (4 mg/ml) in 2% $NaHCO_3$ with an equal vol. dansyl chloride in acetone (2 mg/ml)	340	520
2 Incubate at 37°C for 2 h in the dark; shake intermittently		
3 Precipitate protein with 2 vol. acetone		
4 Centrifuge; wash precipitate with 2 vol. acetone		
5 Dissolve precipitate in buffer and pretreat with SDS as usual		
Fluorescamine		
1 50–100 μg protein dissolved in 100 μl 15 mM Na phosphate buffer (pH 8.5) in 5% SDS and 5% sucrose	390	475
2 Heat at 100°C for 5 min		
3 Cool; add 5 μl fluorescamine (1 mg/ml) in acetone and shake		
4 Add 5 μl bromophenol blue tracking dye (5 mg/ml) if needed		
5 Apply 10 μl sample to gel		
2-methoxy-2,4-diphenyl-3 (2H)-furanone (MDPF)		
As for fluorescamine, but at (3) use MDPF (2 mg/ml) in dimethyl sulphoxide	390	480
o-Phthaldialdehyde		
1 Mix 1 ml protein solution (0.1–50 μg) in 0.05 M Na phosphate, pH 8.5, with 2.5 μl β-mercaptoethanol	340	460
2 Let stand for 10 min		
3 Add 2.5 μl of 1% *o*-phthaldialdehyde in methanol		
4 Place in the dark for 2 h		
5 Pretreat with SDS in the usual way and apply to gel		

[a] Excitation wavelength.
[b] Emission wavelength.

13.4.4. Post-electrophoretic detection

Although the general stain protocols described in the previous chapter (see Section 12.2.5) in general apply, it should be remembered that the presence of SDS might interfere, so that some procedures might have to be adapted to this technique. For example, the standard Coomassie Brilliant Blue R-250 procedure would work here, but it would be preferable to perform the staining step by incubating at 50°C, since at this temperature the SDS micelle will disaggregate more quickly and diffuse more rapidly outside the gel matrix. Of course, silver staining protocols fully apply (a number of them is described by Merril and Washart [70]), with the proviso that, if the protein has to be eluted for further analysis, such staining protocols are in general incompatible with subsequent MS analysis. Below, a classical silver nitrate staining procedure is described.

13.4.4.1. Non-diamine, silver nitrate stain

In this type of stains, silver ions released from silver nitrate under acidic conditions are reduced to metallic silver when the gels are placed in an alkaline solution containing a reducing agent. In the reduction reaction formaldehyde is oxidised to formic acid which is buffered by sodium carbonate [70]. The procedure below is optimised for gels of <1 mm in thickness and the volumes given apply for gels of 16 × 16 cm in size. All steps should be performed in glass trays, with only one gel per tray. The trays should be placed on a shaker at a gentle speed. All steps run at room temperature.

(1) After electrophoresis, place each gel into 500 ml of fixing solution (methanol 50% v/v, acetic acid 10% v/v). Soak the gels for 1 h. Staining can occur just after 1 h; however, the gels may remain in this solution overnight, if needed.

(2) Transfer each gel from the fixing solution into 500 ml of rehydration solution (methanol 10% v/v, acetic acid 5% v/v). Soak the gels in this solution for 10 min.

(3) Discard the rehydration solution and add 200 ml of glutaraldehyde solution (1% w/v) to each tray. Equilibrate the gels in this solution for 30 min.

(4) Discard the glutaraldehyde solution and wash each gel with 500 ml of deionised water for 15 min.

(5) Eliminate the wash and add to each gel 200 ml of dichromate solution (34 mM potassium dichromate, 32 mM nitric acid). Equilibrate the gels in this solution for 5 min.

(6) Discard the dichromate solution and add to each gel 200 ml of staining solution (0.118 mM silver nitrate).

(7) Eliminate the staining solution.

(8) Wash each gel briefly (approximately 5 s) with 50 ml of reducing solution (0.283 M sodium carbonate, 7 mM formaldehyde) and then discard this wash. Follow this initial rinse with another 500 ml of the same reducing solution. If colloidal particles collect in the solution prior to full image development, replace the reducing solution with fresh solution. Continue the development in this solution until the proteins are sufficiently stained. Proteins should become visible within 1–3 min.

(9) In order to block the image development, dispense with the reducing solution and add 500 ml of stop-bath solution (3% v/v acetic acid). Equilibrate the gels in this solution for 5 min.

(10) Dispose of the stop-bath solution. Wash each gel twice with 500 ml of distilled water with 10% methanol or ethanol for 10 min each wash.

(11) Store the gels moist in sealed clear plastic bags.

Caution. Artefactual bands in SDS-PAGE gels are often observed when using highly sensitive silver staining protocols. Skin keratins contaminating the protein samples or the buffers are believed to be a major cause of these bands, appearing in the 50- to 68-kDa region [71–73]. A recent report [74] suggests some remedies to it, based on the observation that such spurious bands seem to originate from the polyacrylamide gels themselves, rather than from the protein samples, sample buffer or electrode buffer. In addition, the gel region responsible for such spurious keratin bands seems to be

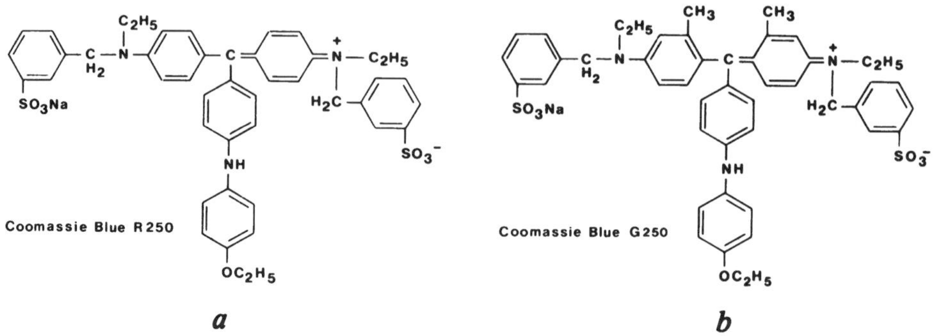

Fig. 13.5. Chemical formulas of Coomassie Brilliant Blue R-250 (a) and of Coomassie Brilliant Blue G-250 (b).

located at or near the sample application wells in the polyacrylamide gels (contaminated combs?). On these grounds, the remedy proposed is quite simple: prior to the actual run and to real sample application, fill the wells with only sample buffer, devoid of protein sample. Conduct electrophoresis with reversed polarity (5 min with a constant current of 40 mA, in a standard Laemmli buffer); remove the sample buffer (into which the keratins will have collected). At this point resume the normal gel polarity, add your sample into the wells and run your SDS-PAGE separation. Upon silver staining, the spurious keratin bands will have almost completely disappeared [74].

13.4.4.2. Colloidal staining

In the following sections, methods will be presented exploiting staining with the Coomassie dyes, a family of dyes extremely popular in electrophoresis, due to their relatively high sensitivity and simplicity of use. Today, they are particularly exploited in their micellar form, which imparts to them a much higher sensitivity. Fig. 13.5 gives the formulas of the two most common ones, the R-250 (R stands for its reddish hue) and G-250 (G stands for its greenish hue) forms. The latter has been exploited, in a colloidal form, by Neuhoff et al. [75,76] for a staining method quite popular today for staining the second dimension of a 2-D gel. This method has a high sensitivity (ca. 30 ng protein/band) but requires an overnight period for development. The procedure is as follows.

(1) Slowly add 100 g of ammonium sulphate to 980 ml of a 2% phosphoric acid solution till it has completely dissolved. Bring it to a 1-l volume by adding 20 ml of a solution of 1 g of Coomassie Brilliant Blue G-250 in water. Shake before use. This staining solution can be used several times.

(2) Fix the gel for 1 h in 12% (w/v) TCA.

(3) Stain overnight with 160 ml of staining solution (0.1% w/v Coomassie G-250 in 2% phosphoric acid, 10% ammonium sulphate) plus 40 ml of methanol (added during staining).

(4) Wash for 1 to 3 min in 0.1 mol/l Tris–phosphate buffer, pH 6.5.

(5) Rinse briefly in 25% v/v aqueous methanol.

(6) Stabilise the protein–dye complex in 20% ammonium sulphate.

13.4.4.3. 'Hot' Coomassie staining

A number of recent reports have appeared, recommending high temperatures for quick staining processes, while guaranteeing high sensitivity. The Kurien and Scofield [77] method involves a 5-min staining step at 70°C, followed by a 20-min destaining step at the same temperature and a rinsing step with distilled water at room temperature for a further 20 min (total processing time 45 min; sensitivity not greater than 100 ng/band). In another approach [78], 2.5% bleach at 55°C, followed by rinsing of the gel in distilled water, has been recommended as a method for speeding up the destaining process (90 min total processing time; 100 ng protein/band as upper detection limit). In a later report, the evolution of a protocol published by Fairbanks et al. [79], who suggested three staining solutions of progressively lower concentrations of Coomassie and isopropanol for increased sensitivity, but which requires up to two working days, has been proposed. This method [80] comprises the following steps.

(1) After SDS-PAGE, place the gels in plastic boxes (Nalgene, Tupperware), with a hole in the lid. Add 100 ml of staining solution (0.05% Coomassie, 25% isopropanol, 10% acetic acid).

(2) Heat the gel in staining solution in a conventional 100 W output microwave oven on full power till reaching the boiling point (ca. 2 min).

(3) Cool at room temperature for 5 min with gentle shaking. After this step, bands with ca. 100 ng protein could be visualised despite the blue background.

(4) Discard the staining solution, rinse with distilled water and immediately discard.

(5) Add 100 ml of a new staining solution (0.005% Coomassie, 10% isopropanol, 10% acetic acid) and microwave again to the boiling point (ca. 1 min and 20 s).

(6) Discard the hot staining solution, rinse with distilled water and discard. At this step, bands with ca. 50 ng protein could be observed.

(7) Add 100 ml of a third staining solution (0.002% Coomassie, 10% acetic acid) and again microwave to the boiling point for 1 min and 20 s.

(8) Discard this third stain and again rinse with distilled water and discard. At this time, bands with ca. 25 ng protein are visible.

(9) Finally, place the gels in 100 ml of destaining solution (10% acetic acid) and microwave for 1 min and 20 s. Allow to cool at room temperature for ca. 5 min. At this point, bands with at least 5 ng protein became visible against a light blue background. If this last step is repeated 2 or 3 times (or if the gel is left on a shaker for 15 more min), as little as 2.5 ng protein can be seen, this time against a clear background.

This method is claimed to be faster and more sensitive than any other Coomassie procedure so far reported. It appears that bringing the solution containing the gel to the boiling point is a prerequisite for the improved sensitivity and speed of this method.

13.4.4.4. Turbidimetric protein detection (negative stain)

Although this staining procedure has low sensitivity, it might be useful when extracting proteins to be analysed by mass spectrometry (MS), since often dye-molecules tend to stick to the protein surface and give an erroneous signal by MS. The method is based on the observation that, when an SDS-containing gel is chilled at 0° to 4°C, free SDS precipitates in the gel, forming an opaque background, whereas the SDS bound to protein does not precipitate, leaving a clear protein band [81]. The method presented below involves the use of high-ionic-strength solutions for precipitating SDS and producing this effect more reliably, instead of inducing SDS precipitation by chilling [66].

(1) Immediately after the SDS step, place the polyacrylamide gel containing the separated proteins into a covered basin with 10 to 12 gel volumes of 4 M Na acetate.

(2) Place on a platform shaker at low speed for 40–50 min for 1-mm-thick gels or longer for thicker gels.

(3) View the gel over a black, non-reflective surface using a fluorescent desk lamp to illuminate the gel from the side. If the bands are not yet visible, allow longer times for precipitation. The incubation should not proceed for too long, though, otherwise the bands will diffuse and resolution will be lost.

(4) For recovery of proteins by electroelution, soak gel slices in water to reduce the ionic strength.

13.4.4.5. Negative metal stains

Just as the method reported in the previous section, there exist a family of negative stains which are based on the formation of insoluble metal salts in the presence of SDS, leaving protein bands unstained when viewed against a dark background. The hypothesis is that all metal cations (e.g. Co, Cu, Zn, Ni) capable of forming sparingly soluble metal salts such as hydroxides, carbonates and chlorides, can be used to detect proteins in SDS-gels. Protein visualisation will occur through deposition of insoluble complexes of the type Me^{2+}–Tris–SDS to form a semi-opaque background. Most of these negative stains act rapidly (within a few minutes), do not require any protein fixation within the gel matrix, and the proteins stained in this manner are reported to be easily recovered from gels for further analysis. Some commonly used negative stains and their protein detection limits include: (a) copper chloride (5 ng/mm) [82]; (b) zinc chloride and zinc sulphate (10–12 ng) [83,84]; (c) potassium acetate (0.12–1.5 μg) [85]; and (d) sodium acetate (0.1 $\mu g/mm^3$) [86].

Detection limits for these negative stains fall between the sensitivity levels of Coomassie Blue and silver stains. While these negative stains require the presence of SDS in the gel, Candiano et al. [87] have described a negative staining technique for proteins in polyacrylamide gels, operating with and without SDS, based on precipitation of methyltrichloroacetate and reported to have a sensitivity of barely 0.5 ng protein. Table 13.3 gives the protocol for the negative zinc/imidazole staining.

While there are a number of advantages of this negative stain protocol, such as reversibility [88], compatibility with Western blotting and subsequent immunodetection [89,90] and compatibility with MS procedures [91], there are also some problems, notably the difficulty of creating a photographic documentation of the gel. These gels, in fact, inherently exhibit low contrast and are usually photographed against a black background with side illumination. The light-scattering in the background gives usually a poor image quality. Bricker et al. [92] have now proposed an easy documentation protocol, which consists in using a low-cost flatbed digital scanner. This device utilises reflected rather than transmitted light to acquire an image. The milky white background of negatively stained gels efficiently back-scatters light, while the clear protein bands do not. When scanned with a black background, the protein bands appear dark against a

TABLE 13.3

NEGATIVE ZINC/IMIDAZOLE STAINING FOR SDS-PAGE

- Prepare 100 ml of 0.5 M imidazole (solution A; 10× conc.); 100 ml of 0.5 M zinc sulphate (solution B; 10× conc.) and 100 ml of 0.5 M Tris/glycinate, pH 8.8 (destaining solution, 10× conc.).
- In separate containers, dilute 1 part of solution A and one part of solution B with nine parts of water. Mix the solutions thoroughly.
- Remove the SDS-gel from the electrophoresis cell and place it in a small basin.
- Pour the imidazole solution (50 ml for a Bio-Rad Mini Protean II, 1-mm thick gel will suffice) on the gel on the basin and mix gently on a platform for 10 min.
- Transfer the gel to the diluted zinc sulphate solution on a rocking platform. Allow 45 s for the gel to develop.
- Transfer the gel to a container filled with doubly distilled water and rinse for 3–5 min. Discard and replace with fresh water solution. The gel can be stored for weeks in water.
- For visualising the protein bands, place the gel against a black background and shine light on it at a shallow angle. The protein bands will be visible as transparent bands (thus black) against an opaque, milky background.
- For recovering the protein zones, the bands can be excised and the proteins eluted by passive diffusion in 30% acetonitrile, 30% isopropanol, 30 formic acid and 10% water.
- For destaining the gel, place it in a 50 mM (1 : 1 dilution) Tris/glycinate, pH 8.8 buffer and shake for 5 min. Replace with fresh solution and again agitate for 5 min.
- The gel is now ready for Coomassie Blue or silver staining, blotting or other manipulations.

white background. After scanning, the image is imported as a TIF file under Corel draw and labelled as needed. The image contrast is good and such pictures are suitable for publication.

13.4.4.6. Fluorescent detection with SYPRO dyes

A large number of staining procedures are based on Coomassie stains, a family of non-polar, sulphonated aromatic dyes, typically utilised in methanol–acetic acid or methanol–trichloroacetic acid solutions, in general belonging to the category of toxic solvents. In addition, it has been demonstrated that Coomassie G solutions, containing trichloroacetic acid and alcohols, lead to irreversible acid-catalysed esterification of Glu side chain carboxyl groups [93], which complicates interpretation of peptide mapping data from mass spectrometry. A recent fluorescent stain, of the SYPRO family, described by Steinberg et al. [94], seems to have solved all the above problems, while considerably enhancing the detection sensitivity (down to about 4 ng per protein band). This novel compound, SYPRO Tangerine stain, does not require solvents such as methanol or acetic/trichloracetic acids; instead, staining can be performed in a wide range of buffers, or simply in 150 mM NaCl, by a simple one-step procedure, in the absence of a destaining step. Stained proteins can be excited by UV light at ca. 300 nm or with visible light at ca. 490 nm, the fluorescence emission of the dye being 640 nm. Non-covalent binding of SYPRO Tangerine dye is mediated by SDS and, to a lesser extent, by hydrophobic amino acid residues in proteins. This is in contrast to acidic silver nitrate staining, which interacts primarily with Arg and Lys residues. Although SYPRO Red and SYPRO Orange stains [95–97] have been already described for sensitive fluorescent detection of proteins in SDS gels, the novel Tangerine dye appears to be an improvement in that it does not require any harsh solvent condition. An example of such staining is

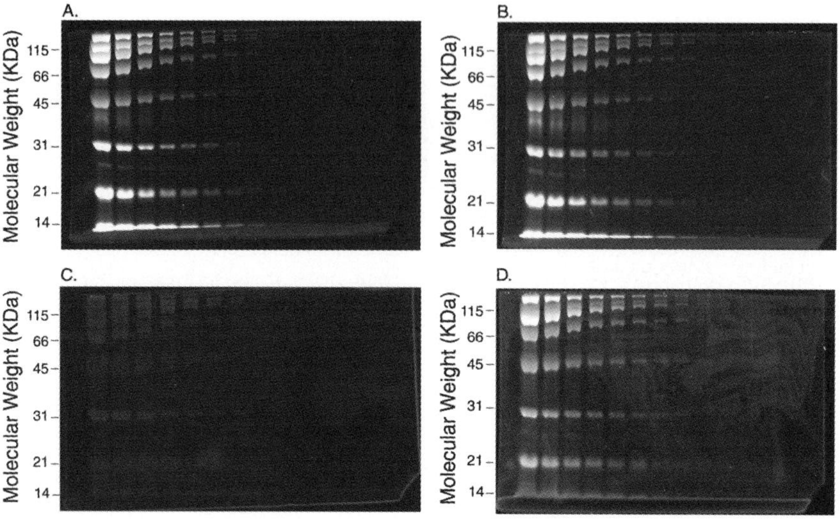

Fig. 13.6. Staining of broad-range molecular mass protein standards with SYPRO Orange and SYPRO Tangerine dye, after an SDS-PAGE run. (A) SYPRO Orange dye staining in 7% acetic acid solution. (B) SYPRO Tangerine dye staining in 7% acetic acid solution. (C) SYPRO Orange dye staining in phosphate-buffered saline solution. (D) SYPRO Tangerine dye staining in phosphate-buffered saline solution. Two-fold dilution series of proteins were prepared in standard SDS sample buffer and the amount of protein loaded ranged from 2 μg per band (1st lane) to 1 ng per band (12th lane). (From [94].)

given in Fig. 13.6, which compares the sensitivities of SYPRO Orange and Tangerine dyes in different solvents. All these SYPRO stains, in addition, have the advantage that, since staining appears to be due to intercalation of dye in the SDS micelle, little protein to protein variability is observed in SDS gels, compared with amino-directed stains such as the Coomassie stains. In case of blotting, either from 1-D (SDS-PAGE) or 2-D (IEF-SDS) gels, an interesting stain appears to be SYPRO Ruby protein blot stain [98,99], which provides a sensitive, gentle, fluorescence-based method for detecting proteins on nitrocellulose or polyvinylidene difluoride (PVDF) membranes. SYPRO Ruby is a permanent stain composed of ruthenium as part of an organic complex that interacts non-covalently with proteins. Stained proteins can be excited by UV light of about 302 nm or with visible light of ca. 470 nm. Fluorescence emission of the dye is approximately 618 nm. The stain can be visualised using a wide range of excitation sources utilised in image analysis systems, including a UV-B transilluminator, a 488-nm argon-ion laser, a 532-nm yttrium–aluminum–garnet (YAG) laser, a blue-fluorescent light bulb, or a blue light-emitting diode (LED). The detection sensitivity of SYPRO Ruby (0.25–1 ng protein/mm^2) is superior to that of Amido Black, Coomassie Blue and india ink staining and nearly matches colloidal gold staining. An additional advantage of SYPRO Ruby blot stain is that it is fully compatible with subsequent protein analysis, such as Edman-based sequencing and mass spectrometry. Some inherent disadvantages, though, could be the cost of such stains and the difficulties in providing proper documentation, since present-day scanners (and photographic equipments) are not yet set-up for fluorescent observation.

13.4.5. Possible sources of artefactual protein modification

If a protein extracted from an SDS-gel has to be subjected to analysis by high-resolution techniques, such as MALDI-TOF, one should be aware of a number of potential modifications which proteins in SDS-mediated techniques can experience. Mass-spectrometry-based methods have successfully identified a number of such modifications including cysteine-acrylamide adducts [100,101], cysteine β-mercaptoethanol adducts [102], methionine oxidation [103], formylation of Ser and Thr [104,105] and aggregates with certain staining agents [56]. It is worth noting that most of these modifications are detected for molecular mass values below 30 kDa, possibly because mass resolution is lost above this size for proteins extracted from diverse matrices. These problems are aggravated in SDS-gels, because typically, in most users laboratories, such gels are not 'washed', i.e. are still full of unreacted monomers. On the contrary, today, in all isoelectric focussing techniques, the gel matrix is 'clean', in that the matrix is fully washed so as to remove unreacted products and catalysts, dried and reswollen when used. This is a normal procedure, since IEF gels are usually run in horizontal chambers, with an open face, whereas SDS gels, in general, are run vertically, enclosed in between two glass plates, which would render such a washing step highly unpractical. As demonstrated by Caglio et al. [106], polymerisation in detergents can rarely be driven above 70% conversion (especially at high surfactant levels), which means that a huge excess of unreacted acrylamide is available in the gel for reacting with proteins. A number of reports have appeared on the use of scavengers, which include 3-mercaptopropanoic acid for Tris-Tricine gels [107,108], glutathione or sodium thioglycolate for modified SDS-PAGE gels [109], or free cysteine for IPG gels [110]. These modifications, together with the removal of SDS from eluted proteins prior to their analysis by MALDI-TOF [111] are improving the mass accuracy and diminishing the risk of protein modification. However, in untreated gels, it has recently been found that severe alkylation by free, excess acrylamide in a gel, occurs for a number of proteins (e.g. ubiquitin, α-lactalbumin, β-lactoglobulin B, LG-B) [112]. Moreover, other modifications can take place during the blotting procedure, as quite often performed for transferring proteins out of SDS-gel matrices. It is quite typical, in this last case, to use formic acid in the transfer solution: this results in massive formylation of proteins too. A case in point is illustrated in Fig. 13.7: in the case of LG-B (M_r 18,285), run in, and extracted from, an SDS gel, MALDI-TOF revealed an impressive number of peaks, a series due to alkylation by acrylamide (spaced at intervals of 71 Da, centred at m/z ratios of 18,356, 18,428, 18,499 and 18,568) and a second series due to formylation (on Ser and/or Thr residues; spaced at intervals of 28 Da, observed at m/z values of 18,312, 18,383, 18,455 and 18,527), for a total of >10 peaks, extending in mass from 18,285 (control LG-B, which had become by far the least abundant species!) up to 18,639. In the case of ubiquitin (M_r 8564), the unmodified protein was observed at m/z 8566, while a series of peaks (up to 10) between m/z values 8593 up to 8835 were attributed to multiple formylation events (ubiquitin contains 3 Ser and 7 Thr residues, which are described as potential formylation sites) (see Fig. 13.8). The fact that no acrylamide adducts can be seen here is not surprising, since ubiquitin does not contain any Cys residue. Remedies to this will be discussed below.

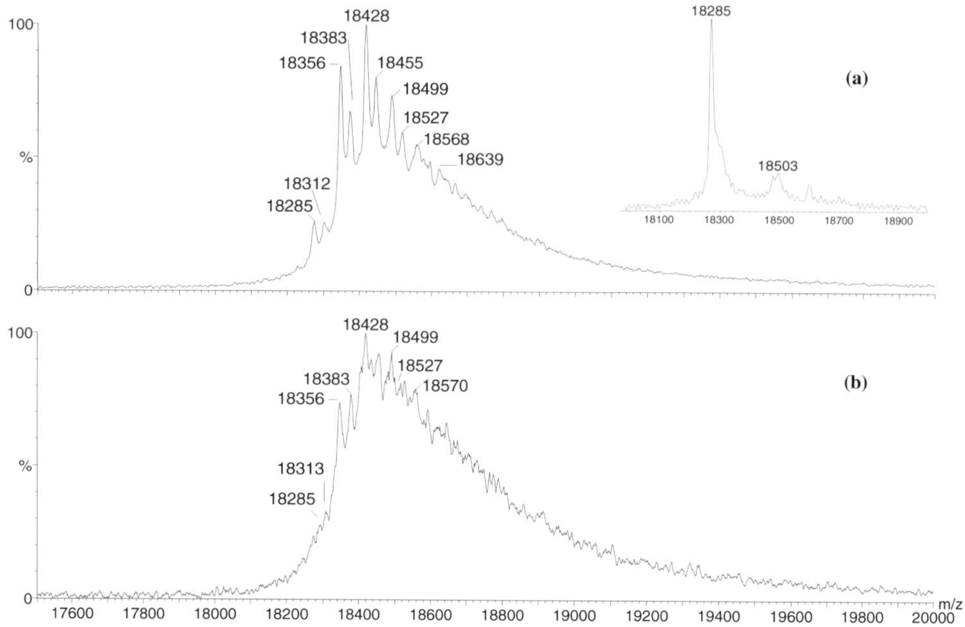

Fig. 13.7. MALDI-TOF mass spectra of LG-B (β-lactoglobulin B) extracted from an SDS-PAGE gel after (a) 10 min and (b) 2 h of passive elution in formic acid–acetonitrile–2-isopropanol–water (50 : 25 : 15 : 10 by volume). The insert refers to an aqueous solution of the same reduced standard protein. (From [112], with permission.)

13.4.6. On the use and properties of surfactants

Four major classes of surfactants are commonly used in biological work: anionic, cationic, zwittcrionic and non-ionic. The structural formulas, chemical and trade names and general properties of a number of them have been given by Hjelmeland and Chrambach [113], Hjelmeland [114], Helenius et al. [115], Neugebauer [116] and Helenius and Simons [117]. Table 13.4 lists some of the most popular non-ionic compounds, whereas Table 13.5 catalogues some of the zwitterionic ones. In addition to a polar group, each surfactant possesses a hydrophobic moiety, and the combination of hydrophobic and hydrophilic sections in the molecule (which is thus said to be amphiphilic) provides the basis for detergent action. Non-ionic detergents are a special class since they are virtually never a single species of molecules, but are instead a group of structurally related compounds. This variability is due to a statistical distribution of chain lengths formed in the manufacture of these detergents by the polymerisation of ethylene oxide. The result, at best, is a single hydrophobic moiety to which polyoxyethylene chains of variable length are attached. Due to their amphiphilic properties all detergents, when dispersed in water, tend to form aggregates in which the hydrophobic section of the molecule is protected from the solvent. These occur as monolayers at the solvent surface and as micelles in solution. Of these two, the micelle is our primary concern since this is the functional unit in protein solubilisation. In

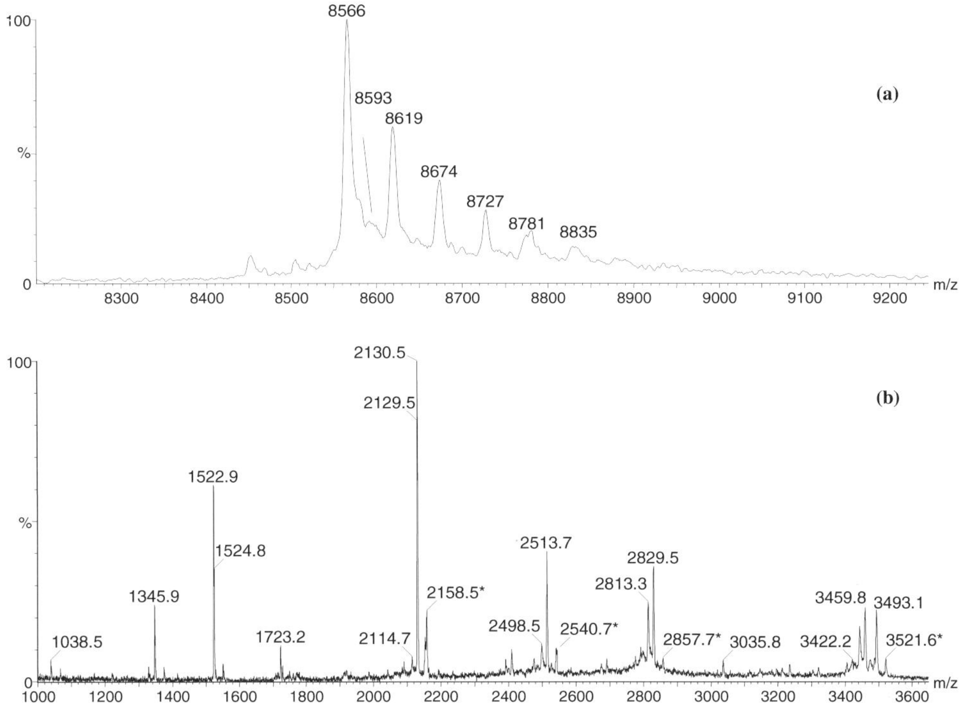

Fig. 13.8. MALDI-TOF mass spectrum of bovine ubiquitin extracted from an SDS-PAGE gel after 10 min of passive elution in formic acid–acetonitrile–2-isopropanol–water (50 : 25 : 15 : 10 by volume). Lower panel: reflectron MALDI-TOF mass spectrum of the tryptic digest of the same ubiquitin. The asterisk refers to formylated fragments. (From [112], with permission.)

TABLE 13.4
NON-IONIC SURFACTANTS

Detergent	Formula	CMC (mM)	MW
Triton X-100 (polyethylene glycol *tert*-octylphenyl ether)	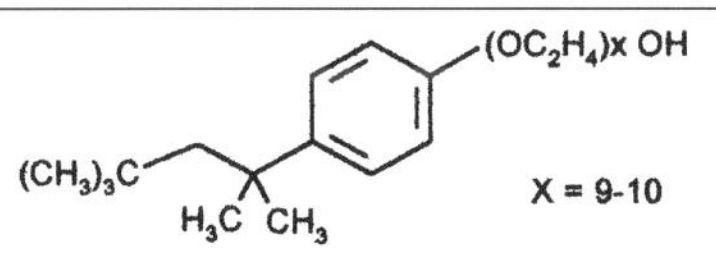	0.02–0.09	625
Brij 35 (polyoxyethylene lauryl ether)		0.05–0.1	1225
Tween 20 (polyoxyethylene sorbitanmonolaurate)	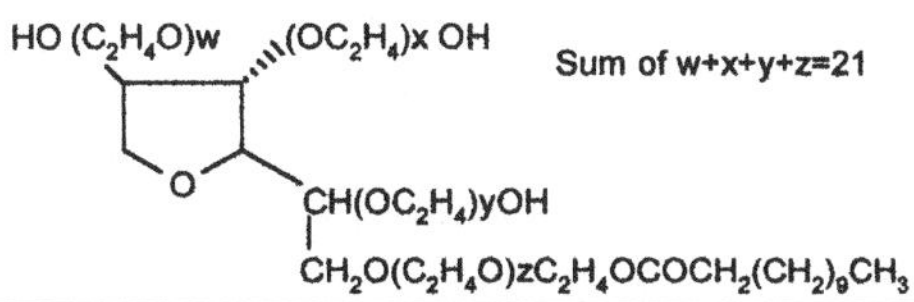	0.06	1228

TABLE 13.5

ZWITTERIONIC SURFACTANTS

Detergent	Formula	CMC (mM)	MW
CHAPS (3-[(3-Cholamidopropyl) Dimethylammonio]-1-propanesulfonate)	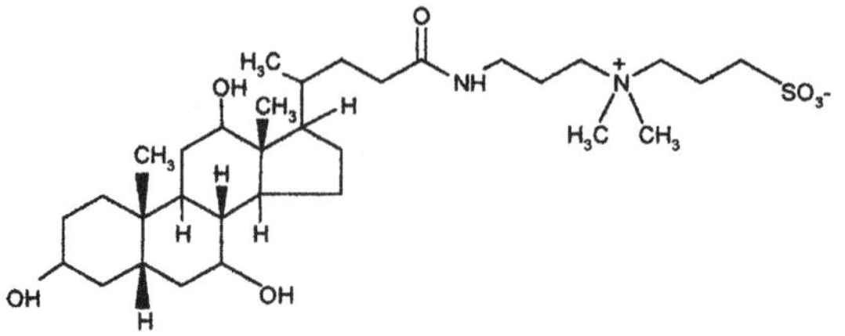	6–10	614.9
Caprylyl sulfobetaine SB-10 (*N*-decyl-*N*,*N*-dimethyl-3-ammonio-1-propanesulfonate)	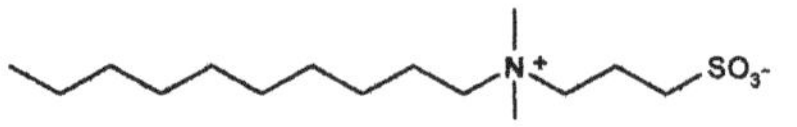	25–40	307.6
Lauryl sulfobetaine SB-12 (*N*-dodecyl-*N*,*N*-dimethyl-3-ammonio-1-propanesulfonate)	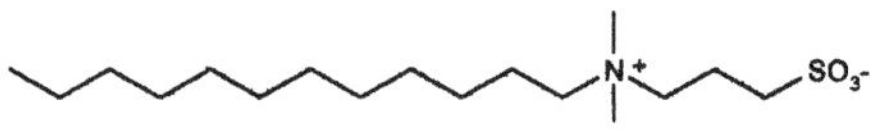	2–4	335.6
Palmityl sulfobetaine SB-16 (*N*-hexadecyl-*N*,*N*-dimethyl-3-ammonio-1-propanesulfonate)	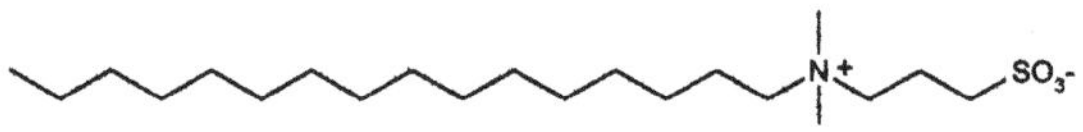	0.01–0.06	391.6

the micelle (so called because the monomers enter into it at variable stoichiometries), molecules are arranged in such a way that polar groups are out and are thus exposed to the solvent, while the hydrophobic segments are buried in the interior, shielded from water. According to Carey and Small [118], clusters of SDS are actually ellipsoids, being some 2.4 nm across the minor axis and composed of approximately 100 SDS residues, depending on the ionic strength and solvent composition.

Surfactants in solution are characterised by an important parameter, called critical micellar concentration (CMC), which is defined as the concentration of detergent at which micelles form. Below the CMC, monomer concentration naturally increases as a function of total detergent, as no micelle exists. Above the CMC value, however, added detergent raises the concentration of micelles, while leaving constant the monomer concentration, since the CMC is an effective solubility limit of detergent monomers [119].

Each detergent is further characterised by two other functional parameters, in addition to the CMC value: an aggregation number (i.e. the number of monomers in the cluster) and the micellar molecular mass (Table 13.6). Two phenomena are readily apparent: (a) a single surfactant can exist in different micellar sizes; (b) among the different classes of detergents, the micellar mass can vary greatly. Thus, increasing ionic strengths

TABLE 13.6

AGGREGATION NUMBER, MICELLAR MASS AND CMC FOR SOME SURFACTANTS (FROM [117], WITH PERMISSION)

Surfactant	Aggregation number	Micellar mass	CMC [a] (mM)	Conditions
Sodium dodecyl sulphate	62	18000	8.2	H_2O
	126	36000	0.52	0.5 M NaCl
CTAB	169	62000	–	13 mM KBr
Triton X-100	140	90000	0.240	H_2O
Triton N-100	100	66000	0.075	H_2O
Lubrol PX	106	64000	–	H_2O
SB_{3-14} [b]	80	~30000	0.6	H_2O
Sodium cholate	2–4	900–1800	13–15	H_2O
Sodium deoxycholate	4–10	1700–4200	4–6	H_2O

[a] CMC = critical micellar concentration.

[b] SB_{3-14} = sulfobetaine with a tetradecyl carbon tail.

drive the SDS micelle towards a double mass and double aggregation number, while lowering drastically the CMC value from 8.2 mM to 0.52 mM. In general, ionic detergents have higher CMCs than non-ionic species, with the CMC of SDS (8.2 mM in plain water) being some two orders of magnitude above the CMC of Triton N-101 (0.075 mM). On the other hand, non-ionic detergents (such as Tritons) tend to have a micellar mass considerably higher than that of ionic species, and this has important practical consequences. Due to their larger micellar size, when used in electrophoresis, non-ionic detergents greatly increase the solvent viscosity, so that protein migration rates will be considerably decreased. Proteins themselves will bind sizeable amounts of non-ionic detergents, in proportion to their number of hydrophobic groups, so that their electrophoretic mobility in sieving matrices will be a function of the new Stoke's radius of the complex.

Use of neutral surfactants, such as Triton X-100 or Nonidet P-40, enables one to add another fractionation parameter to electrophoretic separations, namely the ability to resolve two macromolecular species of identical size and charge but differing in hydrophobicity. We found that out when performing IEF of human globin chains for thalassemia screening [120–122]. Upon IEF in 8 M urea, the separation between β and γ chains was very poor, but when 1–3% NP-40 was added to the IEF gel, two remarkable results were obtained: not only β and γ globins were amply resolved, but γ chains were split into two zones, which were found to be the phenotypes produced by two different genes, called Aγ and Gγ, bearing an Ala to Gly substitution in residue No. 136 of the γ chains. It must be emphasised that, when electrophoresis is performed in 8 M urea, higher amounts of surfactants (e.g. 1–2%, as compared with 0.1–0.2% in absence of urea) should be used, since detergent binding is reduced by these levels of urea, and also the CMC value of surfactants might be markedly altered. This IEF separation of Aγ and Gγ could be reproduced also by free-solution CZE, which suggests that this is a particular case of the vast class of electrophoretic separations called by Terabe et al. [123,124] "micellar electrokinetic chromatography"

(MEKC). MECK is a unique mode of CZE, in that it is capable of separating uncharged compounds. This is due to the fact that uncharged solutes exhibit different micelle–water partition coefficients, according to their relative hydrophobicities. As the charged surfactant micelles migrate in the electric field, with a velocity proportional to their mass/charge ratio, uncharged solutes will migrate too, at a velocity proportional to their residence time in the micelles. Thus, MECK can be viewed as a chromatographic technique, in which the migrating charged micelles act as pseudostationary phases. MECK can be viewed as a hybrid of reversed-phase liquid chromatography (RPLC) and CZE, as the separation process incorporates hydrophobic and polar interactions, a partitioning mechanism and electromigration. In MEKC, anionic alkyl chain surfactants, especially SDS, have been the most widely used species. The popularity of SDS can be attributed to its high aqueous solubility, low CMC, low Kraft point, small ultraviolet molar absorptivity, even at low wavelengths, availability and cost. Serendipitously, SDS has provided the right type of selectivity for many solute mixtures. SDS is a stronger hydrogen bond donor as compared to most other surfactant systems studied so far, such as bile salts, cationic CTAB surfactants and methacrylate-based copolymers. Consequently SDS should be a better surfactant type in many situations considering that the great majority of small solutes that are separated by MEKC contain a hydrogen bond acceptor group, such as nitro, carbonyl and cyano [125,126].

13.4.7. The use of surfactants other than SDS

Williams and Gratzer [127] reported that highly basic proteins such as protamines precipitate in SDS and that highly acidic proteins, such as ferredoxins, behave anomalously during SDS-PAGE, probably owing to poor SDS binding. They therefore replaced SDS with the cationic detergent cetyltrimethylammonium bromide (CTAB). In all other respects, the method is the same as with SDS, except that migration is towards the cathode, which should thus be the lower electrode in a vertical apparatus. With a number of proteins of known size on a gel at 10% T, 5% C, a linear plot of R_f vs. log M_r was obtained over a range of about 10 to 40 kDa. Cationic surfactants are endowed with many desirable properties. They solubilise and denature proteins efficiently, as long as synergistic denaturation by urea or heating is carried out, but they also precipitate large polyanions, including nucleic acids and charged polysaccharides, thereby facilitating the removal of these interfering substances and also the extraction of bound proteins [128].

However, a simple substitution of CTAB for SDS in this way results in the formation of a dense precipitate within the gel, which has been attributed to cetyltrimethylammonium persulphate [129]. Williams and Gratzer [127] overcame the problem of visualising the separated bands in opaque gels by using dansyl pre-labelling and fluorescent detection rather than dye staining. This does not circumvent the objections that the formation of the precipitate results in a non-uniform distribution of ammonium persulphate, leading to non-uniform gel polymerisation, and a non-uniform distribution of CTAB, both of which might affect the protein–surfactant interaction. In addition, the presence of precipitate particles themselves may influence protein migration. There might also be formation of detergent–dye precipitates during any staining/destaining process. Panyim et al. [130] overcame this problem by staining and destaining at 80–100ºC.

There are two ways by which these difficulties can be circumvented. Firstly, if ammonium persulphate is to be used as catalyst, the gels should be prepared in the absence of surfactant. The usual sample preparation and electrophoretic procedures are then followed, with surfactant included in the sample and in the upper electrode reservoir. Residual persulphate in the gel moves ahead of the CTAB during electrophoresis, so that no precipitate is formed. The second approach is to use different polymerisation catalysts.

Marjanen and Ryrie [131] used 10% *T*, 2.67% *C* gels containing 0.1% CTAB photopolymerised with 2 mg/ml riboflavin (see also [132]). The buffer used throughout in both gels and electrode chambers was 0.1 M Na phosphate pH 6.0, added with 0.1% CTAB, but cacodylate (pH 6.0), citrate (pH 6.0) and succinate (pH 5.0) buffers were also satisfactory. Samples were pre-treated by heating for 30 min at 70ºC in 10 mM Na phosphate buffer, pH 6.0, in the presence of 1% CTAB and 1% dithiothreitol. Methylene blue was used as the tracking dye. When 18 standard proteins were examined, a linear plot of R_f vs. log M_r was obtained, with an accuracy very similar to that of SDS-PAGE, but a number of membrane proteins failed to migrate, probably owing to lack of CTAB binding. Other workers used for catalysts a mixture of ascorbic acid, hydrogen peroxide and ferrous ion (Fenton's reagent) [133] or even light and uranium [134].

In a thorough investigation on the factors affecting CTAB-PAGE, Eley et al. [129] adopted gels of 2.5–15% *T*, with a low %*C* (1.33%) made up in 100 mM Na phosphate buffer, pH 7.0, and 0.1% CTAB. The gels were photopolymerised with flavin mononucleotide/TEMED. Samples were pre-treated by heating on a boiling water bath for ca. 4 min in a buffer containing 1% CTAB and 10% β-mercaptoethanol; after cooling, a small amount of malachite green was added as tracking dye. Staining with Coomassie Blue R-250 showed that, although plots of R_f vs. log M_r were sigmoidal, in 7.5% *T* gels linearity was ensured in the 36–96 kDa M_r range. Gels with $T < 7.5\%$ were recommended for proteins with $M_r < 70$ kDa and $T > 7.5\%$ for those with M_r below about 30 kDa. More recently, a discontinuous CTAB-PAGE system has been described by Akins et al. [135].

N-cetylpyridinium chloride (CPC) is another cationic surfactant that has been used for PAGE applications. Schick [136] reported linear plots of R_f vs. log M_r over the range 17–160 kDa for gels with 10 or 12% *T* (at 2.7% *C*), made up in K acetate buffer, pH 3.7, containing 0.01% CPC in both gel and reservoir buffer, which was 0.3 M glycinium acetate, pH 3.7. A ferric-sulphate/ascorbic acid catalyst system had to be employed, due to the low pH in the gel phase.

It would thus appear that both CTAB and CPC can be used to solubilise proteins, including membranaceous structures. Thus, it is reasonable to assume that, in most cases, enough surfactant molecules are bound to the protein to produce complexes with a reasonably constant charge-to-mass ratio and with similar hydrodynamic shapes. Such complexes resemble SDS–protein mixed-micelles in electrophoretic behaviour and thus quite accurate M_r values can be obtained.

13.4.8. Anomalous behaviour

There are a number of exceptions to the binding rule of 1.4 g SDS/g protein, thus leading to possible anomalous migration and erroneous M_r determination (see also Section 13.3). First of all, high ionic strengths reduce the amount of SDS bound to proteins [9], thus interfering with M_r assessments by SDS-PAGE. However, it was also found [137] that the M_r of proteins in the M_r 20–66 kDa interval can be estimated with precision by SDS-PAGE even in presence of NaCl in samples up to 0.8 M. Samples containing either up to 0.5 M KCl or ammonium sulphate up to 10% saturation also seem to give reliable results in SDS-PAGE [138]. Although in the vast majority of cases the M_r of polypeptides, as estimated by SDS-PAGE, are reasonably accurate, there are some exceptions worth mentioning, as described below.

(1) Glycoproteins often exhibit abnormal migration during SDS-PAGE, giving different apparent M_r values when determined in different gel concentrations. This is caused by their hydrophilic glycan moiety, which reduces the hydrophobic interactions between the protein and SDS, thus preventing binding of SDS in the correct ratio. A more accurate procedure appears to be the use of an SDS gradient gel [41]. As an alternative, Tris-borate EDTA buffers allow complexation to the sugars to such an extent that the increased negative charge due to the borate–diol complexes often offsets the loss of charge due to decreased SDS binding [43].

(2) Proteins with high acidic residue contents, such as caldesmon and tropomyosin, also migrate anomalously in SDS-PAGE [139]. This may be due to the repulsion of the negatively charged SDS by the acidic residues. Neutralisation of the negative charge of the acidic groups (e.g. by titration at appropriate pH) restores normal electrophoretic mobility of these proteins in SDS-PAGE [140].

(3) Highly basic proteins, such as histones and troponin I, typically give abnormally higher M_r values by SDS-PAGE. The lower electrophoretic migration of such basic proteins is presumably due to the reduction of the charge/mass ratio of the SDS–polypeptide complex as a result of the high proportion of basic amino acids [141].

(4) Proteins with a high proline content and other unusual amino acid sequences such as ventricular myosin light chain 1 and collagenous polypeptides may also have abnormally high M_r values as determined by SDS-PAGE. The anomaly might be due to the alteration of conformation of the SDS–protein complex.

(5) Finally, polypeptides with M_r below 10 kDa are not properly resolved in uniform concentration SDS-gel and their mass cannot be assessed with precision. Better results can be obtained in pore gradient gels or in 8 M urea SDS-gels containing a higher percentage of cross-linker [44]. These aspects will be treated in more detail below.

13.5. GEL CASTING AND BUFFER SYSTEMS

The following will be a detailed treatment of various methods for performing SDS-PAGE. We will start with the standard method, i.e. the original one in conventional buffers, only to proceed into multiphasic buffers and then in porosity gradients, techniques which have much more to offer in terms of resolving power, although with the added burden of some extra experimental work. Prior to that, however, we will have an excursus on sample treatment in preparation for SDS-PAGE, since this is common to all procedures adopted.

13.5.1. Sample pretreatment

Standard proteins and the unknowns must all be treated prior to electrophoresis, in order to ensure reproducible binding of SDS. Pretreatment also has the scope of inactivating any proteolytic enzyme, which may be present in the sample and give rise to spurious bands by generating breakdown products. It is also customary, in order to minimise proteolysis, to add to the sample cocktails of inhibitors, consisting of natural and synthetic protease inactivators. The former class comprises trypsin inhibitors, such as α_1-antitrypsin, Kazal- or Kunitz-type inhibitors, whereas the last class is made of irreversible inhibitors for serine, cysteine and aspartyl proteases. Among them, we can recall phenylmethylsulfonyl fluoride (PMSF), diisopropyl fluorophosphate (mustard gas, used in Belgium during World War One as nerve gas to kill troops, thus a very toxic compound!) and aminoethylbenzylsulfonyl fluoride [142]. The problem of protease action is not abolished, unfortunately, when denaturing solubilisation is performed. Evidence for proteolysis after solubilisation in 9 M urea [143] or SDS [144] has in fact been described and is also documented by the fact that peptide mapping can be carried out in dilute SDS [145]. It is possible that the kinetics of denaturation of proteases in urea or SDS solutions are slower than those of normal proteins, so that they have some period of time for working on already fully unfolded proteins. Fortunately, this problem seems to be more important in plant tissues than in other biological samples [143] and is minimised by boiling the sample in SDS, so that thermal and SDS denaturation act synergistically.

Another important aspect, in SDS-electrophoresis, is the breaking of disulphide bonds, which is in general achieved by an equilibrium displacement process in a large excess of free thiols. Typically, β-mercapthoethanol, thioglycerol or cysteamine are utilised, although rather high concentrations are needed (e.g. 0.2 M) for rupturing disulphide bonds in proteins. On the contrary, dithiothreitol (DTT) or dithioerythritol (DTE) act at much lower concentrations (e.g. 20 to 50 mM), because of the intramolecular, cyclic condensation process during their oxidation, which will drive the equilibrium more efficiently toward protein reduction (see Fig. 13.9). Problems with any of these reducing agents are due to the fact that dissolved oxygen can reoxidise thiols into disulphides, with the concomitant risk of reoxidation of protein thiols. This reoxidation process can also occur during SDS-electrophoresis, because polyacrylamide gels can provide an oxidising environment due to residual persulphate. A drastic way to prevent reoxidation of reduced thiols in proteins is via nucleophilic substitution, with iodoacetamide [146], or nucleophilic addition with maleimides [147], vinylpiridine [148] or even acrylamide [149]. These are all valid alternatives, with the proviso that utmost care should be exerted when eluted polypeptides have to be further characterised by mass spectrometry. In this last case, erroneous mass determination could ensue if: (a) mixed populations of reacted sulphide groups are present; (b) alkylation proceeds also onto other residues (e.g. ϵ-amino group of Lys). In both cases, erroneous M_r values

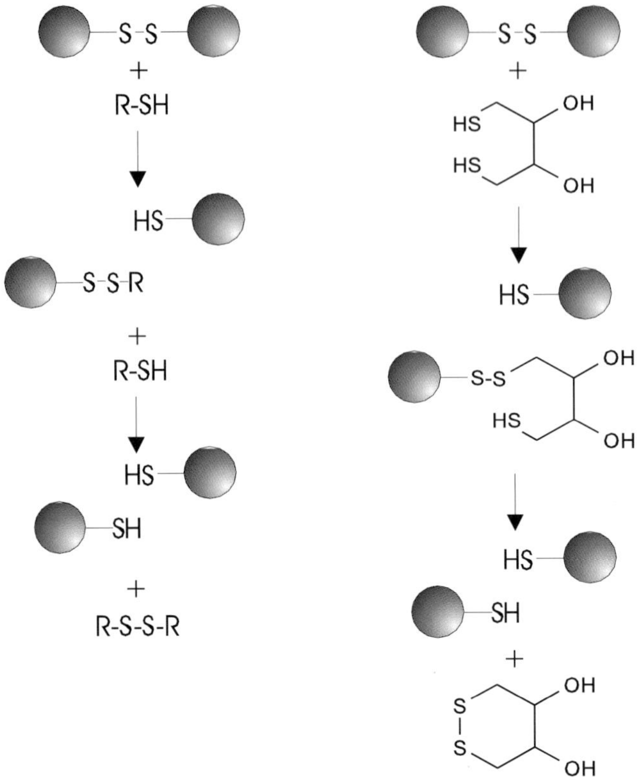

Fig. 13.9. Mechanism of disulfide reduction with free thiols. Left column: reduction with free monothiols (e.g. β-mercaptoethanol). Right column: reduction with cyclisable dithiols (e.g. DTT). The shaded spheres represent protein surfaces. (From [142], with permission.)

will be obtained, which would hamper search for polypeptide identity on data bases. An alternative would be to use tributyl phosphine (Bu_3P), which was in fact the first phosphine species used for disulphide reduction in biochemistry [150]. This compound, which reacts with disulphides as shown in Fig. 13.10, has been claimed to have a number of advantages. First of all, the reaction is stoichiometric, which, in turn, allows the use of very low concentrations (a few mM). Second, phosphines are not as sensitive as other thiols to dissolved oxygen. Third, because of the limited concentration of the agent, subsequent alkylation is much easier to perform. In addition, due to the fact that phosphines do not add to double bonds, as thiols do, one-step protocols, confronting the protein simultaneously with the reducing (e.g. 2 mM phosphine) and alkylating (e.g. 10 mM vinylpyridine) agents, can be adopted [151,142]. The main drawback of tributyl phosphine is that it is volatile, toxic and has a rather unpleasant odour. Another curious aspect of tributyl phosphine, at present not reported in the literature, is that, when this reagent is analysed by MS, it is shown to be extensively oxidised to Bu_3PO, an event which should occur only when it reacts with disulphide bridges for reducing them, as shown in Fig. 13.10. This occurs not only in old bottles of Bu_3P, but also in fresh

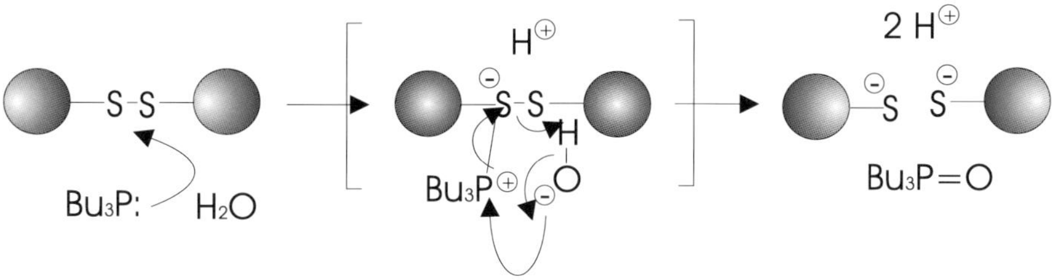

Fig. 13.10. Mechanism of disulphide reduction with phosphines (e.g. tributylphosphine). Although the overall mechanism is known, the degree of concurrence in the electron transfer process between water, the disulphide and the phosphine is speculative (intermediate between brackets). The shaded spheres represent protein surfaces. (From [142], with permission.)

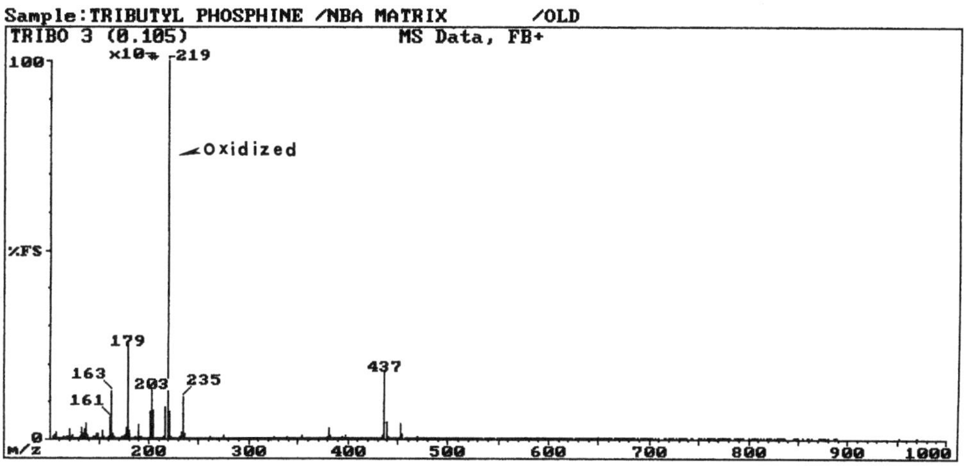

Fig. 13.11. MS analysis of fresh Bu_3P, harvested from a bottle kept under nitrogen. It is seen that, in 10 mM Tris buffer, pH 8.3, it is immediately oxidised to Bu_3PO (M_r 219), with only ca. 2% remaining in the reduced form (M_r 203). (After Hamdan and Righetti, unpublished.)

preparations and when such bottles are rigorously kept under nitrogen. It is just enough to pipette Bu_3P in the typical buffer used for reduction (10 mM Tris, pH 8.3) and within a few minutes (just the time to bring the solution to the MS instrument and perform the analysis) it can be seen that ca. 98% of it is already present as Bu_3PO, with only ca. 2% remaining in the reduced form (see Fig. 13.11). However, curiously, such preparations are just as effective in bringing (and maintaining) reduction of –S–S– bridges in proteins, which suggests that the amount needed for reduction is not in the milli-molar, but rather in the micro-molar range.

13.5.2. The standard method using continuous buffers

The simple standard method of Weber et al. [152] and of Weber and Osborn [138] is usually adequate for a rapid assessment of a purified protein M_r. It is rapid and easy to perform and is very suitable for use with inexpensive and unsophisticated instrumentation. This is why it will be treated here, although it is clear that the use of multiphasic buffers and/or porosity gradient gels give higher resolution. Obviously, SDS of high purity should be used, especially if subsequent removal of SDS and detection of enzyme activity is envisaged [153,154]. It should be remembered that higher-order impurities, such as C_{14} and C_{16} tails, in addition to binding tenaciously to the protein backbone, could give rise to extra bands [155].

Although there are exceptional circumstances when undissociated protein may persist [156], the following pretreatment procedure is generally found satisfactory (standard protocol of [152]).

To a 10 mM phosphate buffer, pH 7.0, containing 1% SDS and 1% β-mercaptoethanol, add protein to a final concentration of 1 mg/ml. Heat to 100°C for 2 min and cool. A small amount (2–10%) of sucrose or glycerol is then added to increase the sample density. Bromophenol blue is a suitable tracking dye and can be added either to the sample or to the cathode (upper) buffer chamber. Another common dye for tracking is Pyronin Y, although both dyes migrate with the discontinuity front (in multiphasic buffers) only over a limited range of gel concentrations. For this reason, Neff et al. [157] have proposed thymol blue, since this dye travels with the true discontinuity front up to 20% *T* gel concentrations. Apply typically 20 μl sample solution to pockets precast in the gel (when staining with Coomassie Blue 2–4 μg protein/band suffice for visualisation).

13.5.2.1. The composition of gels and buffers

Gel concentrations of 3 to 20% or over can be used in any type of PAGE apparatus, the choice of %*T* depending upon the M_r range under investigation. Dunker and Rueckert [13] suggest a 15% *T* gel for 10–60 kDa proteins, a 10% *T* gel for 10–100 kDa species and 5% *T* gels for 20–350 kDa polypeptides. The compositions of solutions for making suitable gels are shown in Table 13.7 [12]. If gels are to be stored before use, cold storage should be avoided because of the poor solubility of SDS at 4°C. Gels can be kept at room temperature for 1–2 weeks since SDS is a powerful bacteriostatic. The apparatus buffer used in the standard method is the same as that used in the gel, i.e. 0.1 M Na phosphate, pH 7.2, containing 0.1% SDS, so that the buffer system is homogeneous throughout. Depending on pocket size, volumes of ca. 20–30 μl of sample can be loaded, with concentrations for each protein species of ca. 5 μg being

TABLE 13.7

COMPOSITION OF STANDARD SDS-GELS OF VARIOUS POROSITIES

Component	$T = 5\%$	$T = 7.5\%$	$T = 10\%$	$T = 15\%$
Acrylamide (g)	4.85	7.28	9.70	14.55
Bis (g)	0.15	0.22	0.30	0.45
Gel buffer (0.2 M phosphate, pH 7.2) (ml)	50	50	50	50
SDS (g)	0.1	0.1	0.1	0.1
H_2O to 95 ml, then add TEMED (ml)	0.15	0.15	0.15	0.15
Ammonium persulphate (1.5%) (ml)	5	5	5	5

adequate for detection. Since SDS–polypeptide complexes have a net negative charge at pH 7.2, they migrate towards the anode, which should thus be the lower electrode in vertical gel apparatuses. The typical voltage gradients given are of the order of 30 V/cm, with running times of 2–3 h, depending on gel length.

An interesting variant of this procedure is the high-temperature SDS-PAGE of Haeberle [158]. It is known that one of the drawbacks of gel slabs run without temperature control is the 'smiling effect', by which the protein bands in the centre lines migrate faster than those in the outer lines, presumably because the temperature at the centre of the gel is higher than at the edges, due to uneven heat dissipation and/or uneven current densities. Cooling the gels and running them at lower voltages minimises this temperature gradient, thus resulting in a more uniform migration rate. However, this remedy slows the rate of protein migration and increases the electrophoretic time to several hours. A valid alternative is to use high-temperature runs. Haeberle [158] has proposed runs in 8% *T*, 5% *C* gels at 70°C, in a modified Porzio and Pearson [159] buffer, having the following composition: (a) electrode buffer 50 mM Tris, 150 mM Gly, 0.1% SDS; (b) gel buffer 100 mM Tris, 300 mM Gly, 0.1% SDS, 5% glycerol, 0.5 mM NaN_3; (c) sample buffer 62.5 mM Tris free base, 3% SDS, 20% glycerol, 6 mg/ml DTT, traces of bromophenol blue, mixed 1 : 1 (v/v) with sample.

Uniform heating of the gel was accomplished by completely submerging the gel in the lower (anodal) buffer pre-heated to 70°C. Fig. 13.12(A,B) illustrates the nearly identical resolution of M_r standards after 20 min at 69°C compared to 2.5 h at 12°C. Both gels were run at 500 V until the dye marker reached the bottom of the gel (13 cm). In order to simulate mini-gels, the 13-cm gels were run till the marker reached 6 cm, with a total running time of only 5 min. Even in this last case (Fig. 13.12C) resolution was still acceptable. The advantages claimed are: (a) the rate of cross-linking is greatly accelerated; (b) compatibility with high-salt loads (up to 700 mM KCl); (c) more uniform migration of dye and proteins. The first point is quite important, since Righetti and Caglio [160] have demonstrated that polyacrylamide gels contain a large number of pendant, unreacted double bonds due to the cross-linker. In turn, such unreacted bonds can covalently affix migrating polypeptides to the gel matrix [161]. High-temperature SDS-PAGE is stated to work equally well for 1- to 5-mm and 0.75-mm-thick gels. However, the temperature should not be increased above 70°C, since above this critical value rubber gaskets leak and bubbles begin to form between the gel and the glass plates. The present protocol is particularly well suited for continuous buffer systems, like the one reported above. In discontinuous buffers, such as those of Laemmli [162]), one should remember to correct for the temperature effect by titrating the various buffers at appropriate pH values, measured at 70°C, due to the rather high temperature dependence of Tris buffers ($\Delta pK/°C = 0.031$).

13.5.3. Use of discontinuous buffers

As in detergent-free systems, the use of multiphasic buffers in SDS-PAGE gives much sharper bands and better resolution, thus is it almost universally used nowadays. The theory of zone stacking and of the other processes involved in SDS-containing

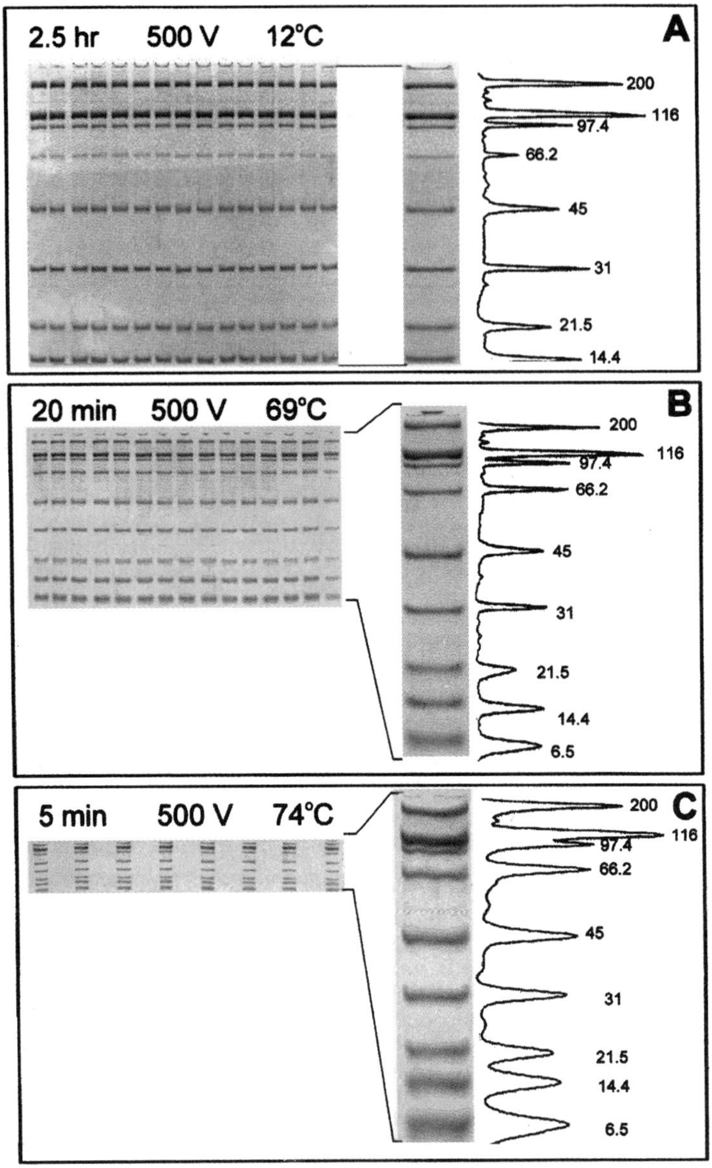

Fig. 13.12. Comparison of standard and high-temperature SDS-PAGE. All gels were 8% *T*, 5% *C*, with the modified Porzio and Pearson [159] buffer described in the text. Gels were run in a Hoefer SE-600 electrophoresis apparatus at 500 V and at the temperatures indicated in the 3 panels. Gels were completely submerged in buffer pre-equilibrated at either 10°C or 70°C. In panels A and B, the temperature was maintained while the gel was run with a circulating water bath. Broad-range M_r markers were run in all lanes (0.25 μl/lane): 200 kDa myosin, 116 kDa β-galactosidase, 97.4 kDa phosphorylase B, 66.2 kDa serum albumin, 45 kDa ovalbumin, 31 kDa carbonic anhydrase, 21.5 kDa trypsin inhibitor, 14.4 kDa lysozyme, 6.5 kDa aprotinin. Staining with Coomassie Blue. (From [158], with permission.)

gels has been comprehensively discussed by Wyckoff et al. [163]. It is important here to recall that the nature of stacking is somewhat altered in the presence of SDS, since SDS-coated proteins have a constant charge-to-mass ratio. As a consequence, they will migrate with the same mobility and thus will automatically stack. Moreover, as the net charge of SDS–protein micelles does not vary in the pH interval 7–10, mobility is not affected within this pH range. It is, therefore, not strictly necessary to have a discontinuity in pH and unstacking of the proteins will simply occur by the change in gel concentration as these analytes leave the stacking gel to enter the running gel. Also the observation that SDS migrates with a mobility higher than SDS–proteins micelles in a restrictive gel, means that SDS will overtake zones of proteins in the resolving gel provided that it is included in the sample and upper buffer reservoir. Thus, in principle, SDS could be omitted from both the stacking and separation gels [164], although few investigators seem to have taken advantage of this option. This type of electrophoresis is typically run at alkaline (pH 8.8 to 9.2) pH values, although an acidic system operating at about pH 4 has also been reported [165], offering comparable results (although, in this last case, lithium, rather than sodium, dodecyl sulphate was required, on account of its better solubility at acidic pH values). Below we will present the two most popular discontinuous systems, those of Neville [31] and of Laemmli [162], which have been particularly widely used, especially the latter, which retains its popularity in spite of the introduction of modified procedures [166].

13.5.3.1. The method of Neville

Typical compositions of buffers and gels used by Neville [31] are shown in Table 13.8. The composition given here for the running gel buffer results in an actual running pH of 9.50 which, together with gels of 11.1% *T*, seems suitable for the majority of separations. Neville [31] also describes alternative compositions, for both the running gel (7.5% *T*) and for buffers giving running pHs of 8.64–10. The sample preparation is as described in Section 13.5.1, except that the stacking gel buffer is used in place of 0.1 M phosphate buffer. As in those days gel slabs were not so popular, the settings given below refer to gel tubes, as commonly adopted in disc electrophore-

TABLE 13.8

COMPOSITION OF BUFFERS AND GELS EMPLOYED IN THE METHOD OF NEVILLE [31]

	pH	Composition
Upper buffer reservoir (cathode)	8.64	40 mM boric acid; 41 mM Tris; 0.1% SDS
Stacking gel buffer	6.10	26.7 mM H_2SO_4; 54.1 mM Tris
Running gel buffer	9.18	30.8 N HCl; 424.4 mM Tris
Lower buffer reservoir (anode)	9.18	Same as running gel buffer
Stacking gel (3.2% *T*, 6.25% *C*)	6.10	3.0 g acrylamide; 0.2 g Bis; buffer to 100 ml; 0.15 ml TEMED[a]; 0.05 g APS[b]
Running gel (11.1% *T*, 0.9% *C*)	9.18	11.0 g acrylamide; 0.1 g Bis; buffer to 100 ml; 0.15 ml TEMED; 0.05 ml APS

[a] *N,N,N′,N′*-Tetramethylethylene diamine.
[b] Ammonium persulphate.

sis. Gel cylinders of 5 mm diameter were run at 1.5 mm per tube at 25ºC until the bromophenol blue marker dye reached the bottom of the tube. The exact position of the dye, which migrates at the borate–sulphate ion interface, was marked by inserting a thin piece of stainless-steel surgical wire into each gel before staining, so that the mobilities of protein bands relative to the dye could be calculated after staining. Using this procedure Neville and Glossmann [167] resolved 35–40 individual protein bands from cell plasma membranes on a single, one-dimension gel. It should be noted that, although the %*T* composition of both stacking and running gels in the present method and in the one of Laemmli [162] are nearly identical, the %*C* levels in the two methods are substantially different. Whereas Laemmli uses the same amount of cross linker in both gel phases (2.6% *C*), Neville uses rather high levels in the stacking gel (6.25% *C*) and very low levels in the running gel (0.9% *C*). There might be good reasons for this: high %*C* in stacking gels were originally proposed by Ornstein [168] and Davis [169] for disc electrophoresis, on the assumption that gels of high %*C* would be much more porous than analogous gels of the same %*T* but with low %*C* (an assumption fully justified, see [170]). On the other hand, running gels of very low %*C* are recommended (especially in a gel slab format) since they can be dried without cracking, due to their better elasticity. The system of discontinuous buffers of Neville [31], based on a borate–sulphate boundary, was selected from the series of 4269 multiphasic buffers calculated by Jovin [171] (this catalogue of buffers was circulating among the adepts well before its publication!) and it was designed to stack all the SDS-saturated proteins, while leaving behind unstacked or partially saturated species.

13.5.3.2. The method of Laemmli

Table 13.9 shows the composition of the gels and buffers described by Laemmli [162]. In the original method, Laemmli poured 10 cm of running gel into tubes 15 cm long and 6 mm inner diameter. After polymerisation, this was overlaid with 1 cm of stacking gel. Protein samples of 0.2–0.3 ml volume were made up in a

TABLE 13.9

COMPOSITION OF BUFFERS AND GELS EMPLOYED IN THE METHOD OF LAEMMLI [162]

	pH	Composition
Upper buffer reservoir (cathode)	8.3	25 mM Tris; 192 mM glycine; 0.1% SDS
Stacking gel buffer	6.8	Tris adjusted to pH 6.8 with HCl and diluted to 125 mM; 0.1% SDS
Running gel buffer	8.8	Tris adjusted to pH 8.8 with HCl and diluted to 375 mM; 0.1% SDS
Lower buffer reservoir (anode)	8.3	Same as upper buffer reservoir
Stacking gel (3.1% *T*, 2.6% *C*)	6.8	3.0 g acrylamide; 0.08 g Bis; buffer to 100 ml; 0.025 ml TEMED[a]; 0.025 g APS[b]
Running gel (10.3% *T*, 2.6% *C*)	8.8	10.0 g acrylamide; 0.27 g Bis; buffer to 100 ml; 0.025 ml TEMED[a]; 0.025 g APS[b]

[a] *N,N,N′,N′*-Tetramethylethylene diamine.

[b] Ammonium persulphate.

solvent containing 62.5 mM Tris–HCl buffer, pH 6.8, 2% SDS, 10% glycerol, 5% β-mercaptoethanol and 0.001% bromophenol blue tracking dye and were heated for 1.5 min in a boiling water bath to ensure complete dissociation and optimum SDS binding. Electrophoresis was run at 3 mA per gel cylinder until the tracking dye reached the bottom of the tube (about 7 h). This method was soon adapted to slab gels. For example, by using 35 cm long gel slabs, Ames [172] separated approximately 60 protein bands in a single run from samples of bacterial cells, the limit of detection being well below 0.2 μg per band.

An interesting point to the users of the SDS-PAGE technique, in all of its possible variants, is the possibility of recovery of biological activity after the run. This would permit in situ zymogramming, i.e. the detection of enzyme-driven reaction directly in the gel slab, as originally reported by Hunter and Markert [173] and subsequently refined by Markert and Moller [174], who suggested the word isozyme to describe the enzyme bands which appeared after specific staining. This method (for reviews see [175,176]) is what made disc electrophoresis extremely popular, since it allowed, for instance, population genetics and refined biochemical studies in even plain tissue homogenates. Of course, in all types of zone electrophoresis, this was quite simple, since the proteins were separated under native conditions, but it would appear to be anathema in SDS-PAGE, where the process of saturating the polypeptide chain with SDS leads to strong denaturation. A recent report by Wang [177]) suggests that this could be possible, at least for some classes of macromolecules, such as protein kinases. After the SDS-PAGE step, the gel is rinsed for 1 h at room temperature in 100 mM Tris–HCl buffer, pH 8.0, containing 5 mM β-mercaptoethanol and 20% isopropanol. For renaturation, the gel is then incubated at 4°C for 16 h in the same buffer, devoid of isopropanol but with 0.04% Tween 40 added. At this point, assays for kinase activity can be performed in presence of the appropriate substrates. Presumably, isopropanol in the first step is used for disaggregating the SDS micelles; by the same token, the Tween 40 present in the second step might be needed for replacing the SDS moiety and preventing massive protein precipitates, although micro-aggregates might still be present, since there does not seem to be a massive leach out of proteins after such a long period of incubation.

13.5.4. Porosity gradient gels

Since the migration of SDS–protein complexes in a gel matrix is mostly determined by its size, it is not surprising that an improved fractionation can be achieved on gels of graded porosity (gels with increasing concentration gradients). In such gradients the migration velocity of a band is inversely related to time and varies exponentially with the distance travelled. Experimentally it is found that not only does the rate of migration of components through a gel gradient vary inversely with time, but also that, after a sufficient time, a stable pattern will develop, in which the different components continue to move slowly but their relative positions remain constant. This has led to the concept of pore limit, which is defined by Slater [178] as the distance migrated from the origin in a specific gradient after which further migration occurs at a slow rate directly proportional to time. Thus the technique of pore gradient electrophoresis is often referred to as pore-limit electrophoresis, although this latter term is perhaps an unfortunate one, since it has led to the common misconception that, once the pore limit is reached, migration stops altogether. Since this is not so (see [179]), it seems preferable to refer to the method as gel gradient electrophoresis.

For a complex mixture there seems to be little doubt that under appropriate conditions a gradient gel can give a resolution superior to a gel of single concentration and it is not necessary for this that all components should have migrated as far as their pore limits. Part of this high resolution results from the fact that, throughout the run, the leading edge of any particular band is moving through more concentrated gel than the trailing edge and hence encounters greater resistance, resulting in a band-sharpening effect [180]. There are two further practical advantages of this method. Firstly, since after the initial 'sorting-out' process a relatively stable band pattern is formed, it is not necessary to control the electrophoretic conditions so precisely as in other electrokinetic procedures, at least for qualitative work. Secondly, once bands have migrated well into the gel and approached their pore limits, diffusion is greatly reduced, so that the gels can be kept unfixed for long periods of time with little loss of resolution [181].

Gels can be prepared with any shape of concentration gradient to suit the particular requirements of the separation. However, the bulk of published work refers to the use of linear or simple concave gradients. Since the pore size is a function of $1/\sqrt{\%T}$, in a linear gradient the pore size changes more gradually at high, rather than at low, values of $\%T$. Thus small molecules have to migrate through a considerable length of gel with a porosity close to the pore limit before finally reaching that limit. Convex concentration gradients would naturally aggravate this situation, but with concave gradients the pore limit is reached more rapidly than with linear gradients. Virtually any device for preparing solution gradients can be applied for the formation of gradient gels. The simplest apparatus for preparing linear gradients requires only two identical vessels, a stirrer and some tubing, as shown in Fig. 13.13. If they are identical, the shape of the

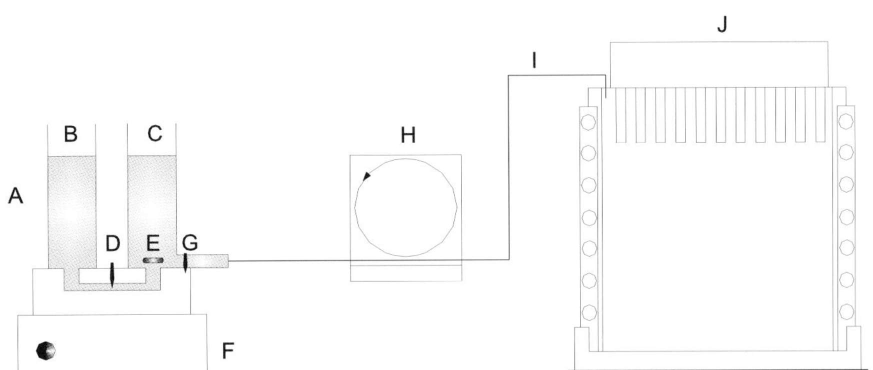

Fig. 13.13. Apparatus for casting linear gradient polyacrylamide gels: A, two-chamber gradient mixer; B, reservoir; C, mixing chamber; D, valve; E, stirring bar; F, magnetic stirrer; G, stopcock; H, peristaltic pump; I, tubing; J, gel slab cassette. (From [182], with permission.)

vessels is immaterial since, as long as they are at the same height, when one volume of solution is withdrawn from flask C, half a volume will flow from reservoir B into C to maintain hydrostatic equilibrium. If C (the mixing chamber) is filled with the high $\%T$ solution and B with the dilute solution and C is continuously stirred, then the $\%T$ of the solution harvested from C and filling the gel slab cassette J will decrease linearly. Cassette J, upon polymerisation, will contain a linearly decreasing porosity gradient from top to bottom. Many published methods are basically only variants of this simple concept (see e.g. [183–185]; and the review in [186]). Small corrections to the volumes of solutions in B and C are made to allow for the different densities of the two solutions, the volume displaced by the stirring bar and the differences in level produced by the dynamic forces resulting from the stirring action. For example, Pharmacia Biotec provides, with their gradient mixer, a compensating rod, taken from the compensating cone designed by Svensson [187] for pouring sucrose density gradients in preparative isoelectric focussing. The quality of the generated gradient can be checked by adding a dye, such as *p*-nitrophenol [188] or simply bromophenol blue [189] to one of the vessels and then scanning the gel with a densitometer.

When having to cast simultaneously a number of gel slabs, a widely used procedure for gradient gel preparation is that of Margolis and Kenrick [180], as exemplified in Fig. 13.14, in which the components are fed into the bottom of a rectangular mould in which the gel plates are supported on a plastic net platform. The apparatus has the advantage that a number of slabs are prepared at the same time, so the gradient

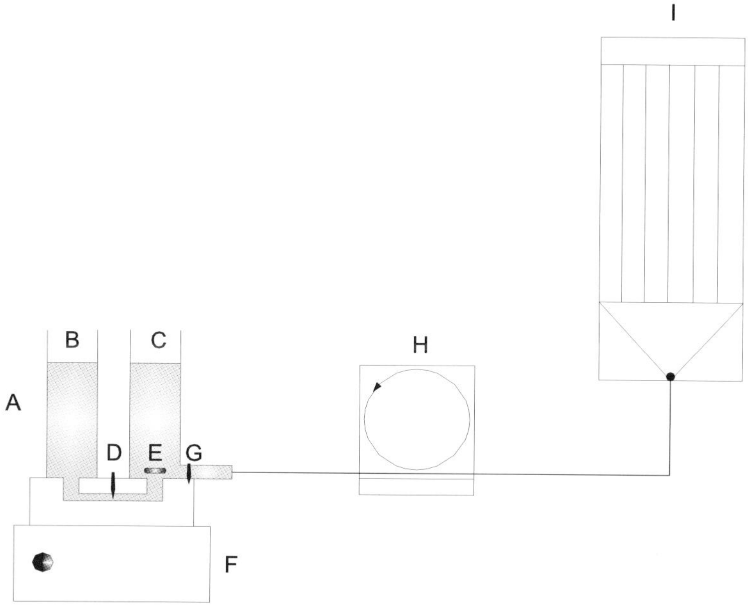

Fig. 13.14. Apparatus for casting batches of linear gradient polyacrylamide gels: A, two-chamber gradient mixer; B, reservoir (high $\%T$); C, mixing chamber (low $\%T$); D, valve; E, stirring bar; F, magnetic stirrer; G, stopcock; H, peristaltic pump; I, gel slab cassette. (From [182], with permission.)

reproducibility within a batch should be very high. Obviously, as the gradient solution is delivered from the bottom of the tower, the low $\%T$ solution should enter first and be displaced upward by the more dense liquid elements being pumped in. The same two authors stress that the most important precaution is to avoid convection currents, including those generated by the heat produced during the polymerisation reaction. This is helped both by the inclusion of a sucrose (or glycerol) density gradient co-linear with the gradient gel mixture itself and also by having a gradient in the concentration of catalysts (TEMED and APS) so that polymerisation will occur first in the top (most dilute) part of the gel gradient and proceed downwards. Using gradients of catalysts is too cumbersome and rarely adopted today; in addition, Righetti et al. [190] have demonstrated that, in the presence of density gradients and in thin gels (as much in vogue today) such convective flows (which would render the gel highly inhomogeneous) are essentially abolished.

If desired, a stacking gel can be applied above a gradient slab, merely by leaving sufficient space between the plates when adding the gradient mixture, and then by pouring the stacking gel mixture in the usual way once the gradient has polymerised. In turn, on top of the stacking layer a further portion of gel mixture can be poured and a slot former inserted for seeding the samples. The concentration limits chosen for the gradient depend upon the samples to be separated. Gradients of 4–24, 4–26, 4–30, 2–30% etc. are generally considered to be suitable for use with most protein samples (e.g. serum, urine, etc.), although, if electrophoresis is unduly prolonged, it is possible for proteins with M_r below 30,000 Da to migrate off the end of such gels. Felgenhauer [191] used 20–50% gradients for proteins below 30,000 Da. Gradients of 2.5–12% or 2–16% [183] are well suited for mixtures of proteins within the M_r 100,000–5,000,000 Da. It is best to maximise the range of pore sizes for any particular gel concentration range, and since this is achieved with 5% cross-linking [179], it is common to use $C = 5\%$ in both gradient making solutions at any value of $\%T$. However, Campbell et al. [192] found that, while pore size is minimal with about $C = 5\%$ for gels with T values below 20%, the proportion of cross-linker, required for minimum pore size, increases with $\%T$ at higher values of T. They reported that gels with maximum sieving effect were obtained with gels of 40% T and 12.5% C and suggested that these should be the limiting conditions on gradient gels. Using gradient gels with 3% T, 4% C at the top, increasing to 40% T, 12.5% C at the bottom, these authors achieved excellent resolution of proteins covering a very broad size range from 670,000 down to 14,000 Da and even of peptides produced by trypsin digestion of bovine serum albumin. However, the concept of gradienting both, acrylamide and cross-linker, is seldom practised today, due to the extra burden of preparing different stock solutions of acrylamide/Bis. All the above steps can be performed with greater versatility and much higher control on the gradient reproducibility if a microcomputer is used for handling the gradient, as shown in Fig. 13.15. In this case, an electromagnetic two-way valve is activated for drawing appropriate liquid amounts from vessels D or E, but then a small-volume mixing chamber (G) has to be placed between the valve and the peristaltic pump, to allow for thorough mixing of the two limiting solutions.

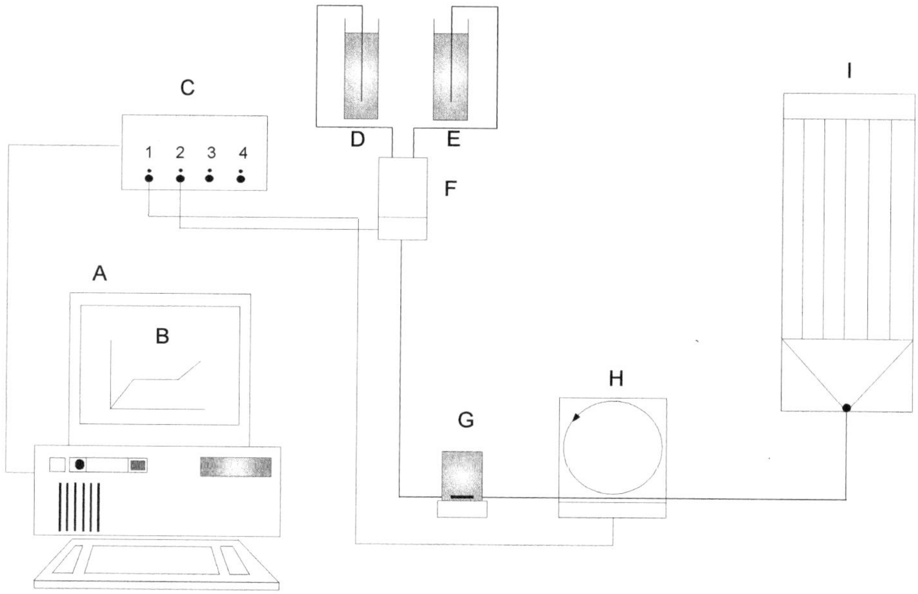

Fig. 13.15. Microcomputer-controlled apparatus for preparation of gradients of polyacrylamide gels in a slab format: A, microcomputer; B, shape of the gradient to be generated; C, switch controller; D, reservoir for high %T acrylamide solution; E, reservoir for low %T acrylamide solution; F, electromagnetic two-way valve; G, mixing chamber; H, peristaltic pump; I, tower containing gel cassettes. (From [182], with permission.)

13.5.5. Peptide mapping by SDS-PAGE

In many cases relationships between proteins cannot be proved unequivocally by electrophoretic mobility alone, especially if they are closely related; thus further analysis is required. If a protein has been already extensively purified, it can be digested in vitro and then the peptides thus produced analysed by SDS-PAGE. This is at the heart of the method of Cleveland et al. [145], who digest such purified proteins in a buffer already containing some SDS (0.5%), via a number of proteolytic enzymes, such as papain, chymotrypsin or *Staphylococcus aureus* protease and then analyse the peptides thus produced by SDS-PAGE. However, this method is of limited applicability, since in many cases purified protein samples are not available, yet the investigator requires information on the similarities between individual protein bands obtained by gel electrophoresis. In order to solve this problem, techniques for in situ peptide mapping have been developed and the methodology extensively described [193,194]. In these cases, the sample proteins are best separated by a preliminary electrophoretic step using virtually any of the usual methods (native electrophoresis, SDS-PAGE, isoelectric focussing, 2D-PAGE). The gels are then briefly stained and destained and placed on a transparent plastic sheet over a light box. Individual protein zones are then cut out with a scalpel or razor blade and trimmed to a size small enough to fit easily into the sample wells of an SDS-PAGE gel to be used for the peptide mapping stage. Before doing

TABLE 13.11

CHEMICAL METHODS FOR CLEAVAGE OF PEPTIDE BONDS IN PROTEINS

Reagent	Bonds cleaved
Cyanogen bromide	–Met–X–
Hydroxylamine	–Asn–Gly–
BNPS skatole	–Trp–X–
N-chlorosuccinimide (*N*-bromosuccinimide)	–Trp–X–
Partial acid hydrolysis	–Asp–Pro–
Partial basic cleavage	–Ser–X–
Heat (110°C, pH 6.8, 1–2 h)	–Asp–Pro–
2-Nitro-*S*-thiocyanobenzoate	–X–Cys–

13.5.6. SDS-PAGE in photopolymerised gels

There are two main problems with the standard polymerisation procedure in the presence of TEMED and APS: due to excess, unreacted acrylamide left behind in the gel matrix, alkylation of Cys residues in proteins is an ever present hazard [110]; additionally, due to the strong oxidation power of APS, Cys may be destroyed or Met altered [102]. Photopolymerisation systems are theoretically well suitable for solving these problems, since polymerisation could be driven to $> 95\%$, because radicals are continuously produced as long as light is absorbed by the dye, as opposed to a single burst of radicals in the case of chemical initiators such as persulphate. Lyubimova et al. [197] have recently described a photopolymerisation in the presence of a triad of novel catalysts: 100 μM methylene blue (MB) combined with a redox couple, 1 mM sodium toluenesulfinate (STS), a reducer, and 50 μM diphenyliodonium chloride (DPIC), an oxidiser (see Fig. 13.17 for their formulas). The results were spectacular: onset of polymerisation could be induced in a matter of seconds and proceeded at a terrific speed till completion, typically within 30 min. The viscoelastic properties of these gels were even better than those of persulphate-activated gels, suggesting a more homogeneous gel structure. The elastic modulus exhibited a maximum in correspondence of a minimum of permeability, both situated at a 5% cross-linker. A theoretical study confirmed the very high reaction kinetics of this system, together with the extraordinary conversion efficiency [198]. The 'methylene blue saga' continued in a number of additional reports. In a first study, we discovered another point of advantage of MB catalysis: its insensitivity to pH in the gelling solution. Persulphate-TEMED polymerisation has pH optima in the pH range 7–10; riboflavin-TEMED only in the pH interval 4–7; ascorbic acid, ferrous sulphate and hydrogen peroxide only at pH 4; TEMED and hydrosulfite a reasonable conversion only in the pH interval 4–6; on the contrary, MB-driven reactions work at full speed in the pH range 4–9 and still very well up to pH 10, a real record [199]! Other, interesting data on matrix structure came with further studies on the MB-driven system: topological data could be derived by starting polymerisation at 2°C and continuing, after the gel point, at 50°C. The data suggested that, at 2°C, the nascent chains formed clusters held together by hydrogen bonds (melting point at 28°C); such clusters were subsequently

Diphenyliodonium chloride (oxidizer) M_r 316.57

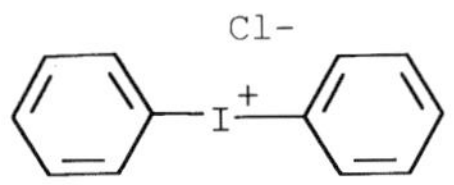

Sodium toluene sulfinate (reducer) M_r 178.18

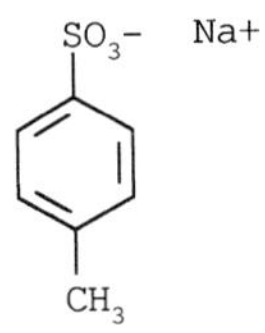

Methylene blue M_r 319.86

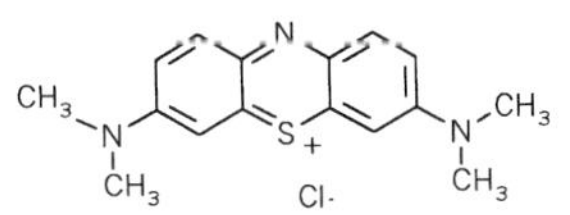

Riboflavin-5'-Phosphate M_r 514.4

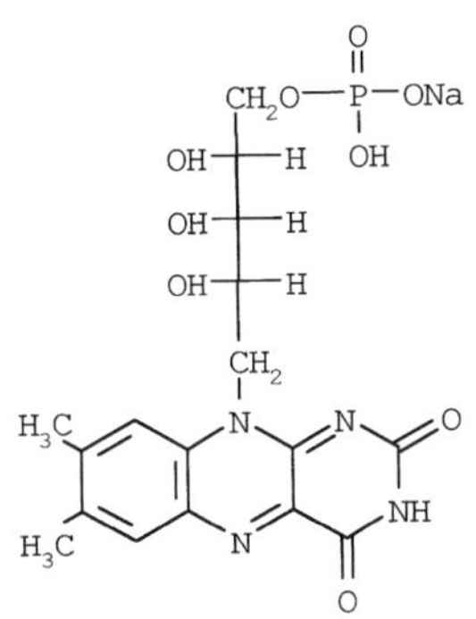

Fig. 13.17. Formulas of the triad of catalysts for photopolymerisation: methylene blue (MB), sodium toluenesulfinate (STS), the reducer, and 50 μM diphenyliodonium chloride (DPIC), the oxidiser. Alternatively (or in combination with MB) riboflavin-5′-phosphate can be used (bottom formula).

'frozen' into the 3-D space as the pendant double bonds of the cross-linkers kept reacting [160]. This would lead to turbid, highly porous and unelastic gels. However, if polymerisation was continued, at the gel point, at 50ºC, clear and elastic gels would be obtained. This was an additional proof of the widely held belief that turbid gels would indicate large-pore structures. Other remarkable features were then discovered: e.g. the insensitivity of MB-driven catalysis to oxygen, against a large inhibition in persulphate reactions. Conversely, whereas 8 M urea would substantially accelerate persulphate catalysis, it left undisturbed the MB system. However, when polymerisation occurred in a number of hydroorganic solvents (all in a 50 : 50 v/v ratio; dimethyl sulphoxide, DMSO, tetramethyl urea, formamide, dimethyl formamide) persulphate-catalysis was severely inhibited, whereas MB-photopolymerisation was essentially unaffected [200]. The same applied to polymerisation in detergents [106]. It thus appeared that MB-catalysis was a unique process, proceeding at optimum rate under the most adverse conditions, and able to ensure nearly complete monomer conversion into the growing polymer.

Given all the above advantages, why has MB-driven photopolymerisation not become popular? A reason could be the one reported by Rabilloud et al. [201] who claimed poor transfer of proteins in 2-D maps and potential adsorption of such analytes into one (or more) of the triad of catalysts grafted at the chain termini. A partial remedy to these problems was drastically reducing the amounts of the three initiators in the gelling solution: 500 μM STS, 20 μM DPIC and 30 μM MB. Such lower levels were found not to affect the final properties of the gels. This new, reduced-level, photopolymerisation system was in fact found to give excellent results in the separation of histones at acidic pH values (pH ca. 4) in the classic urea–acetic acid–Triton system [202]. On the basis of these findings, we believe that one of the main reasons of such a poor performance of MB-catalysed gels could be the very presence of MB, grafted covalently into part of the population of polyacrylamide chains in the matrix. Since MB is positively charged up to ca. pH 10, it could very well interact with the negatively charged SDS–protein micelles, producing smears during migration and poor transfer from the gel. It is a fact that this does not occur with histones, which are highly positively charged at the pH 4 of this electrophoretic system, thus repelled by the like charge of MB. On these assumptions, we have made 'zwitterionic' gel, i.e. gels photopolymerised with 30 μM of MB (positively charged) in presence of 30 μM of riboflavin-5′-phosphate (a negatively charged dye). On the assumption that there would be equal chances for these two catalysts to be incorporated into the gel matrix, said matrix would then be zwitterionic and have minimal tendency to adsorb protein–SDS micelles by ion-pairing. Excellent gels were obtained in such a zwitterionic polymerisation system, not only for marker proteins, but also for total cell lysates, such as *E. coli* total proteins (Olivieri, Castagna and Righetti, unpublished data). Blotting onto nitrocellulose membranes was also shown to be essentially quantitative. As an extra bonus, no oxidation of proteins by the catalysts occurred, as in the case of persulphate-driven polymerisation. Notwithstanding these advantages, alkylation of proteins by free acrylamide (even when driving the reaction to 95%, high %*T* gels would still contain a large excess of unreacted monomers over the potential alkylation sites of proteins) still occurred, suggesting that there is no simple cure to this problem. Thus, if one wants to avoid protein alkylation by free acrylamide during SDS-PAGE, there appear to be only two possible remedies: (a) use washed matrices, as suggested by Westermeier [203]; (b) use pre-alkylated proteins, i.e. protein samples which are not only reduced, but alkylated with iodoacetamide, as suggested by any text book in protein chemistry (see also the chapter on two-dimensional maps). After all, the site of attack has been demonstrated to be almost exclusively the free –SH group of Cys residues (at least during the normal times of electrophoresis); thus, their alkylation efficiently suppresses further reaction with acrylamide during the SDS-PAGE step.

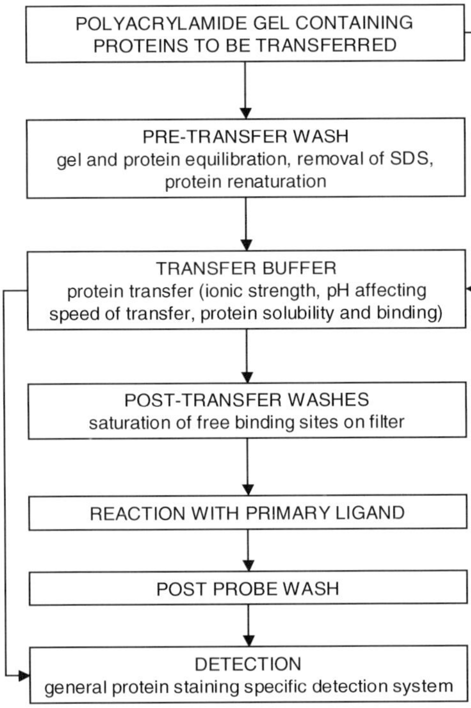

Fig. 13.18. Flow sheet for performing protein transfer from a gel and subsequent detection steps on the membrane.

13.6. BLOTTING PROCEDURES

When analyzing complex samples by SDS-PAGE, it is often desirable to use a method allowing specific detection of some components in the mixture. One of the most sensitive methods is immunodetection, by which specific antibodies (either polyclonal or monoclonal) are used for precipitating components following an electrophoretic separation. This procedure, known as immunofixation, works well with cellulose acetate membranes and agarose gels (it was in fact first reported by Grabar and Williams [204], precisely on agarose matrices), but it can hardly be adopted when running polyacrylamide gels, due to their restrictive nature (i.e. small pore size). This last problem was overcome by the development of blotting techniques, in which the separated proteins are transferred from the polyacrylamide gel onto the surface of a thin matrix such as nitrocellulose. As a result, the proteins are immobilised onto the open face of this thin sheet (via hydrophobic bonds in case of nitrocellulose) and are thus readily accessible to interaction with antibodies and other ligands. This procedure originated from a technique developed by [205] for transferring DNA (Southern blotting) and subsequently adapted for RNA transfer (Northern blotting). Almost inevitably, when this procedure was subsequently adapted to proteins [206], it became popularly known as Western blotting. The major steps in Western blotting are shown in Fig. 13.18. After terminating the electrophoretic step (in general, but not exclusively, an SDS-PAGE, either 1-D or 2-D), the gel should be briefly incubated in the appropriate transfer buffer for removal of SDS and of other gel constituents which could cause problems during transfer. This also minimises shrinking or swelling of the gel during transfer and allows potential protein renaturation (if possible at all). After transfer by a suitable procedure, the protein pattern can be visualised using a total protein stain. Several general detection methods for proteins on Western blots have been described by Merril and Washart [70], to whom the readers are referred for further details (see also a review by [207], related to blotting and immunoblotting in 2-D map analysis). A variety of membranes have been described for implementing protein transfer from polyacrylamide matrices. Nitrocellulose is the most commonly used blotting membrane at this time. Interaction of denatured polypeptides with nitrocellulose is probably mediated mostly by hydrophobic interaction, although the complete mechanisms is not yet fully understood. However, nitrocellulose has low affinity for some proteins, particularly those of low M_r [208]. Given this problem, some authors have suggested covalent fixation of proteins to nitrocellulose via glutaraldehyde [209] or *N*-hydroxysuccinimidyl-*p*-azidobenzoate [208]. Other matrices that have been developed include diazobenzyloxymethyl (DBM) modified cellulose paper [210] and diazophenylthioether (DPT) paper [211]. Negatively charged proteins interact electrostatically with the positively charged diazonium groups of these matrices, followed by irreversible covalent linkage via azo derivatives [210]. DPT paper has been shown to be more stable than DBM paper, but equally efficient. Diazo paper is reported to show less resolution of separated proteins compared to nitrocellulose or nylon membranes [212]. In addition, glycine may interfere with DBM protein blotting [208]. Nylon membranes are also used for protein blotting. Two such commercially manufactured membranes are GeneScreen (from NEN, Boston, MA, USA) and Zeta-bind from AMF/Cuno (Meriden, CT, USA). Zeta-bind has a very high protein binding capacity [213] and requires extensive blocking for eliminating non-specific binding. Nylon membranes consistently provide a higher efficiency of protein transfer from SDS gels than nitrocellulose [213,214]. However, one disadvantage of nylon matrices is that they bind the common anionic dyes such as Coomassie Blue and Amido Black, which may result in background staining that could obscure the detection of transferred proteins. Polyvinylidene difluoride (PVDF) membranes can also be utilised for protein blotting [215]; they are hydrophobic in nature and proteins bound to them can be detected by most anionic dyes as well as by immunodetection protocols.

13.6.1. Capillary and electrophoretic transfer

When originally described [205] blotting was performed by capillary forces, in a set-up similar to that shown in Fig. 13.19. The gel is overlaid with the membrane (nitrocellulose in this case) and then with a large stack of dry filter paper sheets. The assembly is then covered with a rigid plate (glass or plastic) on top of which a weight (0.5 to 1 kg) is placed. The dry stack of filter papers pulls liquid (with the dissolved proteins) from the gel layer, allowing the proteins to be eluted from the gel matrix and be trapped by the membrane. Although this method is not much in vogue today for

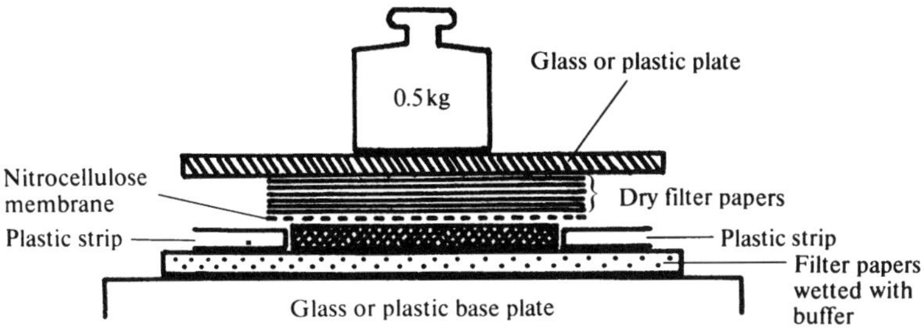

Fig. 13.19. Set-up for capillary blotting. The polyacrylamide gel is represented by the hatched area between the two plastic strips.

SDS-PAGE, it can have remarkable efficiency when adopted for eluting proteins from agarose gels [216], with transfer times of less than 1 min! Pressure blotting can also be efficiently adopted for thin polyacrylamide IEF gels, with a contact time of about 1 h. An alternative procedure is vacuum blotting [217], by which efficient transfers (30–40 min, instead of overnight for a typical pressure blot) can be achieved. The other advantages of vacuum blotting are: (a) it is quantitative, there are no back transfers; (b) it leads to sharper zones and better resolution; and (c) it reduces the mechanical stress on the gel. By far, however, the preferred method today is electrophoretic blotting, performed either by the tank or the semi-dry procedures. A typical set-up is shown in Fig. 13.20: it is noticed that the trapping membrane is typically positioned on the anodic side of the polyacrylamide gel, since after SDS-PAGE the protein–SDS complex has a high net negative charge. Originally, vertical buffer tanks with coiled platinum wire electrodes fixed on the two sides were used. For this technique, the gel and blotting membrane are clamped in grids between filter papers and sponge pads and suspended in the tank filled with buffer. The transfers usually occur overnight. Semi-dry blotting between two

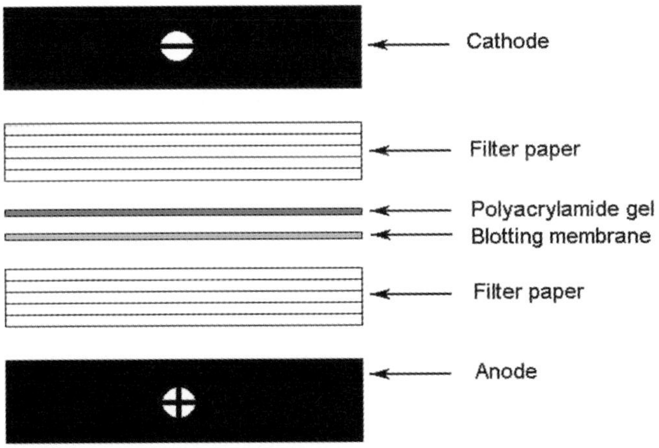

Fig. 13.20. Experimental set-up for wet or semi-dry blotting from SDS-PAGE gels. Note that the blotting membrane is placed on the anodic side of the gel, due to the very high net-negative charge of the protein–SDS complex.

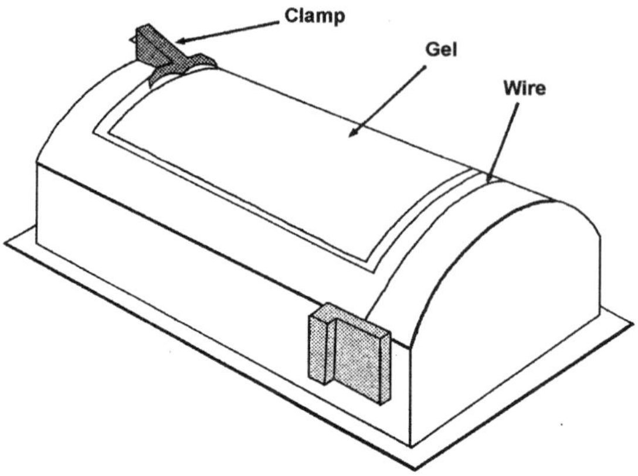

Fig. 13.21. Guillotine-like apparatus for detaching the gel matrix from the backing plastic support. The gel is clamped on the curved surface of the shield-like instrument, where the steel wire will slide at the interface between gel and plastic backing.

horizontal graphite plates has gained more and more acceptance in the last few years. Only a limited volume of buffer, in which a couple of sheets of filter paper are soaked, is necessary. This technique is simpler, cheaper and faster and allows use of discontinuous buffer systems [218,219]. Graphite seems to be the best electrode material in semi-dry blotting, since it conducts well, does not overheat and does not catalyse oxidation products. A kind of isotachophoretic effect takes place in this system: the anions migrate at the same speed, so that a regular transfer takes place. A current no higher than 0.8 to 1 mM/cm^2 of blotting surface is recommended. If higher currents are used, the gel can overheat and the proteins can coagulate. The transfer time is approximately 1 h and depends on the thickness and concentration of the gel. When longer transfer times are required as for thick (1 mm) or highly concentrated gels, a weight is placed on the upper plate so that the electrolyte gas is expelled out of the sides. Johansson [220] has also described a double-replica blotting: an alternating electric field is applied on a blotting sandwich with a membrane on each side of the gel with increasing pulse time, so that two symmetrical blots result. This could be quite useful if the membranes have to be probed with different detection systems, such as different antibodies.

Ready-made or home-made gels backed by support films are today quite common for zone electrophoresis and IEF of proteins. These films are impermeable to current and buffer flow, so that they must be separated from the gels prior to performing electrophoretic transfers or capillary blotting. In order to separate the gel from the film without damage an apparatus exists (Fig. 13.21) with a taut thin steel wire which is pulled between them.

13.6.2. Detection systems after blotting

These systems comprise general staining protocols as well as specific detection methods. In the first category, besides staining with Amido black or Coomassie Brilliant

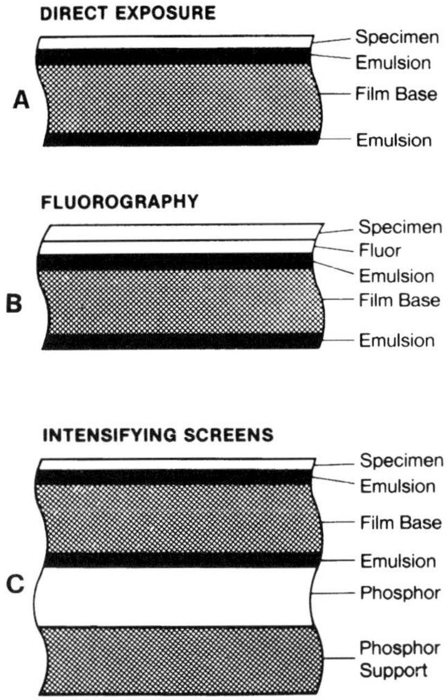

Fig. 13.22. The three classical methods of autoradiography: (A) by direct exposure; (B) by fluorography; and (C) with the use of intensifying screens. (By courtesy of Eastman Kodak Company.)

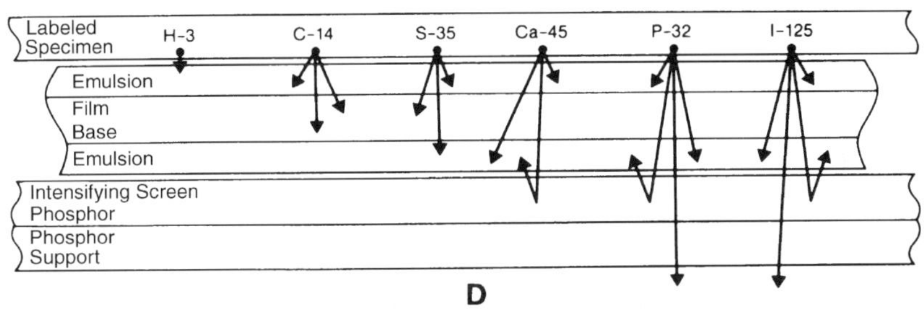

Fig. 13.23. Penetration of different radiations from radio-isotopes into X-ray film. Note that the intensifying screen phosphor method can only be used with high-energy emitters (e.g. gamma radiations) since they can penetrate through the film thickness and excite the phosphor underneath. (By courtesy of Eastman Kodak Company.)

Blue, mild staining methods such as the very sensitive Indian Ink method [221] or reversible ones, such as with Ponceau S [222] or Fast Green FCF have been described. In many cases, one can also use: (a) a general immunostain [223]; (b) colloidal gold or colloidal iron [224,225]; (c) autoradiography; or (d) fluorography [212].

The principle of autoradiography and fluorography (coupled often to the use of intensifying screens) is illustrated in Fig. 13.22, whereas Fig. 13.23 shows how the radiation of different isotopes can penetrate to a given depth through the top and bottom emulsions, which coat both sides of the film base. More recent developments call for new types of intensifying screens, in which phosphor imaging plates are composed of a thin (ca. 500 μm) layer of very small crystals of $BaFBr:Eu^{2+}$ in a plastic binder [226]. In practice, this phosphor imaging plate is exposed to a dried 1-D or 2-D gel containing separated radiolabelled protein zones. After exposure, the plate is transferred to a scanner where light from a HeNe laser (633 nm) is absorbed by the metastable F centres activated by the radiolabel in the phosphor screen. This in turn results in the emission of a blue (390 nm) luminescence proportional to the original amount of radiation incident on the plate. The image of the bands or spots in the original gel is thus reconstructed by the scanner and stored electronically. These techniques were extremely popular in the seventies and eighties, especially for revealing spots on 2-D maps, due to their extremely high sensitivities and were typically performed on dried polyacrylamide gel slabs [227]. However, due to risks involved with the use of penetrating radiations, such as the gamma emitter ^{125}I, and the risk of environmental pollution, there is now a trend to avoid radioactivity in the laboratory.

Whenever possible, and available, specific immunodetection methods are used, since they often reach sensitivities close to those of radiolabelling, without generating environmental problems. In this procedure, after the proteins have been transferred to the membrane, but before activating any immunochemical protocol, the membrane must be deactivated, i.e. all its unoccupied binding sites must be blocked. Failure to do that would result in the adsorption of antibodies all over the membrane surface, obliterating then any specific signal generated by potential antigens blotted onto the membrane. Bovine serum albumin (3–5% w/v) in phosphate buffered saline, PBS, 1 h incubation, is the most commonly used blocking agent, although other proteins could be used (e.g. other animal sera, ovalbumin, casein, gelatin). A solution of non-fat dried milk (3% w/v in PBS) has become popular as a blocking agent and usually results in very low background staining. At this point, washing the excess of blocking agent, the membrane is ready for incubation with antibodies. Although it would be possible to label the primary antibody (or other ligand) with a suitable reporter group for a direct visualisation of reactive proteins on the blot, this approach is not popular, since it requires derivatisation of the primary antibody, a process which could adversely affect its specificity or affinity. Therefore, it is usual to use an indirect, or sandwich, approach by utilising labelled secondary (or tertiary) antibody reagents. Thus, following blocking, the membrane is incubated with a solution containing the appropriately diluted specific primary antibody. The membrane is then washed and reacted with a solution containing a secondary antibody specific for the species and immunoglobulin class of the primary antibody. The secondary antibody can be fluorescently labelled (e.g. with fluorescein isothiocyanate, FITC), radiolabelled (usually with ^{125}I) or conjugated to an enzyme (e.g. horseradish peroxidase, alkaline phosphatase, β-galactosidase or glucose oxidase). Alternatively, the secondary antibody can be replaced by appropriately labelled staphylococcal protein A or streptococcal protein G. These reagents bind specifically to the F_c region of immunoglobulins (Ig), with protein G reacting with a broader range of species and classes than does protein A. Methods employing fluorescently labelled secondary antibodies require the use of UV illumination, while radiolabelled antibody procedures depend on time-consuming autoradiographic steps. Thus, methods based on the use of enzyme-conjugated secondary antibody reagents

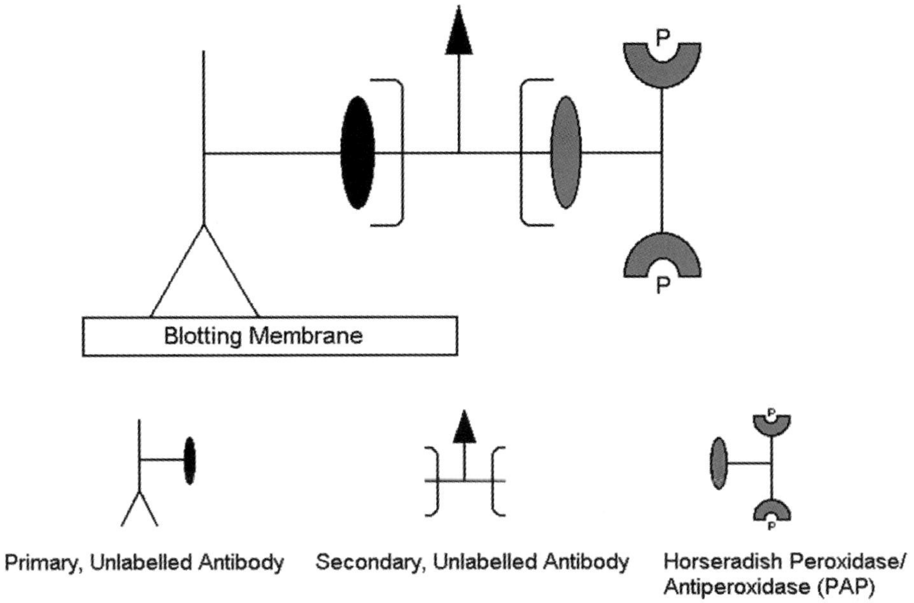

Fig. 13.24. Scheme of the system of antibodies building up onto the spot containing the specific antigen blotted onto a membrane, after a 1-D or 2-D electrophoretic separation in a polyacrylamide gel. In this particular example a cascade of three antibodies is used. Note that the last antibody in this cascade is covalently reacted with a reporter/amplifier, in this case the enzyme peroxidase from horseradish. Revelation will take place when the membrane will be confronted with appropriate substrates and dyes which, upon reaction, will form insoluble, stable, coloured reaction products on the site.

have become the most popular and kits of these reagents are commercially available. Fig. 13.24 shows how this system of antibodies builds up onto the spot containing the specific antigen blotted onto the membrane, although in this particular example a cascade of three antibodies is used. It is seen that the last antibody in this cascade carries an enzyme, which will be used for detection and signal amplification. Detection is in fact achieved by activating an enzyme reaction, by using appropriate substrates which form insoluble, stable coloured reaction products at the sites on the blot where the enzyme-conjugated secondary (or tertiary, as in Fig. 13.24) antibody is bound. The two most popular substrates for use with peroxidase-conjugated antibodies are diamino benzidine, which gives brown bands, and 4-chloro-1-naphthol, which produces purple bands. The best substrate system for the visualisation of alkaline phosphatase-conjugated antibodies is a mixture of 5-bromo-4-chloro indoxyl phosphate and nitroblue tetrazolium.

Other detection systems have been developed in order to increase he sensitivity of detection of proteins on blots. One approach is the use of colloidal gold-labelled antibodies [228] or protein A. These reagents have the advantage that the stain is visible, due to its reddish-pink colour, without further development; their sensitivity can be further increased by silver enhancement. Triple antibody probing methods can also be used for increasing the sensitivity of detection (see Fig. 13.24). In this case, preformed complexes of enzyme and antibodies are linked by secondary antibodies to the primary antibodies, examples being the peroxidase–antiperoxidase (PAP) and alkaline phosphatase–anti-alkaline phosphatase (APAP) procedures. Secondary antibodies can also be conjugated with the steroid hapten, digoxigenin, which can then be detected by using an enzyme-conjugated anti-digoxigenin antibody [229].

Another popular method for increasing the sensitivity of detection on blots exploits the specificity of the reaction between the vitamin biotin (M_r 224 Da) and the protein avidin. Antibodies can be readily conjugated with biotin and the resulting conjugates used as the secondary reagents for probing blot transfers. A third step must then be used for visualisation using avidin conjugated with a suitable reporter enzyme (e.g. peroxidase, β-galactosidase, alkaline phosphatase or glucose oxidase). Even greater sensitivity can be achieved at this stage by using preformed complexes of biotinylated enzyme with avidin, since many enzyme molecules are present in these complexes, thus generating an enhanced signal. Egg white avidin (M_r 68,000) is often used, but streptavidin (M_r 60,000, from *Streptomyces avidinii*) is today preferred, since it has a p*I* close to neutrality and is not glycosylated.

13.7. CONCLUSIONS

SDS-PAGE seems to have reached a plateau in terms of method development. It is hard to conceive that the technique could be further improved, since just about all possible methodological advances known in electrokinetic methods have been applied here. Also detection techniques seem to have been pushed to the maximum, so it is hard to guess, by present-day knowledge, what new developments could possibly be made. SDS-PAGE is a fundamental method for 2-D map analysis, since it represents the second dimension run and the run that will finally remain in the record, since it is at the end of this step that proteins are stained, or blotted and extracted and further analysed with the powerful tools today availablc in proteomics, such as MALDI-TOF mass spectrometry of both the intact polypeptide chain and of its peptides generated by proteolytic digestion. It might be asked which, of all possible techniques here described, is best suited for 2-D mapping. In the early days (e.g. in the Human Anatomy Program of the Anderson's) it was customary to perform the SDS-PAGE in concave porosity gradients. Today this method has been largely abandoned, in favour of linear porosity gradients, which are much more user friendly and easy to be implemented. In addition, in many reports, one could find the use of a double zone-sharpening technique, namely the use of discontinuous buffers (typically the Laemmli) coupled to porosity gradients in the running gel slab. This technique is also of diminishing importance today, again due to too cumbersome manipulations. It has been found that one of the two methods suffice. So, if one adopts discontinuous buffers, the stacking of zones is adequate to give high resolution, rendering unnecessary the simultaneous use of a porosity gradient in the running slab. Conversely, if one does (or cannot) use buffer discontinuities, then a porosity gradient is a must. With this caveat, we think it is high time that we lead the readers into the realm of 2-D map analysis, as expounded in the following chapter.

13.8. REFERENCES

1. A.T. Andrews, Electrophoresis: Theory, Techniques and Biochemical and Clinical Applications, Clarendon Press, Oxford, 1986, pp. 118–141.
2. G.M. Rothe and W.D. Maurer, in M.J. Dunn (Ed.), Gel Electrophoresis of Proteins, Wright, Bristol, 1986, pp. 37–140.
3. G.M. Rothe, in A. Chrambach, M.J. Dunn and B.J. Radola (Eds.), Advances in Electrophoresis, VCH, Weinheim, 1991, pp. 252–358.
4. R. Westermeier, Electrophoresis in Practice, VCH, Weinheim, 1993, pp. 167–188.
5. Q. Shi and G. Jackowski, in B.D. Hames (Ed.), Gel Electrophoresis of Proteins: A Practical Approach, Oxford University Press, Oxford, 1998, pp. 13–34.
6. K.A. Ferguson, Metabolism, 13 (1964) 985–995.
7. R. Pitt-Rivers and F.S.A. Impiombato, Biochem. J., 109 (1968) 825–833.
8. W.W. Fish, J.A. Reynolds and C. Tanford, J. Biol. Chem., 245 (1970) 5166–5168.
9. J.A. Reynolds and C. Tanford, Proc. Natl. Acad. Sci. USA, 66 (1970) 1002–1007.
10. J.A. Reynolds and C. Tanford, J. Biol. Chem., 245 (1970) 5161–5165.
11. A.L. Shapiro, E. Vinuela and J.V. Maizel, Biochem. Biophys. Res. Commun., 28 (1967) 815–820.
12. K. Weber and M. Osborn, J. Biol. Chem., 244 (1969) 4406–4412.
13. A.K. Dunker and R.R. Rueckert, J. Biol. Chem., 244 (1969) 5074–5080.
14. C.A. Nelson, J. Biol. Chem., 246 (1971) 3895–3901.
15. J.P. Segrest, R.L. Jackson, E.P. Andrews and V.T. Marchesi, Biochem. Biophys. Res. Commun., 44 (1971) 390–395.
16. S. Panyim and R. Chalkley, Arch. Biochem. Biophys., 130 (1969) 337–346.
17. T.V. Waehneldt, Bio-Systems, 6 (1975) 176–187.
18. P. Lambin, Anal. Biochem., 85 (1978) 114–125.
19. C. Tanford, The Hydrophobic Effect. Formation of Micelles and Biological Membranes, Wiley, New York, 1980, 2nd ed., pp. 159–164.
20. K. Shirahama, K. Tsujii and T. Takagi, J. Biochem. (Tokyo), 75 (1974) 309–319.
21. T. Takagi, K. Tsujii and K. Shirahama, J. Biochem. (Tokyo), 77 (1975) 939–947.
22. W.L. Mattice, J.M. Riser and D.S. Ckark, Biochemistry, 15 (1976) 4264–4272.
23. P. Lundahl, E. Greijer, M. Sandberg, S. Cardell and K.O. Eriksson, Biochim. Biophys. Acta, 873 (1986) 20–26.
24. E. Mascher and P. Lundahl, J. Chromatogr., 476 (1989) 147–158.
25. M. Wallstén and P. Lundahl, J. Chromatogr., 512 (1990) 3–12.
26. K. Ibel, R.P. May, K. Kirschner, H. Szadkowski, E. Mascher and P. Lundahl, Eur. J. Biochem., 190 (1990) 311–318.
27. P. Lundahl, Y. Watanabe and T. Takagi, J. Chromatogr., 604 (1992) 95–102.
28. K. Ibel, R.P. May, M. Sandberg, E. Mascher, E. Greijer and P. Lundahl, Biophys. Chem., 53 (1994) 77–84.
29. W.H. Westerhuis, J.N. Sturgis and R.A. Niederman, Anal. Biochem. 284 (2000) 143–152.
30. M. Samsò, J.R. Daban, S. Hansen and G.R. Jones, Eur. J. Biochem., 232 (1995) 818–824.
31. D.M. Neville, J. Biol. Chem., 246 (1971) 6328–6334.
32. D. Rodbard and A. Chrambach, Proc. Natl. Acad. Sci. USA, 65 (1970) 970–977.
33. D. Rodbard and A. Chrambach, Anal. Biochem., 40 (1971) 95–134.
34. A. Chrambach and D. Rodbard, Science, 172 (1971) 440–451.
35. G.M. Rothe, Electrophoresis, 3 (1982) 255–262.
36. J.F. Poduslo and D. Rodbard, Anal. Biochem., 101 (1980) 394–406.
37. F. Hashimoto, T. Horigome, M. Kanbayashi, K. Yoshida and H. Sugano, Anal. Biochem., 129 (1983) 192–199.
38. H. Schagger and G. von Jagow, Anal. Biochem., 166 (1987) 368–379.
39. A.L. Burlingame, R.K. Boyd and S.J. Gaskell, Anal. Chem., 70 (1998) 647R–716R.
40. S. Panyim, J. Biol. Chem. 246 (1971) 7557–7560.
41. J.P. Segrest, R.L. Jackson, Methods Enzymol. 28B (1972) 54–60.
42. P. Lambin and J.M. Fine, Anal. Biochem., 98 (1979) 160–168.

219. E.R. Tovey and B.A. Baldo, Electrophoresis, 8 (1987) 384–387.
220. K.E. Johansson, Electrophoresis, 8 (1987) 379–383.
221. K. Hancock and V.C.W. Tsang, Anal. Biochem., 133 (1983) 157–162.
222. O. Salinovich and R.C. Montelaro, Anal. Biochem., 156 (1986) 341–347.
223. J.M. Kittler, N.T. Meisler and D. Viceps-Madore, Anal. Biochem., 137 (1984) 210–216.
224. M. Moeremans, M. De Raeymaeker, G. Daneels and J. De Mey, Anal. Biochem., 153 (1986) 18–22.
225. M. Moeremans, G. Daneels and J. De Mey, Anal. Biochem., 145 (1985) 315–321.
226. R.F. Johnston, S.C. Pickett and D.L. Barker, Electrophoresis, 11 (1990) 355–360.
227. W.M. Bonner, Methods in Enzymology, Vol. 96, Academic Press, New York, 1983, pp. 215–225.
228. G. Daneels, M. Moeremans, M. De Raeymaeker and J. De Mey, J. Immunol. Methods, 89 (1986) 89–91.
229. C. Kessler, Molec. Cell. Probes, 5 (1991) 161–205.

CHAPTER 14

Two-Dimensional Maps

CONTENTS

14.1. INTRODUCTION

Proteomics is an emerging area of research in the post-genomic era that deals with the global analysis of gene expression, via a combination of techniques for resolving, identifying, quantitating and characterising proteins. In addition, a fundamental part of proteomics is bioinformatics, for storing, communicating and interlinking proteomic and genomic data as well as mapping information from genome projects [1]. Although each technique can be applied independently, their impact can be maximised when used in concert for the study of complex biological problems.

For the last 25 years, two-dimensional polyacrylamide gel electrophoresis (2-D PAGE) has been the technique of choice for analysing the protein composition of a given cell type and for monitoring changes in gene activity through the quantitative and qualitative analysis of the thousands of proteins that orchestrate various cellular functions. Proteins are usually the functional molecules and, therefore, the most likely components to reflect qualitative (expression of new proteins, post-translational modifications) and quantitative (up and down regulation, co-ordinated expression) differences in gene expression [2]. Just as an example of the incredible growth in the field, in the journal *Electrophoresis* alone an avalanche of special issues has appeared, over the years, collecting several hundreds of articles dealing with 2-D PAGE and its application in any possible area of biological research) [3–25]. Many more of those issues have appeared since; starting from the year 2000, the publisher has announced that six issues per year of *Electrophoresis* will be devoted to this topic.

Notwithstanding its extraordinary resolving power, even 2-D PAGE seems to have reached a plateau in the number of proteins that can be resolved and detected in a single 2-D map. The advent of immobilised pH gradients (IPG) [26] has certainly made a big improvement, but its resolving power cannot be pushed any further. Also solubilising cocktails have improved, due to the introduction of new surfactants [27] and new reducing agents [28]; however, progress seems to be levelling off. The same applies to staining protocols, which seem not to have produced any increment in detection sensitivity in the last five years [29]. Yet, the enormous complexity of the proteome (it is hypothesised that up to 100,000 genes could be present in the human genome, but the proteins in the proteome could be as many as 1 million, if one considers all the possible post-translational modifications) calls for improvements in resolving power, possibly coupled to increments in detection sensitivity for tracking trace components. It is the aim of the present chapter not only to review the evolution of 2-D PAGE up to the present, but also to suggest new avenues and potential new developments in the field. As it stands today, 2-D PAGE is a complex field, in which perhaps the electrophoretic knowledge has become minoritary: 30% of it belongs to the fine art of electrokinetic methodologies, combining with proper skills an isoelectric focussing step to an orthogonal SDS-PAGE; 30% of it is proper use and knowledge of mass spectrometry (MS), especially in the variant MALDI-TOF (matrix assisted laser induced desorption ionisation, time of flight), for a proper assessment of the precise M_r value of polypeptide chains and their fragments; the remaining 30% is computer science coupled to good skills in interrogating databases and extracting pertinent data. The remaining 10%, perhaps, is fantasy and intuition, sprinkled in the above cocktail. Shake it (do not stir it) and serve it properly chilled.

14.1.1. The early days and the evolution of 2-D PAGE

Any single dimension method (e.g. IEF, SDS-PAGE, and isotachophoresis to name just a few, high-resolution techniques) cannot resolve more than 80–100 different components, and even that only under the most favourable circumstances. Thus, when a sample is highly heterogeneous, other methods have to be sought. One such a procedure is 2-D PAGE, which exploits a combination of two single dimension runs. Two-dimensional maps could be prepared by using virtually any combination of one-dimensional methods (not necessarily electrophoretic!) but the one that has won universal recognition is that combining a charge (typically an IEF protocol) to a size (e.g. SDS-PAGE) fractionation, since this results in a more even distribution of components over the surface of the map. While this combination of separation methods was used at quite an early stage in the development of 2-D macromolecular mapping, it was the elegant work of O'Farrell [30] that really demonstrated the full capabilities of this approach (see also [31,32]). He was able to resolve and detect about 1100 different proteins from lysed *Escherichia coli* cells on a single 2-D map and suggested that the maximum resolution capability might have been as high as 5000 different proteins. Apart from the meticulous attention to detail, major reasons for the advance in resolution obtained by O'Farrell, compared to earlier workers, included the use of samples labelled with ^{14}C or ^{35}S to high specific activity, and the use of thin (0.8 mm) gel slabs for the second dimension, which could then be dried down easily for autoradiography. This detection method was able to reveal protein zones corresponding to one part in 10^7 of the sample (usually 1–20 μg was applied initially, since higher loads caused zone spreading, although up to 100 μg could be loaded). Coomassie Blue, in comparison, was about 3 orders of magnitude less sensitive and could reveal only about 400 spots. For the first dimension, O'Farrell adopted gel rods of 13 cm in length and 2.5 mm in diameter. The idea was to run samples fully denatured, in what became known as the 'O'Farrell lysis buffer' (9 M urea, 2% Nonidet P-40, 2% β-mercaptoethanol and 2% carrier ampholytes, in any desired pH interval). For the second SDS-PAGE dimension, O'Farrell [30] used the discontinuous buffer system of Laemmli [33] and, for improved resolution, a concave exponential gradient of polyacrylamide gel (usually in the intervals 9–15 or 10–14% T, although wide porosity gradients, e.g. 5–22.5% T, were also suggested). It is thus seen that, since its very inception, O'Farrell carefully selected all the best conditions available at the time; it is no wonder that his system was adopted as such in the avalanche of reports that soon followed. He went as far as to recognise that some protein losses could occur during the equilibration of the IEF gel prior to running of the SDS-PAGE. Depending on the identity of the protein and the duration of equilibration, losses were estimated to vary from 5 up to 25%. Even for that O'Farrell proposed a remedy: he reported that, in such cases, the equilibration step could be omitted and, for minimising streaking in the second dimension, due to only partial SDS saturation, one could increases the depth of the stacking gel in the SDS-PAGE run from 2.5 cm up to 5 cm. The increased length of the stacking gel, coupled to lower initial voltage gradients and to a higher amount of SDS in the cathodic reservoir, allowed for proper saturation of the protein species by the SDS moiety. It is no surprise that, with such a thorough methodological development, there were hardly any modifications to this technique, something that in the scientific world is a very rare event (as soon as a method is published, usually a cohort of modifications is immediately reported, in the hope that the 'second discoverer' will be the winner!).

Some minor modifications, with time, were adopted and they certainly helped in further improving the technique. The factors affecting the resolution of 2-D maps were discussed by a number of workers (e.g. [34–36]), as a result of which it appeared that it was beneficial to reduce the concentration of acrylamide in the first dimension IEF gel to as low as 3.5% *T* (a smart idea indeed, as it will be seen ahead) and to run the IEF gel at 800–1000 V for at least 10,000 V × hour (we will see that with IPGs it is not uncommon to deliver much stronger conditions than those, up to 60,000 V × hour). In addition, Perdew et al. [37] proposed to replace the non-ionic detergent Nonidet P-40 with similar proportions of the zwitterionic detergent CHAPS (3-[(3-cholamidopropyl)dimethylammonium]-1-propane sulphonate, which they claimed had superior membrane protein solubilising properties and was particularly effective at disaggregating hydrophobic proteins. By the same token, in order to overcome the difficulties of separating basic proteins in the first, IEF dimension, O'Farrell et al. [38] introduced the technique of non-equilibrium pH gradient isoelectric focussing (NEPHGE), in which the sample was applied at the acidic end of the gel and the IEF terminated prior to reaching steady-state conditions. Elegant micro-versions of O'Farrell's method were reported in [39,40], in which the first dimension IEF run was performed with gels in small capillary tubes followed by a second dimension SDS-PAGE step on postage-stamp-size gel slabs with a 1–40%*T* or 6–25% *T* gradient.

At the opposite end of the scale, instead of miniaturisation, Anderson and Anderson [41–43] started thinking of 'large scale biology' and building instrumentation (called the ISO-DALT system) for preparing and running a large number of O'Farrell type gels together. This approach greatly enhanced reproducibility and comparison between the resulting protein maps while enabling a very large number of samples to be handled in a short time. A variation (termed BASO-DALT), giving enhanced resolution of basic proteins by employing NEPHGE in the first dimension, was also described [44]. These approaches made practical the Molecular Anatomy Program at the Argonne National Laboratory in the USA, the object of which was to be able to fractionate human cells and tissue with the ultimate aim of being able to describe completely the products of human genes and how these vary between individuals and in disease [45,46]. It is also thanks to the Herculean efforts of the two Anderson's if present-day 2-D maps have reached such a high stage of evolution. (It is worth a perusal through two special issues of *Clin. Chem.* [47,48] to see the flurry of papers published and the incredible level of development reached already in the years 1982–1984!)

14.1.2. A glimpse at modern times

Although the power of 2-D electrophoresis as a biochemical separation technique has been well recognised since its introduction, its application, nevertheless, has become particularly significant in the past few years, as a result of a number of developments, outlined below.

(1) The 2-D technique has been tremendously improved to generate 2-D maps that are superior in terms of resolution and reproducibility. This new technique utilises a unique first-dimension, that replaces the carrier ampholyte-generated pH gradients with immobilised pH gradients (IPG) and replaces the tube gels with gel strips supported by a plastic film backing [49].

(2) Methods for the rapid analysis of proteins have been improved to the point that single spots eluted or transferred from single 2-D gels can be rapidly identified. Mass spectroscopic techniques have been developed, that allow analysis of very small quantities of proteins and peptides [50–52]. Chemical microsequencing and amino acid analysis can be performed on increasingly smaller samples [53]. Immunochemical identification is now possible with a wide assortment of available antibodies.

(3) More-powerful, less expensive computers and software are now available, allowing routine computerised evaluations of the highly complex 2-D patterns.

(4) Data about entire genomes (or substantial fractions thereof) for a number of organisms are now available, allowing rapid identification of the genes encoding a protein separated by 2-D electrophoresis.

(5) The World Wide Web (WWW) provides simple, direct access to spot pattern databases for the comparison of electrophoretic results and to genome sequence databases for assignment of sequence information.

A large and continuously expanding application of 2-D electrophoresis is proteome analysis. The proteome is defined as "the PROTEin complement expressed by a genOME", thus it is a fusion word derived from two different terms [54,55]. This analysis involves the systematic separation, identification and quantitation of a large number of proteins from a single sample. Two-dimensional PAGE is not only unique for its ability of simultaneously separating thousands of proteins, but also for detecting post- and co-translational modifications, which cannot be predicted from genome sequences. Other applications of 2-D PAGE include identification (e.g. taxonomy, forensic work etc.), the study of genetic variation and relationships, the detection of stages in cellular differentiation and studies of growth cycles, the examination of pathological states and diagnosis of disease, cancer research, monitoring of drug action. Among the books dedicated to this topic, one could recommend those of Wilkins et al. [1], Kellner et al. [56] and Rabilloud [57], among book chapters, Westermeier [58] and Hanash [59], to name just a few.

Perhaps one of the key points of the success of present 2-D PAGE was the introduction of the IPG technique, as stated above, and in particular the early recognition that wide, non-linear pH gradients could be generated, covering the pH 3.5–10 interval. Such a gradient (see Fig. 12.17 in Chapter 12) was calculated for a general case involving the separation of proteins in complex mixtures, such as cell lysates and applied to a number of 2-D separations [60–62]. The idea of this non-linear gradient came from an earlier study by Gianazza and Righetti [63], computing the statistical distribution of the p*I*s of water-soluble proteins and showing that as much as 2/3 of them would focus in the acidic region (taking as a discriminant pH of 7.0) and only 1/3 in the alkaline pH scale. Given this relative abundance, it was clear that an optimally resolving pH gradient should have a gentler slope in the acidic portion and a steeper course in the alkaline region. Such a general course was calculated by assigning to each 0.5 pH unit interval in the pH 3.5–10 region a slope inversely proportional to the relative abundance of proteins in that interval. In a separation of a crude lysate of *Klebsiella pneumoniae*, a great improvement in resolution of the acidic cluster of bands was in fact obtained, without loss of the basic portion of the pattern [60]. This manuscript was also a cornerstone in IPG technology, since it demonstrated, for the first time, that the pH gradient and the density gradient stabilising it did not need to be co-linear, because the pH could be adjusted by localised flattening while leaving the density gradient unaltered. This non-linear pH 3.5–10 gradient formed the basis (with perhaps minor modifications) of most of the wide non-linear gradients adopted today and sold by commercial companies. The other events that made IPGs so powerful were the recognition that a much wider portion of the pH scale could be explored; thus very acidic pH intervals (down to pH 2.5) were described, for equilibrium fractionation of acidic proteins, such as pepsin [64], as well as very alkaline intervals (e.g. pH 10–12) for analysis of high p*I* proteases (subtilisins [65]) and even of histones [66]. Already in 1990 our group described 2-D maps in what was, at that time, the most extended pH gradient available, spanning the pH 2.5–11 interval) [67]. We went so far as to describe even sigmoidal pH gradients, which have also found applications in 2-D analysis [68,69]. This shows that, already starting from 1990, all the ingredients for success of 2-D maps exploiting IPGs in the first dimension were fully available in the literature.

14.2. SOME BASIC METHODOLOGY PERTAINING TO 2-D PAGE

Fig. 14.1 shows what is perhaps the most popular approach to 2-D map analysis today. The 1st dimension is preferably performed in individual IPG strips, laid side by side on a cooling platform, with the sample often adsorbed into the entire strip during rehydration. At the end of the IEF step, the strips have to be interfaced with the 2nd dimension, almost exclusively performed by mass discrimination via saturation with the anionic surfactant SDS. After this equilibration step, the strip is embedded on top of an SDS-PAGE slab, where the 2nd-D run is carried out perpendicular to the 1st-D migration. The 2-D map displayed at the end of these steps is the stained SDS-PAGE slab, where polypeptides are seen, after staining, as (ideally) round spots, each characterised by an individual set of $\mathrm{p}I/M_\mathrm{r}$ coordinates.

Although most of the relevant protocols for properly performing 2-D maps can be found in the respective chapters devoted to IEF (and IPG) and to SDS-PAGE, it is worth here to recall some important methodologies, especially those involving sample solubilisation and preparation prior to the first dimension IEF/IPG step. Appropriate sample preparation is absolutely essential for good 2-D results. Due to the great diversity of protein sample types and origins, only general guidelines for sample preparation are provided here. The optimal procedure should ideally be determined empirically for each sample type. If sample preparation is performed properly, it should result in complete solubilisation, disaggregation, denaturation and reduction of the proteins in the sample. We recall here that different treatments and conditions are required to solubilise different types of samples: some proteins are naturally found in complexes with membranes, nucleic acids or other proteins; some proteins form various

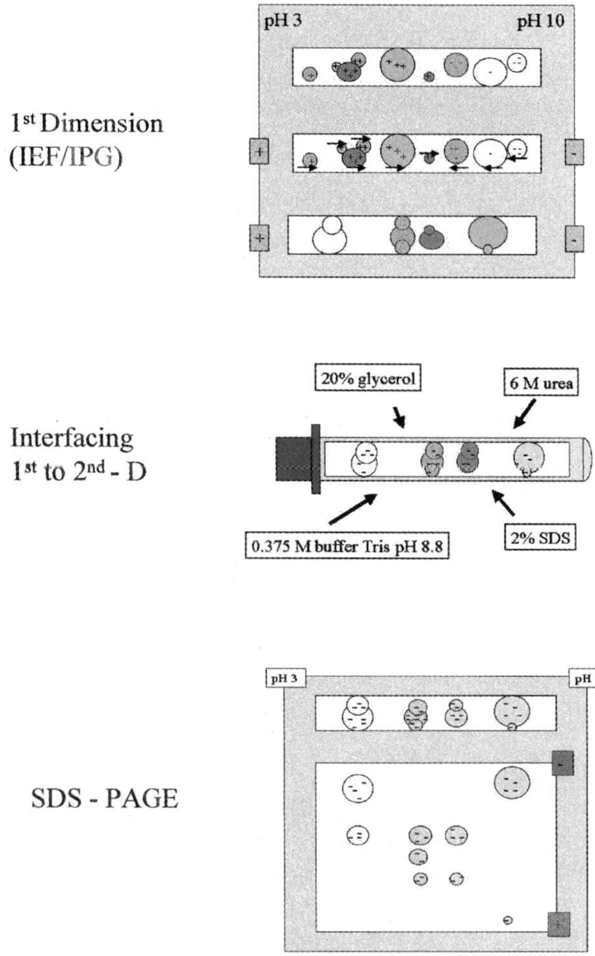

Fig. 14.1. Pictorial representation of the steps required for obtaining a 2-D map, with the first dimension typically done in IPG strips, followed by equilibration of the focussed strips in SDS-interfacing solution and finally by the 2nd dimension run in SDS-PAGE.

non-specific aggregates; some others precipitate when removed form their physiological environment. The effectiveness of solubilisation depends on the choice of cell disruption methods, protein concentration and dissociation methods, choice of detergents and on the overall composition of the sample solution. Lenstra and Bloemendal [70], Molloy et al. [71] and an entire issue of *Methods of Enzymology* [72] can be consulted as general guides to protein purification and/or for specific solubilisation problems, e.g. for membrane proteins and other difficult samples. The following general sample preparation guidelines should always be kept in mind.

(1) Keep the sample preparation strategy as simple as possible, so as to avoid protein losses.

(2) Cells and tissues should be disrupted in such a way as to minimise proteolysis and other modes of protein degradation. Cell disruption should be done at low temperatures and should ideally be carried out directly into strongly denaturing solutions containing protease inhibitors.

(3) Sample preparation solutions should be freshly made or stored frozen as aliquots. High-purity, deionised urea should always be used.

(4) Preserve sample quality by preparing it just prior to IEF or by storing samples in aliquots at −80°C. Samples should not be repeatedly thawed.

(5) Remove all particulate material by appropriate centrifugation steps. Solid particles and lipids should be eliminated because they will block the gel pores.

(6) In presence of urea, samples should never be heated. Elevated temperatures produce higher levels of cyanate from urea, which in turn can carbamylate proteins. A specific technique has been in fact described for producing 'carbamylation trains' as p*I* markers by boiling specific proteins in urea [73–75].

Below we will briefly describe some general methods pertaining to sample preparation.

14.2.1. Methods of cell disruption

These methods can be divided into two categories: 'soft' and harsher methods. Soft methods are in general employed when the sample under investigation consists of cells that can be lysed under mild conditions (e.g. tissue culture cells, red blood cells etc.). Briefly, they are:

Osmotic lysis. When isolated cells are suspended in hypotonic solution, they swell and burst, releasing all the cellular content [76].

Freeze–thaw lysis [77,78]. Many types of cells can be lysed by subjecting them to one or more cycles of quick freezing and subsequent thawing.

Detergent lysis. Most detergents solubilise cellular membranes, thereby lysing cells and liberating their content.

Harsh methods are those employing, in general, mechanical means for cell or tissue disruption. They can be summarised into:

Sonication. Ultrasonic waves generated by a sonicator lyse cells through shear forces. Care should be exerted for minimising heating and foaming [79,80].

French pressure cell. This cell can be pressurised well above 1000 atm. Cells are lysed by shear forces resulting from forcing a cell suspension through a small orifice at high pressure. Particularly useful for microorganisms with cell wall (e.g. bacteria, algae).

Grinding. Some types of cells can be broken by hand grinding with a mortar and pestle, often admixed with quartzite or alumina powders. Well suited for solid tissues and micro-organisms

Mechanical homogenisation. One of the most popular for soft, solid tissues. Classical devices are the Dounce and the Potter–Elvehjem homogenisers. Blenders and other electrically driven devices are often used for large samples [81–83].

Glass-bead homogenisers. The abrasive force of the vortexed beads is able to break cell walls, liberating the cell contents. Useful for cell suspensions and micro-organisms [84,85].

14.2.2. Proteolytic attack during cell disruption

When cells are lysed, hydrolases (phosphatases, glycosidases and especially proteases) are in general liberated or activated. In presence of chaotropes, glycosidases and phosphatases are quickly denatured, but the denaturation of proteases might have slower kinetics. For instance, Cleveland et al. [86] performed peptide mapping in dilute SDS solutions, suggesting that proteases had the ability to work in such denaturing media. This often results in much altered spots in the subsequent 2-D analysis, to the point that no band recognition is any longer possible. Proteases are in general inhibited when the tissue is disrupted directly in strong denaturants, such as 8 M urea, 10% TCA or 2% SDS [87–91]. But there are still two major problems: (a) some proteases might be more resistant to denaturation than the majority of the other cellular proteins, so that, while the former have been unfolded, proteases might have some time for attack before being denatured too; (b) in other cases (e.g. red blood cells, RBCs) the cells might have to be lysed under native conditions, in order to eliminate some cellular components, so that there is ample time for strong proteolytic aggression. In any event, proteases are less active at lower temperature, that is why, during cell disruption, low temperatures (2–4°C) are in general recommended. In addition, most tissue proteases are inactive above pH 9.5, so that proteolysis can be often inhibited by preparing the sample in Tris-free base, sodium bicarbonate or basic carrier ampholytes at pH values close to 10. Some proteases, however, might retain activity even when all the above precautions are taken. Thus, as a safety precaution, it is advisable to use a cocktail of protease inhibitors. Such combinations are available from a number of commercial sources and in general comprise both chemical and proteinaceous inhibitors. Below, a list of the most common inhibitors is given.

(1) Phenylmethylsulphonyl fluoride (PMSF). It is used at concentrations up to 1 mM. It is an irreversible inhibitor that binds covalently into the active site of serine proteases and some cysteine proteases. Since PMSF is quickly inactivated in aqueous solutions, it should be prepared just prior to use. Since it is also less active in presence of DTT or thiol reagents, the latter species should preferably be added at a later stage.

(2) Aminoethyl benzylsulphonyl fluoride or Pefabloc SC (AEBSF). It is similar to PMSF but it is more soluble and less toxic. Used at concentrations up to 4 mM.

(3) EDTA or EGTA. Generally used at 1 mM concentrations. They inhibit metalloproteases by chelating the metal ions required for activity.

(4) Peptide protease inhibitors. They are: (a) reversible inhibitors; (b) active in presence of DTT; (c) active at low concentrations under widely different conditions. Among them: leupeptin, active against serine and cysteine proteases; pepstatin, inhibiting aspartyl proteases, such as pepsin; aprotinin, active against many serine proteases; bestatin, an inhibitor of aminopeptidase. In general, they are used at a level of 2–20 μg/ml.

(5) Tosyl lysine chloromethyl ketone (TLCK) and tosyl phenyl chloromethyl ketone

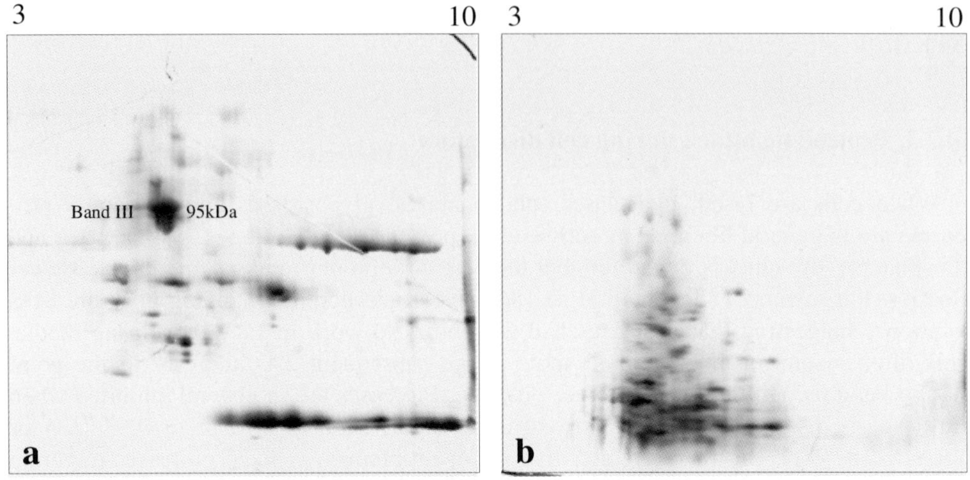

Fig. 14.2. 2-D map of RBC membranes: (a) treated with protease inhibitors during the lysis and washing steps; (b) control, untreated with protease inhibitors. Note, in this last case, the massive proteolytic action. The numbers at the gel top indicate the pH interval (pH 3–10, non-linear). (From [92], with permission.)

(TPCK). These rather similar compounds irreversibly inhibit many serine and cysteine proteases. Both used at levels of 0.1–0.5 mM.

(6) Benzamidine. It inhibits serine proteases and is used at concentrations of 1–3 mM.

The importance of safeguarding your sample against accidental proteolytic attack during preparation should never be underestimated. A case in point is the dramatic example of 2-D maps of RBCs membranes, given here, when processed in the absence or presence of proper inhibitor cocktails (Fig. 14.2). The primary concern when preparing RBC membranes for electrophoresis is to remove haemoglobin, which is present at extremely high levels in the RBC and can cause severe streaking in 2-D maps. Most of the procedures in the literature involve washing the RBCs in physiological buffered saline followed by lysis and washing in hypotonic buffer. However, there has been a lack of attention for the activation of proteases upon RBC lysis. Indeed Heegaard and Poglod [93] stated that no differences in the protein patterns were observed between protease-treated and standard, non-protease-treated, RBC membranes. Interestingly, the same authors also reported that there was a large difference in the number of RBC membrane spots and in their distribution in the 2-D maps shown by seven previous papers on RBC membranes. The number of RBC membrane 2-D spots reported in the seven papers quoted by Heegaard and Poglod [93] varied from less than 100 to more than 600, which should have been a strong indication that there were major discrepancies in the sample preparation methodologies presented. Rabilloud et al. [94] have shown 2-D maps of RBC membranes using a conventional protocol, protease inhibitors and a highly solubilising cocktail containing urea, thiourea and amidosulfobetaine 14, a novel zwitterionic surfactant which has improved the solubility of membrane proteins from a variety of sources [27]. The 2-D map shown in Fig. 14.2a is a silver stained

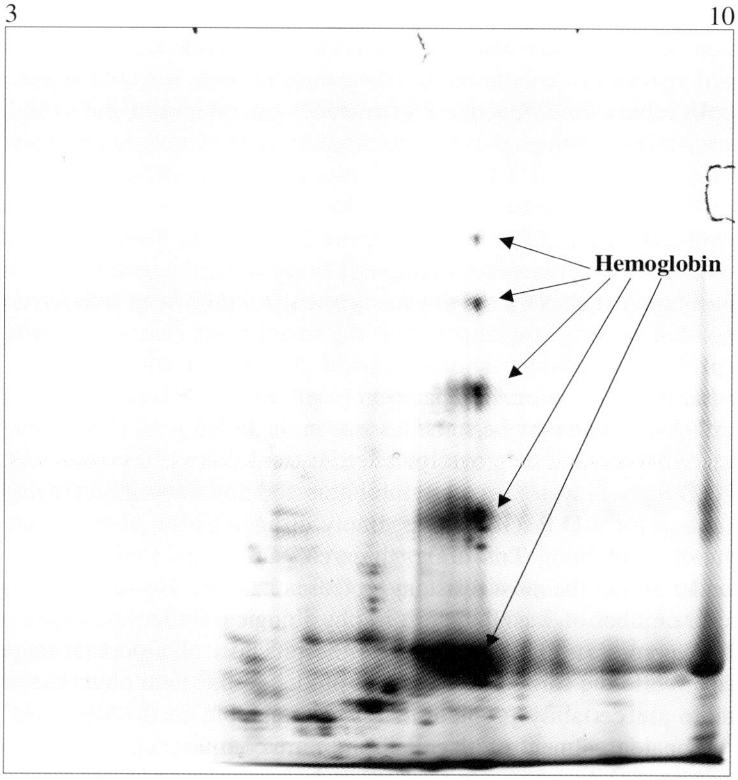

Fig. 14.3. 2-D map of a RBC membrane preparation extracted with sodium carbonate pH 11. Note the presence of large amounts of globin chains and its oligomers, possibly formed via mixed disulphide bridges. The numbers at the gel top indicate the pH interval (pH 3–10, non-linear). (From [92], with permission.)

10–20% *T* gel of RBC membranes prepared using protease inhibitors [92]. The pattern is quite similar to the RBC profiles obtained by Rabilloud et al. [94] and major landmark membrane proteins such as Band III (95 kDa) are visible. In contrast, the 2-D map in Fig. 14.2b is a silver-stained 10–20% *T* gel of RBC membranes prepared without protease inhibitors. Clearly in the absence of inhibitors there has been massive degradation of the high molecular mass proteins and the resulting mixture of peptides barely extends above a M_r of 50 kDa.

But there is more to it, as demonstrated in the following. Fig. 14.3 shows a 10–20% *T* silver-stained gel of RBC membranes prepared with sodium carbonate. Even though the packed RBCs are lysed in 4°C 100 mM carbonate at pH 11, there appears to be proteolysis, because only relatively low M_r material is visible as compared to Fig. 14.2a. The only high molecular mass species on the gel are multimeric complexes of haemoglobin, which is a contaminant. In contrast to carbonate treatment of bacterial cells, which seems to produce high quality membranes, this procedure is not useful for RBCs as it appears to allow proteolysis and also causes haemoglobin to be irreversibly attached to the membrane. These examples suggest that, unfortunately, no procedure

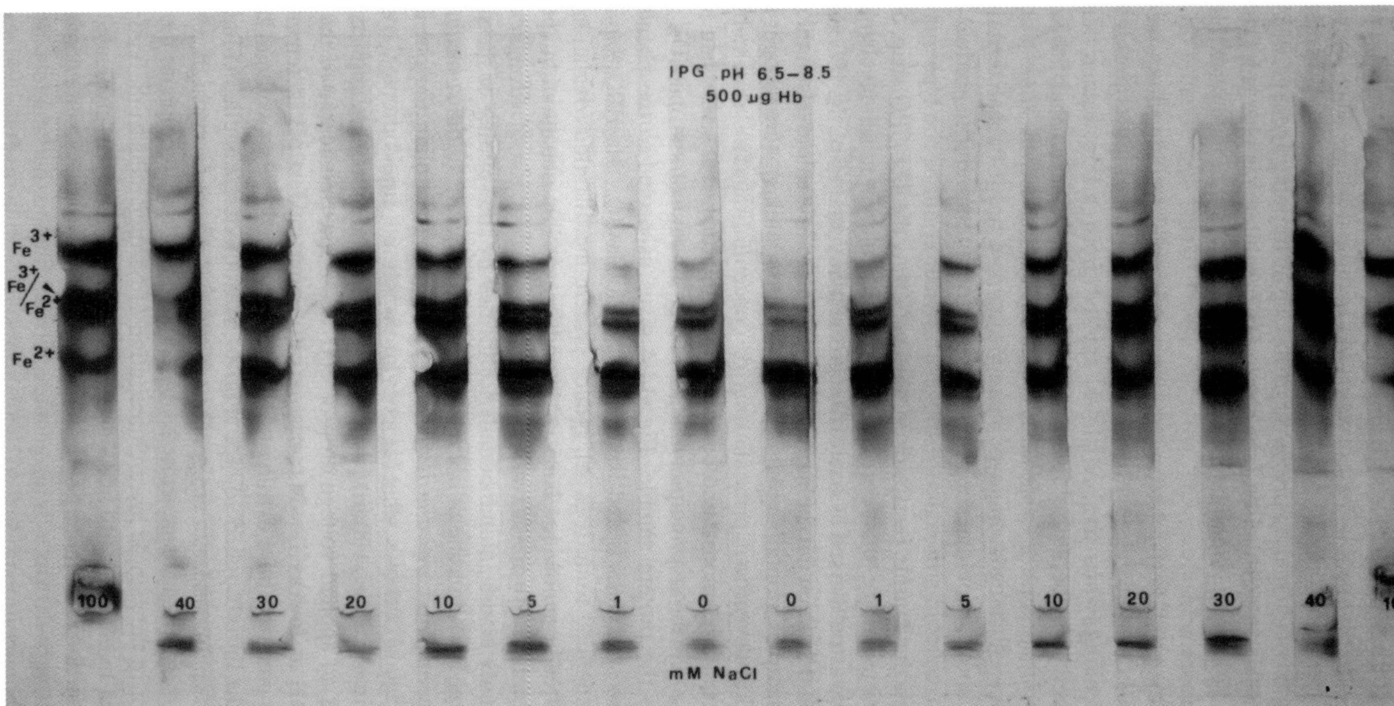

Fig. 14.4. Effects of strong salts on protein patterns in IPGs. Native haemoglobin (Hb) was applied to a pH 6.5–8.5 IPG gel and added with increasing amounts (up to 100 mM) of NaCl. Reduced (Hb-Fe^{2+}), partially oxidised (Fe^{3+}/Fe^{2+}) and fully oxidised (Fe^{3+}) Hb species are marked on the left gel side. Note, at progressively higher salt levels, the higher haeme oxidation and the appearance of sample precipitates in the application pocket. (From [121] with permission.)

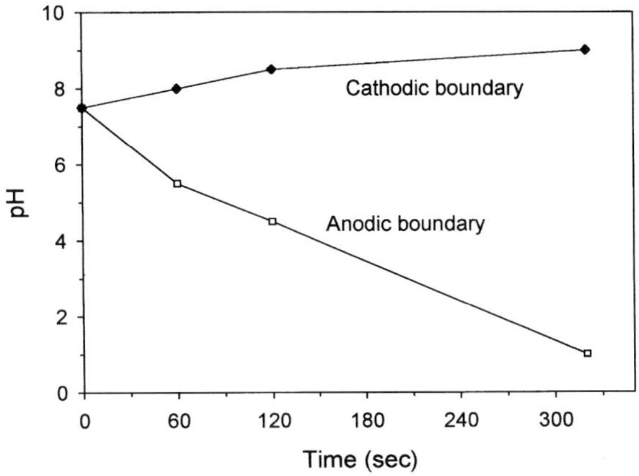

Fig. 14.5. Assessment of pH of salt ion boundaries in the sample zone as a function of time. Pockets cast in the middle of the gel (pH 7.5) were filled with 20 μl of 100 mM NaCl. The cathodic and anodic hedges of the pocket were covered with thin strips of alkaline and acidic pH indicators, respectively. The pH in the two boundaries was assessed by visual inspection of the colour changes at the given time intervals at constant 2000 V. (From [121], with permission.)

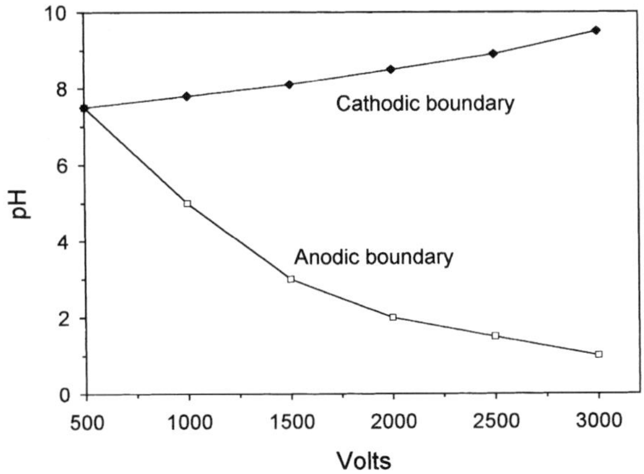

Fig. 14.6. Assessment of pH of salt ion boundaries in the sample zone as a function of applied voltage. The experiment of Fig. 14.5 was repeated, except that pH estimations were made as a function of different voltage gradients applied (from 500 V up to 3000 V) after 10 min from the application of the electric field. Here, as well as in Fig. 14.5, the alkaline pH estimates must be regarded as approximate, since the pH indicators, being negatively charged, move away from the Na^+ boundary towards lower pH values. Conversely, the pH of the anodic boundary is a much better estimate since the pH indicator, when it starts leaching out of the filter paper strip, moves with the Cl^- boundary. (From [121], with permission.)

molar ratio of ca. 1 : 1) (see Figs. 14.7 and 14.8). Low-voltage runs for extended initial periods (e.g. 4 h at 500 V) are also beneficial (see Fig. 14.6). Although the data here presented refer to native protein runs, it must be emphasised that they apply to denatured samples too, especially if these samples are run side by side in a continuous gel slab, rather than in individual strips. In the former case, variable salt levels in the various

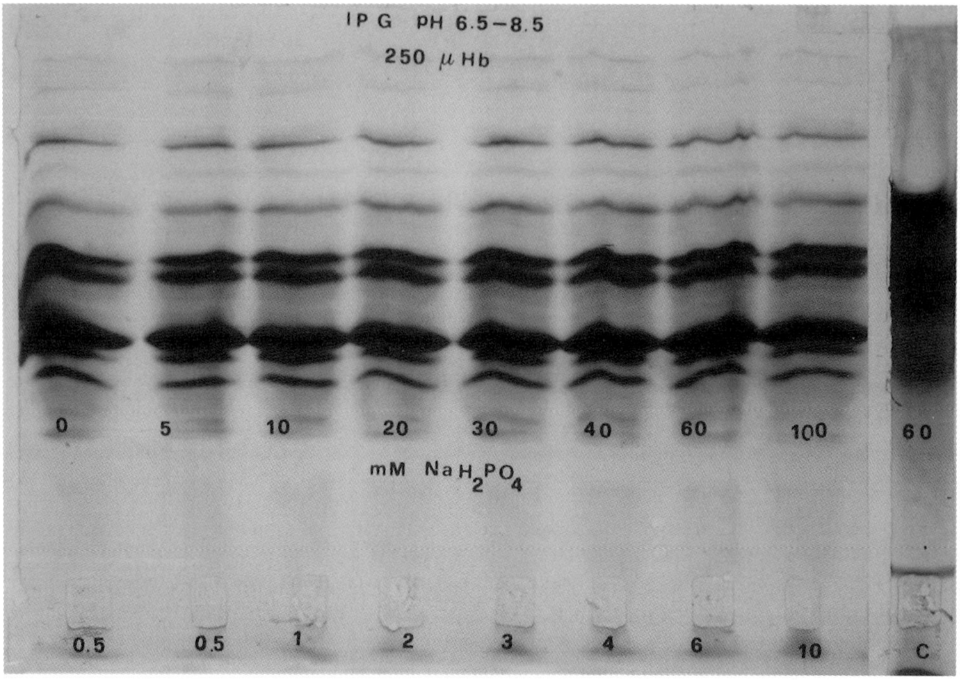

Fig. 14.7. Effects of strong salts on protein patterns in IPGs, in presence of soluble carrier ampholytes. Native haemoglobin (Hb) was applied to a pH 6.5–8.5 IPG gel and added with increasing amounts (up to 100 mM) of sodium phosphate, titrated to pH 7.5, simultaneously added with increasing levels (up to 10%) of pH 6–8 carrier ampholytes (CA). Note, in this case, the pronounced reduction of oxidised species and the abolition of sample precipitation in the application pocket. C = control lane, added with 60 mM phosphate but in the absence of CAs. (From [121], with permission.)

samples would cause fanning out of the protein fronts, which would thus invade laterally all adjacent sample tracks and completely ruin the IPG pattern. Even in individual strips, excess salt in the sample can considerably quench migration of proteins to the p*I* value; the salts will have to migrate out and collect in the electrodic reservoirs before any appreciable protein migration will ensue. As a result, the run might take an incredible length of time, to the point at which proteins (and the gel matrix too) could hydrolyse, especially in runs performed in rather alkaline (or acidic) pH ranges. We would thus like to offer the following 'pentalogue':

(1) Avoid high salt levels in your sample (>40 mM).

(2) Avoid salts formed by strong acids and bases (e.g. NaCl, Na_2SO_4, Na_2HPO_4).

(3) In presence of high salt levels, when running native proteins, add high levels of carrier ampholytes (e.g. 3–4% to 50 mM salt).

(4) If salt is needed, e.g. for sample solubility, use only salts formed from weak acids and bases (e.g. Tris-acetate, Tris-glycinate) or any of the Good's buffers (e.g. ACES, ADA, MOPS, etc.) titrated around the p*K* of their amino group (with an appropriate weak counterion!).

(5) In presence of high salt levels, run your sample at low voltages for several hours

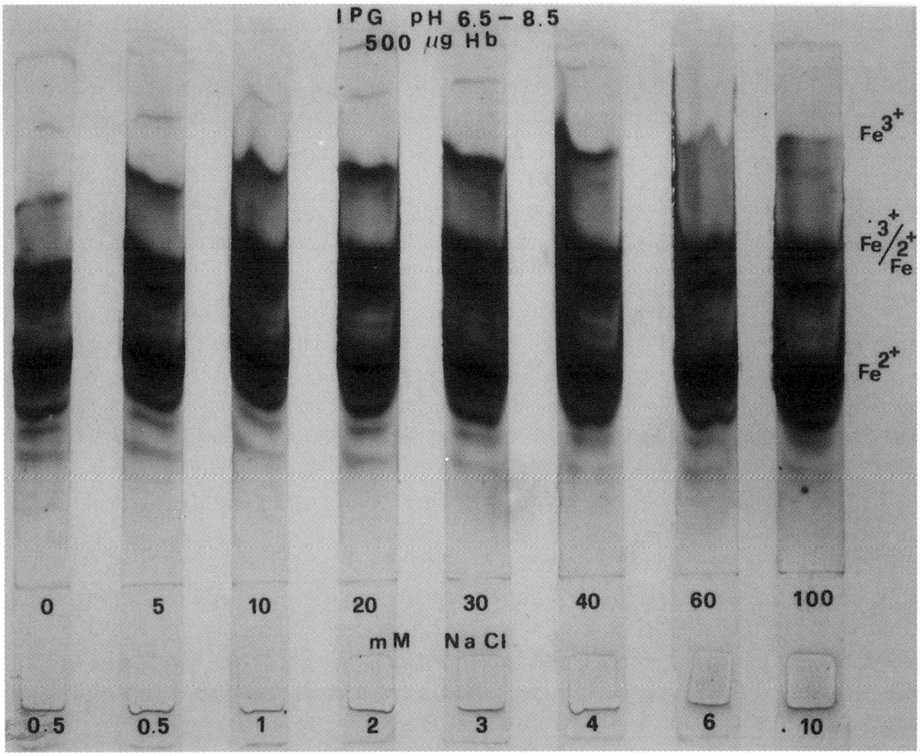

Fig. 14.8. Effects of the simultaneous addition of strong salts and of 'protective' carrier ampholytes on protein patterns in IPGs. Native haemoglobin (Hb) was applied to a pH 6.5–8.5 IPG gel and added with increasing amounts (up to 100 mM) of NaCl and proportionally increasing amounts of pH 6–8 carrier ampholytes (up to 10%). Reduced (Hb-Fe^{2+}), partially oxidised (Fe^{3+}/Fe^{2+}) and fully oxidised (Fe^{3+}) Hb species are marked on the left gel side. Note the pronounced reduction of oxidised species (compare with Fig. 14.4) and the absence of protein precipitation in the anodic application pockets. (From [121], with permission.)

(possibly no greater than 2–300 V) so as to prevent formation of strongly acidic and alkaline boundaries.

Now, like Moses, you can wander in the proximity of Mount Sinai with your 'pentalogue' under your arm and be a winner.

14.2.5. Solubilisation cocktail

We have seen in Section 14.1.1 that for decades the most popular lysis solution has been the O'Farrell cocktail (9 M urea, 2% Nonidet P-40, 2% β-mercaptoethanol and 2% carrier ampholytes, in any desired pH interval). Although much in vogue also in present times, over the years new, even more powerful, solubilising mixtures have been devised. We will discuss below the progress made in sample solubilisation, since this is perhaps the most important step for success in 2-D map analysis. Great efforts were dedicated to such developments, especially in view of the fact that many authors noted that hydrophobic proteins were largely absent from 2-D maps (see e.g. [122]). These authors noted that, quite strikingly, in three different species analysed (*Escherichia coli, Bacillus subtilis, Saccharomyces cerevisiae*), all proteins above a given hydrophobicity value were completely missing, independently from the mode of IEF (soluble CAs or IPG). This suggested that the initial sample solubilisation was the primary cause for loss of such hydrophobic proteins. The progress made in solubilising cocktails can be summarised as follows (see also [123,124]).

(1) Chaotropes. Although urea (up to 9.5 M) has been for decades the only chaotrope used in IEF, recently thiourea has been found to further improve solubilisation, especially of membrane proteins [71,125–128]. The inclusion of thiourea is recommended for use with IPGs, which are prone to adsorptive losses of hydrophobic and isoelectrically neutral proteins. Typically, thiourea is added at concentrations of 2 M in conjunction with 5–7 M urea. The high concentration of urea is essential for solvating thiourea, which is poorly water soluble (only ca. 1 M in plain water [129]). Among all substituted ureas (alkyl ureas, both symmetric and asymmetric) Rabilloud et al. [125] found thiourea to be still the best additive. However, it would appear that not much higher amounts of thiourea can be added to the IPG gel strip, since it seems that at >2 M concentrations, this chaotrope inhibits binding of SDS in the equilibration step between the 1st and 2nd dimension, thus leading to poor transfer of proteins into the 2-D gel. It should also be remembered that urea in water exists in equilibrium with ammonium cyanate, whose level increases with increasing pH and temperature [130]. Since cyanate can react with amino groups in proteins, such as the N-terminus α-amino or the ϵ-amino groups of Lys, these reactions should be avoided since they will produce a false sample heterogeneity and give wrong M_r values upon peptide analysis by MALDI-TOF MS. Thus, fresh solutions of pure grade urea should be used, in concomitance with low temperatures and with the use of scavengers of cyanate (such as the primary amines of carrier ampholytes or other suitable amines). In addition, protein mixtures solubilised in high levels of urea should be subjected to separation in the electric field as soon as possible: in presence of the high voltage gradients typical of the IEF protocol, cyanate ions are quickly removed and no carbamylation can possibly take place (B.R. Herbert and P.G. Righetti, unpublished) whereas it will if the protein/urea solution is left standing on the bench.

(2) Surfactants. These molecules are always included in solubilising cocktails to act synergistically with chaotropes. Surfactants are important in preventing hydrophobic interactions due to exposure of protein hydrophobic domains induced by chaotropes. Both the hydrophobic tails and the polar head groups of detergents play an important role in protein solubilisation [131]. The surfactant tail binds to hydrophobic residues, allowing dispersal of these domains in an aqueous medium, while the polar head groups of detergents can disrupt ionic and hydrogen bonds, aiding in dispersion. Detergents typically used in the past included Nonidet P-40 or Triton X-100, in concentrations ranging from 0.5 to 4%. More and more, zwitterionic surfactants, such as CHAPS, are replacing those neutral detergents [95,132], often in combination with small amounts (0.5%) of Triton X-100. In addition, small amounts of CAs (<1%) are added, since they appear to reduce protein-matrix hydrophobic interactions and overcome detrimental

effects caused by salt boundaries [133]. Linear sulphobetaines are now emerging as perhaps the most powerful surfactants, especially those with at least a 12-carbon tail (SB 3–12). They were in fact already demonstrated to be potent solubilisers of hydrophobic proteins (e.g. plasma membranes), except that they had the serious drawback of being precipitated out of solution due to low urea compatibility (only 4 M urea for SB 3–12) [134]. This drawback of scarce solubility of many non-ionic or zwitterionic surfactants with long, linear hydrophobic tails in urea solution seems to be a general problem [44,135] and thus has prompted the synthesis of more soluble variants [136]. The inclusion of an amido group along the hydrophobic tail greatly improved urea tolerance, up to 8.5 M, and ameliorated separation of some proteins of the erythrocyte membranes [137]. Recently, Chevallet et al. [27] embarked on a project for determining the structural requirements for the best possible sulphobetaine. Three major features were found to be fundamental: (a) an alkyl or aryl tail of 14–16 carbons; (b) a sulphobetaine head of 3 carbons; (c) a 3-carbon spacer between the quaternary ammonium and the amido group along the alkyl chain. The most promising surfactants were found to be ASB 14 (amidosulphobetaine) containing a 14-C linear alkyl tail and C8Ø, which possesses a *p*-phenyloctyl tail (see Fig. 14.9). Both of these reagents have since been used successfully in combination with urea and thiourea to solubilise integral membrane proteins of both *E. coli* [138] and *Arabidopsis thaliana* [139,140]. Recently, Malone and Kramer [141] have re-evaluated the use of different detergents and proposed the following choices: (a) when working with mixtures of hydrophobic and hydrophilic proteins, like those seen in whole cell samples, combinations of non-ionic/zwitterionic surfactants, with hydrophobicity/hydrophilicity indexes at the low and high end of the scale, offer best solubilisation results; (b) when analysing water-soluble samples, such as serum or aqueous extracts, best performances are obtained with a single zwitterionic detergent. In all cases, they claim that, for most samples, best results are obtained when less than 2% total detergent is used for solubilisation, although this seems to apply not to the 1st, but to the 2nd dimension, since such low levels of surfactants would allow better SDS equilibration of the IPG strip, with shorter incubation times.

(3) Reducing agents. Thiol agents are typically used to break intramolecular and intermolecular disulphide bridges. Cyclic reducing agents, such as dithiothreitol (DTT) or dithioerythritol (DTE) are the most common reagents admixed to solubilising cocktails. These chemicals are used in large excess (e.g. 20 to 40 mM) so as to shift the equilibrium toward oxidation of the reducing agent with concomitant reduction of the protein disulphides. Because this is an equilibrium reaction, loss of the reducing agent through migration of proteins away from the sample application zone can permit reoxidation of free Cys to disulphides in proteins, which would result not only in horizontal streaking, but also, possibly, in formation of spurious extra bands due to scrambled –S–S– bridges and their cross-linking different polypeptide chains. Even if sample is directly reswollen in the dried IPG strip, as customary today, the excess DTT or DTE will not remain in the gel at a constant concentration, since, due to their weakly acidic character, both compounds will migrate above pH 7 and be depleted from the alkaline gel region. Thus, this will aggravate focussing of alkaline proteins and be one of the multifactorial factors responsible for poor focussing in the alkaline pH scale. This problem is not trivial at all, and deserves further comments. For example, Righetti

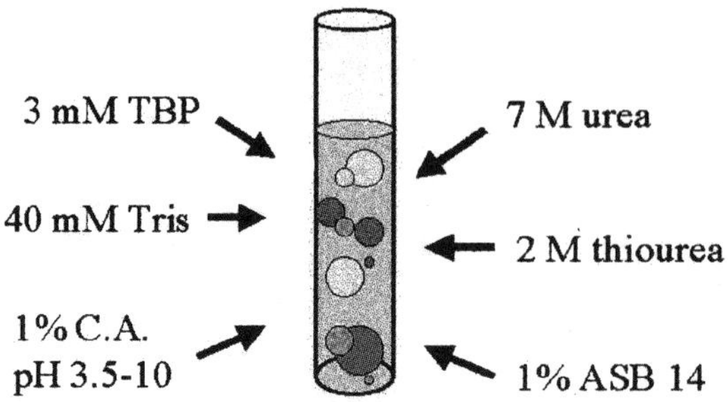

Fig. 14.10. Composition of the cocktail today preferred for best sample solubilisation in preparation of the 1st dimension run. In some cases, the level of urea is reduced from 7 to 5 M. The wide-range carrier ampholyte solution is added to facilitate sample entry in the IPG strip.

and pre-treatment in 2-D map analysis. Reducing and alkylating the resulting free –SH groups is, of course, a well-known procedure and highly recommended in any biochemistry textbook dealing with protein analysis. Alkylation will prevent all the noxious phenomena reported above, like reformation of disulphide bridges, producing smears and even spurious bands due to inter-chain cross-linking, even among unrelated polypeptide chains. However, as strange as it sounds, although reduction and alkylation is performed by anyone working with 2-D maps, this step is not done at the very beginning of the sample treatment, just prior to the 1st dimension IEF/IPG run, but in between the 1st and 2nd dimension, during the interfacing of the IEF/IPG strip with SDS-denaturing solution, in preparation for the SDS-PAGE final step. This probably stems from earlier reports by Goerg et al. [145,146], who recommended alkylation of proteins with iodoacetamide during the interfacing between the 1st and 2nd dimension, on the grounds that this treatment would prevent point streaking and other silver-staining artefacts. As luck goes, this recommendation was followed without any questioning by everyone working with 2-D maps. Clearly, at the light of the above discussion, it appears to be a much smarter move to reduce and alkylate the sample just to start with, i.e. prior to even the 1st IEF/IPG dimension. The drawbacks for alkylation with iodoacetamide in between the two-dimension runs have also been highlighted, recently, by Yan et al. [147], although on different grounds than as discussed above. These authors noted that, in most 2-D protocols, the discontinuous Laemmli buffer [33] is used, which calls for a stacking and sample gels to be equilibrated in a pH 6.8 buffer. In order to prevent pH alterations, most people use a modified stacking buffer with a reducing agent (DTT or DTE) and alkylating agent (iodoacetamide) at pH 6.8, so that the strips can be loaded as such after these two treatments, avoiding any further pH manipulations [148]. However, at this low pH both reduction and alkylation are not so efficient, since the optimal pH for these reactions is usually at pH 8.5–8.9 [149]. As a result of this poor protocol, Yan et al. [147] have reported additional alkylation by free acrylamide during the SDS-PAGE run, with the same protein exhibiting a major peak of Cys- carboxyamidomethyl and a minor one of Cys-propionamide. It should be borne in mind, in fact, that, whereas the risk of acrylamide adduct formation is much reduced in IPG gels (but not in unwashed IEF gels!), it is quite real in SDS-PAGE gels, due to the fact that these gels are not washed and that surfactants, in general, hamper incorporation of monomers into the growing polymer chain ([150]; see also Section 13.5.1). Different alkylating residues on a protein will complicate their recognition by MALDI-TOF analysis, a tool much in use today in proteomics. Here too, however, although Yan et al. [147] clearly identified the problem, they did not give the solution we are proposing here, namely to reduce and alkylate the sample prior to any electrophoretic step. They suggest to retain still the original protocol of alkylation in between the two dimensions, but to increase the pH of the equilibration to pH 8.0, add a much large amount of alkylating agent (125 mM iodoacetamide) and increase the time of incubation to 15 min. We again stress that, although this new protocol is an improvement over previous ones, it still does not cure the problems of the 1st IEF/IPG dimension, namely smears and formation of spurious bands, both due to reoxidation of reduced but non-alkylated Cys residues.

(5) Alkylation with iodoacetamide in presence of thiourea: a conflicting situation. We have seen above that reduction and alkylation prior to any electrophoretic step is a must, if one wants to avoid a number of artefacts, typically encountered in the alkaline pH region. For example, recently, Olivieri et al. [151] have demonstrated, by MALDI-TOF MS, that failure to alkylate prior to the IEF/IPG step would result in a large number of spurious spots in the alkaline pH region, due to 'scrambled' disulphide bridges among like and unlike polypeptide chains. This series of artefactual spots comprises not only dimers, but an impressive series of oligomers (up to nonamers) in the case of simple polypeptides such as the human α- and β-globin chains, which possess only one (α-) or two (β-) –SH groups. As a result, misplaced spots are to be found in the resulting 2-D map, if performed with the wrong protocol. Subsequently, the same group [152], via the same MALDI-TOF MS procedure, have monitored the kinetics of protein alkylation by iodoacetamide over the period 0–24 h at pH 9. Alkylation reached ca. 70% in the first 2 min, yet the remaining 30% required up to 6 h for further reaction (which, however, never quite reached 100%). The use of SDS during the alkylation step resulted in a strong quenching of this reaction (thus further corroborating the notion that alkylation in between the 1st and 2nd dimension is a useless procedure), whereas 2% CHAPS exerted a much reduced inhibition. It was during these investigations that the same authors found a disturbing phenomenon: during alkylation of α- and β-globin chains, substantially different results were obtained when the sample was dissolved in plain 8 M urea or in the mixture 5 M urea and 2 M thiourea. In the former case, almost complete disappearance of all homo- and hetero-oligomers was achieved, whereas in the latter case appreciable amounts of dimers and trimers were still left over, even upon prolonged incubation. A thorough investigation by ESI-MS demonstrated that indeed thiourea was competing with the free –SH groups of proteins for reaction and was scavenging iodoacetamide at an incredible rate (in ca. 5 min of incubation all iodoacetamide present in solution had fully disappeared). Iodoacetamide was found to add to the sulphur atom of thiourea; the reaction was driven in this direction also by the fact that, once this adduct was formed, thiourea was deamidated and the reaction product generated a cyclic compound (a thiazolinone derivative). Thus, those using urea/thiourea solutions should be aware of this side reaction and of the potential risk of not achieving full alkylation of the free, reduced –SH groups in Cys residues. Fortunately, alkylation of proteins is still substantial if iodoacetamide is added as a powder to the protein solution

treated with the reducing agent, instead of as a solution. In addition, due to the fact that all iodoacetamide is scavenged by thiourea, the final concentration of thiourea in the solubilising solution will be reduced by the same molar amounts of iodoacetamide added.

As a conclusion, reduction and alkylation at the very start of any 2-D fractionation would have the following immediate advantages [151,152]: (1) reduce considerably smears in the alkaline region, above pH 8; (2) prevent formation of spurious bands due to mixed disulphide bridges; (3) abolish formation of a mixed population of Cys-propionamide and Cys-carboxyamidomethyl species, due to alkylation with acrylamide and iodoacetamide, respectively.

There remain the aspects of how to perform it. If reduction is done with DTT or DTE, it must be a two-step operation. The sample, brought to pH 8.5–8.9 (with a weak base, like free Tris and/or basic CAs!) is first incubated for ca. 1 h to allow for full reduction of –S–S– bridges. After that, a 2-fold molar excess of alkylating agent (iodoacetamide) is added and the sample left to incubate for an additional hour (or more) to allow for full alkylation (of both, free Cys groups in proteins as well as free thiols in the DTT or DTE additives). Now, due to the fact that DTT or DTE are typically used at the ca. 40 mM level, this means that 80 to 100 mM iodoacetamide will have to be added. This per se does not pose any risk in the subsequent IEF/IPG fractionation, since this molecule is neutral, so large amounts can be tolerated in the focussing step. However, upon prolonged focussing (1–2 days, as often applied in IPG runs) there could be the risk of partial hydrolysis of the amido bond in iodoacetamide, this provoking undue currents and destabilising the focussing process. Here is where the use of TBP might be preferred. Since the levels used of TBP, as a reducing agent, are minute (2 mM), this means that the levels of alkylating agent (iodoacetamide) to be added will also have to be small, typically of the order of 5 to 10 mM, i.e. at least one order of magnitude less than in the case of DTT or DTE reduction. But there is more to it. Due to the fact that TBP does not react with some alkylating agents, such as acrylamide and 4-vinylpyridine [153], it could offer the opportunity of a simplified, one-step reduction and alkylation procedure using TBP and acrylamide (as alkylating agent) simultaneously, as proposed by Molloy [123]. However, it remains to be seen if alkylation with acrylamide is as efficient as with iodoacetamide and, if not, for how long the incubation should be protracted in order to ensure full alkylation of all species present.

14.2.6. Sample application

This ritual is just as important, for a successful 2-D fractionation, as all the steps described above. There are different modes of sample application to the 1st dimension IEF/IPG strip.

(1) Pulse loading. This is the method typically adopted in all fractionation techniques, including chromatography, since due to entropic forces tending to dissipate the sample zone, the latter has in general to be applied as a very thin zone. This, of course, does not hold true for any focussing procedure, which reaches steady-state conditions. Nevertheless, it has been customary, for years, to apply the sample (especially in the

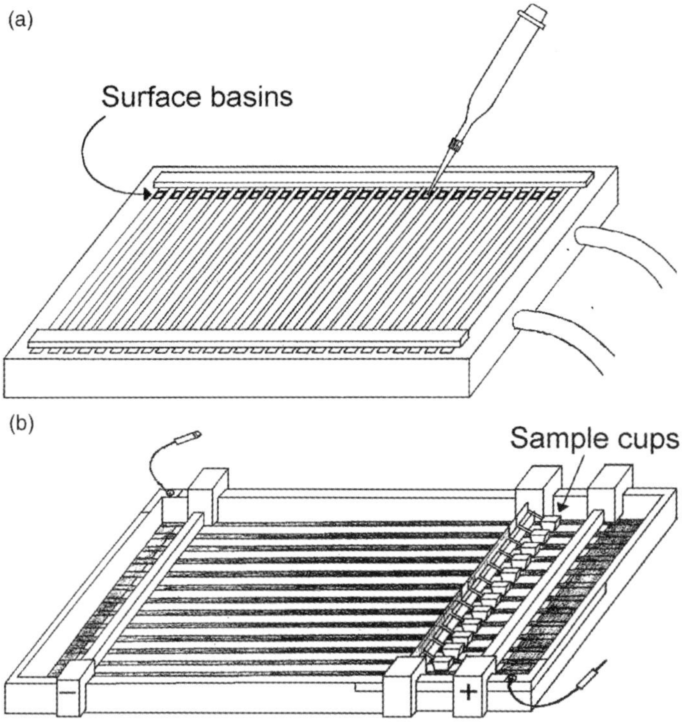

Fig. 14.11. Scheme of different sample application procedures to rehydrated IPG strips. (A) Surface basins (the long strips at the two gel extremities are the electrodic reservoirs, i.e. filter paper strips soaked in anolyte and catholyte, respectively). (B) Battery of cups supported on a holding bar for sample application to rehydrated IPG gel strips.

old procedure of CA-IEF) in pockets precast in the gel, or in surface basins or adsorbed onto granulated materials (e.g. Sephadex, Paratex films and the like; see, as an example of surface basins, Fig. 14.11A). Sample cups did seem to offer a valid alternative to loading, since they permitted higher volume loadings (up to 100 μl) than with any other conventional means. To this purpose, Amersham Pharmacia Biotech devised a bar holder, able to position a battery of cups on a linear array of pre-hydrated IPG strips. This holder would gently press the cups against the gel surface, so as to prevent sample leakage; each cup would then be filled with the desired volume of any sample (see Fig. 14.11B). This method too is being used less and less, since it was found that significant amounts of proteins were lost at the application point, due to protein aggregation and precipitation as they migrated from the free liquid phase into the gel, with the ultimate result being poor resolution and 'underloading' [154]. This effect has been explained as a concentration phenomenon occurring at the gel–sample interface, where proteins massively accumulate as they are dramatically slowed down in their transit from a liquid to a gel phase. At this interface also anomalous ionic boundaries will form and the two concomitant phenomena would favour protein aggregation and precipitation [121,155].

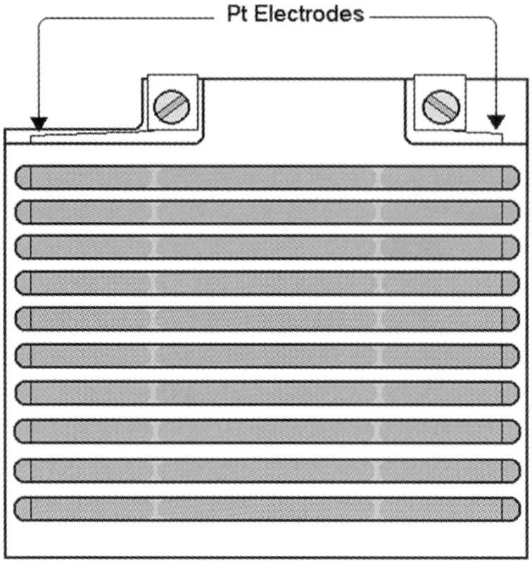

Fig. 14.12. Pictorial representation of the IPG gel strips rehydration tray, for sample passive or active diffusion into the immobilised pH gradient gel. Since the tray is provided with electrodic contacts at the two ends, active sample loading can be implemented, by applying a gentle voltage gradient during rehydration (courtesy of Bio-Rad and Proteome Systems Ltd.).

(2) In gel rehydration loading. This method is today by far preferred [127,155–157] since it facilitates higher protein loads and reduces focussing times. With this technique, the dehydrated IPG strip is directly reswollen with the protein sample dissolved in the rehydration solution (notice that, in this last case, the same solution is used to solubilise the protein and rehydrate the IPG strip) (see Fig. 14.12). After a suitable rehydration time (approximately 5–6 h), the IPG strip is ready for the 1st dimension IEF, with the proteins already uniformly distributed within the gel matrix. The clear advantage of in-gel rehydration is the large volume of sample that can be applied (up to 450 μl for a 18 cm long, 4 mm wide, 0.5 mm thick IPG strip) compared to 100 μl for conventional cup loading. The other major advantage is minimisation of sample aggregation and precipitation: since the sample is diluted through the entire gel strip, no local, accidental build-ups of proteins can occur (except of course, in the focussing zone, but here, even if large sample amounts are collected and precipitate locally, it would not be a problem since separation has already occurred; it must be borne in mind that the real danger of protein aggregation and precipitation, as typical of pulse loading, is the fact that this event occurs among unlike proteins, so that ultimately highly heterogeneous proteins collect into a single precipitin zone, a process lethal to any separation operation). The third advantage of this protocol is that, due to dilution of proteins throughout the gel, much accelerated focussing protocols can be adopted, utilising higher-voltage gradients at the very start of the IEF/IPG run. Thus, with in-gel sample rehydration, it is rare for IEF to exceed a total of 30,000 V × h, even with milligram protein loads [123]. Table 14.1 gives the protocol adopted by Molloy [123] in the case of in-gel hydration loading procedures. The in-gel rehydration, as compared

TABLE 14.1

A TYPICAL IEF REGIME USING THE MULTIPHOR II (AMERSHAM-PHARMACIA) WITH IN-GEL REHYDRATION LOADING FOR SAMPLES CONTAINING MINIMAL SALT (<50 mM) (FROM [123], WITH PERMISSION)

Phase No.	Voltage (V)	Maximum current [a] (mA)	Time (h)
1	150	5	0.1
2	300	3	0.1 [b]
3	1000	3	0.5 [b]
4	2500	1	0.5–1
5	5000	1	3–5

[a] If the maximum current exceeds these limits there is a danger of the IPG strip burning. As a precaution, the voltage should be decreased and the IEF time extended at the lower voltage.
[b] The time may be lengthened depending on the protein load and salt contamination.

to sample cup loading, has been recently quantitatively evaluated by Zuo and Speicher [158], who have reported the following findings. At high sample loads, cup loading can result in as much as 50% sample loss, whereas in the case of in-gel rehydration, even when applying up to 0.5 mg protein, rarely sample losses of >15–20% are observed. Instrumental to good protein recoveries seems to be also the addition of 2% carrier ampholytes in the sample buffer, whereas use of thiourea did not significantly affected protein recoveries, although it did improve resolution in the final 2-D gel. Interestingly, the same authors have demonstrated that another 10% sample loss is experienced during the two-step equilibration procedure prior to the 2nd dimension SDS-PAGE, as recommended by Goerg et al. [146]. Also Kramer and Malone [159] underline the importance of adding carrier ampholytes (CA) to the IPG strip for improved protein entrance and recoveries. They suggest that best results are obtained with blends of CAs from different manufacturers.

(3) In-gel hydration loading under a low voltage. This technique, which we call active sample loading, is distinguished from the previous one (called passive), in that, during IPG strip rehydration, a gentle voltage gradient (typically 50 V) is applied, usually for the entire duration of the reswelling (overnight). It is believed that this procedure would further facilitate sample entry. Olivieri et al. [92] have recently found an interesting aspect of such active sample loading: its ability to facilitate entry of high M_r proteins. If one looks at Fig. 14.2 one can notice the lack of spectrins, large (280 kDa for the α-chains), filamentous proteins, which represent the major constituents of the cytoskeletal network underlying the RBC plasma membrane and are typically associated with band 4.1 and actin to form the cytoskeletal superstructure of the RBC plasma membrane. When Heegaard and Poglod [93] reported 2-D maps of RBCs, spectrins were highly visible as a string of bands penetrating for a short distance into the top (the most porous) part of the SDS-PAGE slab. Curiously, not only these gels are devoid of them, but also, upon a close inspection of the 2-D maps of Rabilloud et al. [94], they seem to be missing there as well and there are no comments from these authors about this, although they characterise and recognise a number of major

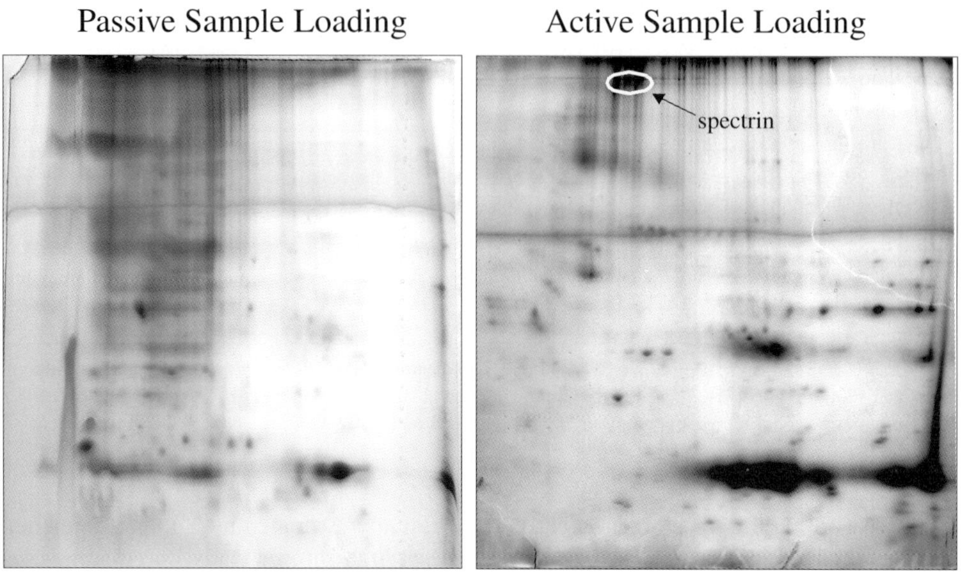

Fig. 14.13. Active vs. passive sample loading. Passive sample loading (left panel): IPG strips, pH range 3–10 linear, 18 cm long, rehydrated with 500 μl protein sample and 10 mM iodoacetamide overnight. IEF run: 80-kV × h. Second dimension gel: 8–18% *T*. Active sample loading (right panel): IPG strips, pH range 3–10 linear, 18 cm long, rehydrated with 500 μl protein sample and 10 mM iodoacetamide on a Protean IEF cell at 50 V overnight. IEF run: 80-kV × h. Second dimension gel: 8–18% *T*. (From [92], with permission.)

spots by MALDI-TOF. However, the reports of Olivieri et al. [92] and that of Rabilloud et al. [94] are the only ones in which RBCs are analysed by IPG-DALT, i.e. by adopting immobilised pH gradients in the first dimension. All the previous published work dealt only with CA-IEF-DALT. In the late eighties, Rabilloud et al. [160] reported the massive loss of all large proteins in Immobiline gels and found out that even the free monomers contained in the bottles would precipitate large macromolecules out of solution. Later on, the problem was found to be the auto-polymerisation of basic Immobilines, fired by the radioactive cloud of Chernobyl, to oligomers and *n*-mers, able to cross-link and aggregate all large proteins (a ferritin precipitation test had been devised for detecting the presence of such homo-oligomers [161]). The problem was finally solved by the introduction of the Immobiline II generation, in which *n*-propanol (an inhibitor of radical polymerisation) was used as a solvent for alkaline Immobilines [162]. Even though today IPGs should be trouble-free, still it is known that quite often large, filamentous proteins have difficulties in penetrating and migrating through the Immobiline strip, for reasons not yet completely understood. Olivieri et al. [92] have therefore tried alternative methods of sample entry: Fig. 14.13 compares passive (left panel) vs. active (right side) sample entry in the IPG strip. In the latter technique, as the total volume of sample (in this case 450 μl) is allowed to reswell into the dry IPG strip, a gentle voltage gradient (barely 50 V over the entire gel length) is applied, the assumption being that the simultaneous passive adsorption and active electrophoretic migration would prevent local sample overloads, which could lead to aggregation and

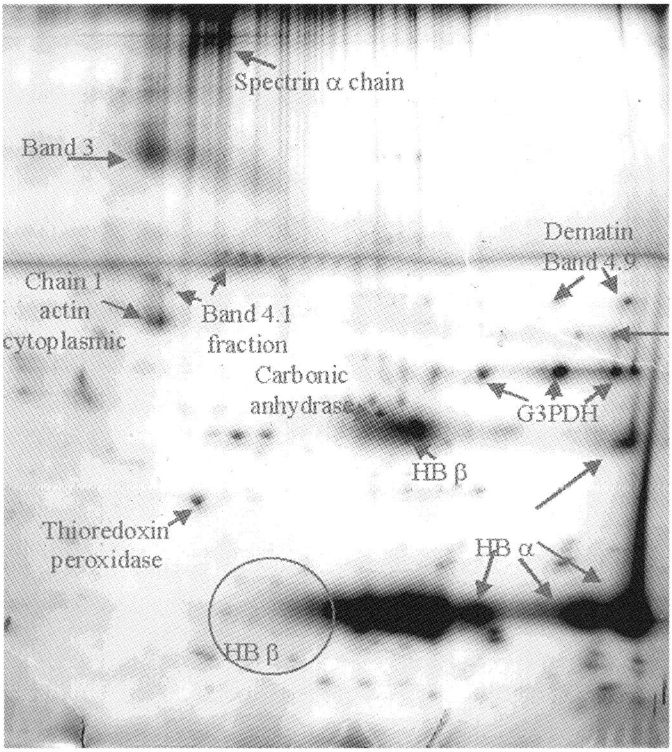

Fig. 14.14. Identification of some RBC membrane proteins performed by using the ExPASy molecular biology server and PeptIdent proteomics tool. SWISS-PROT and TrEMBL databases were searched for human proteins by mass tolerance of 0.1 Da and allowing 0 or 1 missed cleavage. The first dimension run was done by the active sample entry protocol. (From [92], with permission.)

precipitation. It is in fact seen that at least 3 major spectrin bands have entered the gel (right side) whereas nothing is visible in the case of passive sample entry. In order to prove this point, these authors have eluted all major spots and identified them by MALDI-TOF MS of the digested peptides: Fig. 14.14 shows not only the correct identification of spectrin α-chains, but also of a number of major spots belonging to the RBS membrane.

(4) Soft and thicker gels for high-capacity loading. With regular IPG strips (0.5 mm thick, 0.3 mm wide, different lengths from 7 up to 18 cm) it is difficult to load more than 1–2 mg total protein. Although this amount, for an analytical gel, is considered to be a high load, it is often not enough if one aims at detecting and eluting, for further chemical characterisation, a number of minor spots. In this last case, considerably higher sample loads are required, often greater by at least one order of magnitude. It appears to be possible to increase the sample load to 10 mg (and even higher) total protein per IPG strip, provided the following modifications are followed (B.R. Herbert and P.G. Righetti, unpublished). First of all, the polyacrylamide matrix can be made 'softer', i.e. it can be cast more diluted. Typically, all commercial, ready-made IPG strips are made

to contain 4% *T* acrylamide (with various levels of cross-linker, usually 2–5%*C*) and are only 0.5 mm thick. Diluting the gel matrix greatly enhances the gel load ability, as demonstrated long ago by Gelfi and Righetti [163,164]. For example, when decreasing the %*T* from 6% to 3%, the protein load per unit of gel volume can be more than doubled; below 3% *T*, the sample loadability increases exponentially. It was in fact on the basis of these observations that preparative IPGs in gel matrices were abandoned and evolved into multicompartment electrolysers with Immobiline membranes, where preparative separations of proteins occur in a liquid vein and the gel phase is reduced to an isoelectric membrane functioning as a titrating unit [165,166]. We prefer to cast 3.5% *T* (and even lower) IPG gels, which ensures considerably higher protein loads and faster focussing times, due to the higher gel porosity. As a second modification, one can increase the gel thickness from 0.5 to 1 mm; this higher thickness allows too higher sample loads and it is still compatible with the 2nd dimension SDS-PAGE, which is routinely run in 1 to 1.5 mm thick gels. As a third modification, the width of the IPG strip can be increased from 3 mm (as available from commercial sources) to 6 mm, this too allowing doubling of the sample load. Note that this width is also compatible with proper stacking when using discontinuous SDS-PAGE, like the Laemmli buffer [33]. With these three simple modifications it is common to be able to load 10 mg total proteins and even higher levels.

Some final comments are due in this section about two other aspects: one is IPG strips pre-treatment before the run and the other is on a comparative study on the different instruments used for the 1st dimension run.

(1) IPG strips pre-washing. A number of reports have lamented, in the past, horizontal and point streaking of proteins due to the 1st dimension run. In a recent report, Chan et al. [167] have claimed complete elimination of horizontal streaking of basic proteins if the commercial, dried IPG strips were washed prior to use, in 100 mM ascorbic acid, pH 4.5, for 24 h before use and if the subsequent IEF run were performed at 40°C, instead of 15°C, as customary. Interestingly, this procedure of reduction in ascorbic acid had been proposed in 1989 by Righetti et al. [168], who observed that, during gel polymerisation in presence of persulphate, the basic Immobilines would be oxidised to different extents, forming N-oxides. These oxides, in turn, would oxidise proteins migrating through the relevant gel zones (i.e. in the alkaline milieu). These authors had recommended reduction for only 45 min, whereas Chan et al. [167] now propose as much as 24 h washing. Perhaps this elimination of horizontal streaking might be due to prevention of oxidation of reduced (but not blocked) –SH groups in proteins by the N-oxides in the Immobilines. Since re-formation of disulphide bridges is a time-dependent phenomenon, its avoidance might indeed reduce or abolish horizontal streaking. If this is the case, probably the novel procedure we have recommended (reduction and alkylation directly during the sample solubilisation step, prior to any electrophoretic run) should take care of that, thus avoiding such a terribly lengthy washing procedure.

(2) Use of different IEF chambers for the IPG run. The performance of the three most popular chambers for the 1st dimension IEF/IPG run (Multiphor II and IPGphor from Amersham-Pharmacia and the Protean IEF cell from Bio-Rad) has been recently evaluated by Choe and Lee [169], using 18 cm long IPG strips run under various conditions. The Multiphor II consistently resulted in the highest number of spots

detected per gel, independent of sample type, width of the IPG interval and method for calculating the number of spots. The Protean IEF cell had the next highest number of spots detected per gel. The IPGphor afforded good reproducibility in the total number of computer-detected spots, whereas the Protean IEF cell offered better reproducibility in the total number of manually detected spots from gel to gel. As a final conclusion, Choe and Lee [169], as a measure of quantitative reproducibility, suggested that the Protean IEF cell, which was the easiest instrument to use, performed better than the other instruments, although all three of them demonstrated good reproducibility in the experiments performed.

14.2.7. Sequential sample extraction

When a sample is highly complex, or when desiring to extract under widely different conditions, it might pay to apply a protocol of sequential extraction, as described by Molloy et al. [71] in the case of total extracts of *E. coli*. This sequence is composed of three steps, as follows.

(1) The cells are first ruptured in an Aminco French press with two presses at 14,000 psi. After removal of unbroken cells, the total cell lysate is diluted with ice-cold sodium carbonate, pH 11, and stirred slowly on ice for 1 h. This step brings into solution only the most soluble proteins.

(2) The precipitated, insoluble material after the carbonate treatment is recovered by centrifugation. The pellet is then resuspended in the classical reagent for 2-D maps (8 M urea, 4% CHAPS, 100 mM DTT and 0.5% alkaline carrier ampholytes). This step solubilises more hydrophobic proteins.

(3) After centrifugation, the remaining, insoluble pellet is further treated with the enhanced solubilising solution, incorporating 5 M urea, 2 M thiourea, 2% ASB14 and 2 mM TBP.

When the three resulting 2-D maps were compared, it was seen that, except from some spots appearing in more than one gel, the protein patterns were quite specific; in addition, the final gel displayed several abundant protein spots that were missing from the two previous ones. Subsequent identification confirmed these as integral outer-membrane proteins, which had not been detected previously in any 2-D maps reported for *E. coli*, owing to their greater hydrophobicity and thus to their poor solubility with classical sample solubilisation cocktails. This was despite the high copy number of some of the identified proteins (e.g. the *E. coli* integral outer membrane protein, OmpA, was present at 10^5 copies per cell) [170,171]. From this point of view, it is of interest to have an estimate of what copy number is needed before a protein can be detected on a gel. This problem has been addressed by Herbert et al. [172] who have given the following guidelines, as also visualised in Fig. 14.15, which represents the relationship between the quantity of protein loaded onto a 2-D gel and the final concentration of a 20 kDa protein present at 10, 10^3 and 10^5 copies per cell. On a double log plot, this three protein levels are indicated by three diagonal lines correlating the two ordinates, both representing the on-gel concentration of said protein, the one to the left expressed in grams, the one to the right expressed in moles. Assuming that

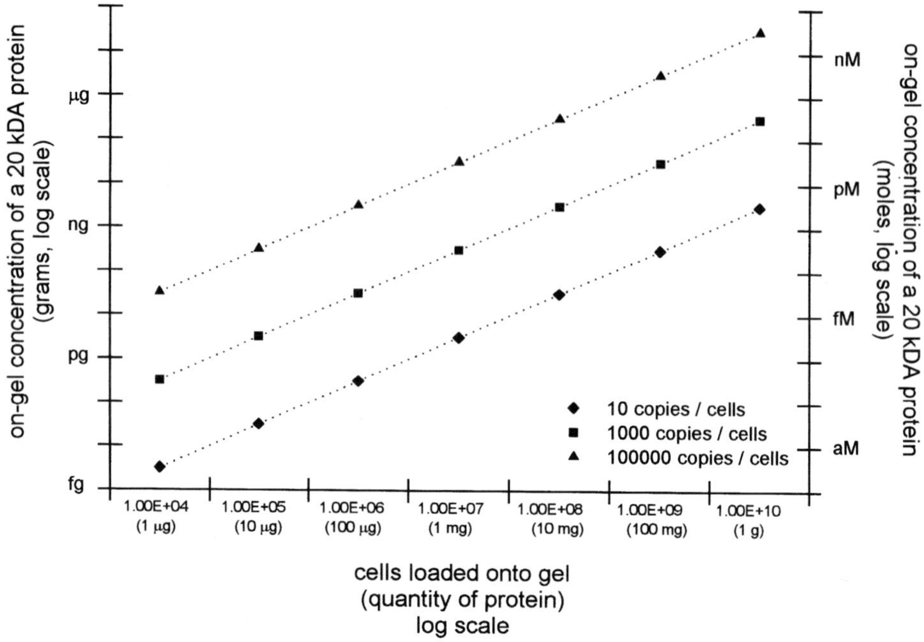

Fig. 14.15. Final concentration of a 20 kDa protein after 2-D gel electrophoresis and how it varies with the quantity of cells or proteins loaded onto the gel, and the copy number of the protein per cell. This is based on 10^9 cells being 1 g fresh weight, 200 mg dry weight and containing 100 mg protein. (From [172], with permission.)

no prefractionation is performed, with a loading capacity of 10 mg per gel, only the proteins present at more than 10^5 copies per cell (17 pmol) can be easily identified using analytical procedures. With a highly sensitive silver stain, proteins present at $>10^3$ copies per cell (0.17 pmol) will be detectable, but difficult to analyse with analytical techniques. Immunoblotting, using a combination of high-affinity monoclonal antibodies and enhanced chemiluminescence, will allow detection of proteins present at >10 copies per cell (1.7 fmol). Therefore, the proportion of total protein complement that can be seen on a gel for any tissue will depend on the copy number, on the quantity of protein loaded on the gel and on the method of detection. For example, on a typical $160 \times 180 \times 1.5$ mm 2-D PAGE gel, not more than 20% of the expressed genes of mammalian cells (assuming 5000 proteins corresponding probably to 10–15,000 polypeptides or isoforms) can be detected, representing only the tip of the proteome 'iceberg' in this case [173]. Since many of the low copy number proteins are likely to have important regulatory functions in cells, it is clear that many of them will go undetected and that the major challenge of today proteomics will be to find ways to make such rare proteins visible. At the end of this chapter, we will present some novel approaches to this vexing problem.

14.3. MASS SPECTROMETRY IN PROTEOMICS

As stated in the introduction, perhaps 30% of the field of proteomics, as practised today, is the proper use and knowledge of mass spectrometry (MS), especially in the variant MALDI-TOF (matrix assisted laser induced desorption ionisation, time of flight), for a proper assessment of the precise M_r value of polypeptide chains and their fragments. We will thus review here the different approaches and the basic concepts of MS, together with the advantages and limitations of the various techniques applied so far. For a deeper insight on these topics, the reader is referred also to a number of recent reviews [52,174–178]. An entire section on mass spectrometry, as hyphenated to separation techniques, can also be found in Vol. 270 of *Methods in Enzymology*, edited by Karger and Hancock [179], with chapters by Seifert and Caprioli [180], Banks and Whitehouse [181], Beavis and Chait [182] and Schwartz and Jardine [183]. Several chapters on MS analysis can also be found in the book of Kellner et al. [56,184–188].

Over the past 10 years, MS has become the technique of election for protein characterisation. There are a number of reasons for this. The first is the development of new methods for protein and peptide ionisation, especially MALDI-TOF and ESI (electrospray ionisation). The second reason is the rapid growth of databases: the human genome sequence is now available (ca. 3 billion bases sequenced) and software exists to search proteins, expressed sequence tags (EST) and genome databases with MS data. The third is the fact that MS provides detailed information on post-translational modifications, such as phosphorylation and glycosylation. It must be emphasised that, in general, the mass of a protein is insufficient for identifying a protein with confidence, so that they are invariably converted to shorter peptides, usually by digestion with trypsin, which cleaves specifically at the Lys and Arg residues. The proteins in databases such as Swiss-Prot are also cleaved in the same manner and databases with tryptic fragments provided. The same can be done with EST databases (translating in all six reading frames) and, by extension with genome databases, peptide sequence tags can be searched directly at the DNA level. Such virtual peptide and DNA databases are at the core of MS searches. Although the current procedure is to try to identify proteins eluted from 2-D maps via assessment of the mass of the resulting peptides, this should not be taken to mean that measurement of the precise mass of an undigested protein should be abandoned. Such value could be vital in helping searching databases when the mass of only a few peptides is available: the knowledge of a couple of fundamental physico-chemical parameters, namely the pI (as assessed with precision via immobilised pH gradients) and of the mass (as derived by MS protocols) is very valuable in narrowing down the search to proteins belonging to a specific pI/M_r region. Thus, we encourage scientists to assess, when possible and when enough material is available, both the native M_r and the M_r of the resulting peptides after digestion.

In general, for protein identification from a 2-D map, the visualised spot should be excised, digested and the derived peptides analysed by MS techniques. This can be done manually for a few spots (Fig. 14.16), but it is not feasible when large-scale studies have to be conducted, involving hundreds of gels per week. Thus, several groups have developed robots for automating this task. The companies involved (with their URL addresses) are: Bio-Rad (http//www.proteomeworks.bio-rad.com/); Genomic Solu-

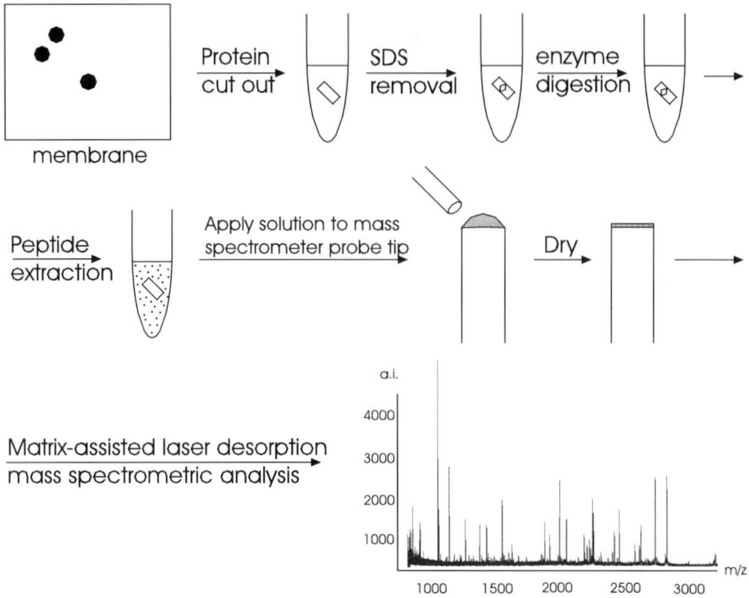

Fig. 14.16. Strategy employed for the MS mapping of peptides generated from proteins electroblotted onto Immobilon CD membranes from 2-D electrophoretic gel slabs. (From [189], with permission.)

tions (http//www.genomicsolutions.com/); Protana (http//www.protana.com/); Proteome Systems (http//www.proteomesystems.com/).

Most of them use the standard 96-well format of microtiter plates and usually rely in cutting the spot of interest from a blot and depositing the cut-out membrane into a well to which the relevant chemistry is added (buffers and proper amount of trypsin) for digestion. However, the new tendency (as developed by Proteome Systems) consists in leaving intact the membrane onto which the spots resulting from the 2-D map have been transferred from the polyacrylamide gel of the 2-D PAGE step. After visualisation of the spots of interest with a suitable stain and acquisition of the image in an electronic format, the blotting membrane is left intact and the spots to be submitted to MS analysis sprayed in situ with the relevant chemicals (buffers, proteases and the reagents needed for the following MS analysis, such as sinapinic acid). This is obtained by utilising miniaturised robotic systems, able to deposit minute droplets of liquids (as little as 20 pl) with the same technology developed for ink-jet printers. After digestion, the entire membrane is then inserted in the MS machine and all the relevant spots analysed. This minimises errors due to mislabelling or misplacing of cut-out membrane areas. In addition, it permits storage of the intact membrane after MS analysis. Since, during MS probing, only a small portion of the protein spots is consumed, this would allow reprobing of the same spots in future analysis, if re-evaluation of the data originally obtained were requested. Binz et al. [190], however, have developed a different approach, which has been termed ‘molecular scanner’. In this system, polypeptide chains from an unstained 2-D gel are electroblotted through a porous membrane, containing immobilised trypsin, whereby they are cleaved to the corresponding peptides. The peptides thus generated

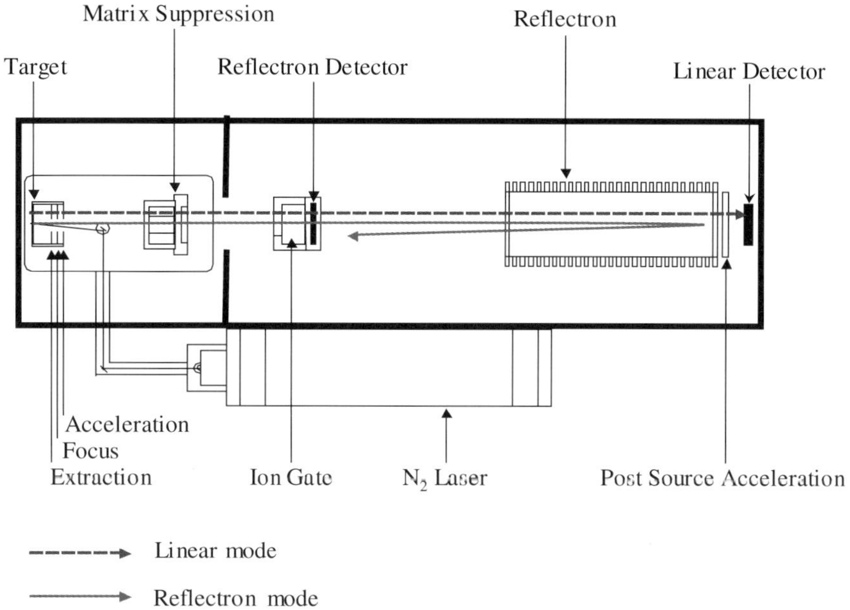

Fig. 14.17. Schematic diagram of a MALDI instrument. Microlitre quantities of liquid samples are mixed with a matrix, such as sinapinic acid, and dried onto a stainless steel or gold-plated target. A pulsing laser, e.g. a N_2 laser of 337 nm wavelength, is used to irradiate the matrix-embedded sample. This creates molecular ions that are accelerated by an electric field existing between the sample target and the grid. Ions then enter a field-free flight tube with a velocity essentially proportional to their mass, and their time of flight in this tube is measured at the linear detector. For increased resolution, modern MALDI instruments have an ion mirror (called reflectron) for reflecting the ions back towards a detector near the ion source.

are trapped onto a second membrane, which is finally probed in a MALDI-TOF mass spectrometer, by stepping the laser spot at 400 μm intervals and collecting the desired spectra. The membrane thus analysed is a facsimile, in peptide form, of the original polypeptide map in the 2-D gel. Software is then used to search databases in the usual way and also to reconstruct a gel image. However, the method is in its infancy and needs to be validated: it remains to be seen, e.g. if the time allowed for the blotting is sufficient to permit proper tryptic cleavage of the relevant proteins and also if blotting is quantitative (e.g. it is known that large proteins need much longer blotting time for adequate transfer). It also needs to be demonstrated that all peptides are indeed collected onto the second membrane. While this is in general the case for intact proteins, some of the peptides obtained by digestion might be too hydrophilic and thus might not be adsorbed by the second membrane. Below will follow a description of the most popular instruments used in proteomic research.

14.3.1. MALDI-TOF mass spectrometry

The essential experimental set-up for MALDI-TOF MS is illustrated in Fig. 14.17.

TABLE 14.2

PROPERTIES OF MALDI MATRICES

Matrix	Analytes [a]		Suggested solvent (water : organic)	Ionisation	Adduct
	peptides	proteins			
Gentisic acid	+	+/−	9 : 1	+	M_r +136
Sinapic acid	+/−	+	2 : 1	+	M_r +206
3-Indoleacrylic acid	+	+	2 : 1	++	M_r +185
4-HCCA	+	+	2 : 1	+++	–

[a] +, matrix may be used for most peptides and proteins; +/−, matrix may (or may not) work.
[b] The more + signs, the more intense the signal and the higher the charge state of the most intense peak.
[c] Expected mass of the most intense satellite peak.

Short-duration (nanosecond) pulses of laser light are directed at matrix–protein sample mixtures (protein-doped matrix crystals) that are inserted into the MS instrument on a mechanical probe. The laser light causes a portion of the matrix–protein sample to be volatilised and ionised. The resulting gas-phase ions are accelerated to a fixed energy in an electrostatic field and directed into a field-free flight tube. At the end of this journey, the ions impact on a 'linear' detector, whereupon the intervals (Δt_m) between the pulse of laser light and the ion impacts (i.e. the times of flights) are measured. The masses (M_r) of the ions passing through the flight tube can then be determined via the approximate relationship:

$$M_r = 2qV(\Delta t_m)^2/l^2 \quad (14.1)$$

where q is the charge on the ion, V is the potential through which the ion is accelerated and l is the length of the flight tube. For improving mass resolution, modern MALDI-TOF instruments have an ion reflector, which turns the ions around in an electric field, sending them towards a second 'reflectron' detector. The reflector allows partial peptide sequence analysis by post-source decay [191]. Sample preparation is crucial for MALDI experiments. The surroundings of a protein molecule must be such that an intense light pulse can propel the intact molecule into a vacuum. To this aim, proteins have to be incorporated into crystals of a second material. The protein-containing crystal absorbs the light pulse and uses its energy to eject material from its surface. Only few compounds can achieve that, and they are generally referred to as 'matrix'. Ideally, the protein and the matrix should not react to form a stable product; additionally, the matrix crystals must remain in vacuo for extended periods of time, thus their sublimation rate should be as low as possible. Table 14.2 lists some of the common matrices for MALDI-TOF, with their relevant properties. It is seen, though, that most matrices produce satellite signals called 'adduct peaks', which result from the photochemical breakdown of the matrix into more reactive species, which ultimately can add to the polypeptide. These four compounds can be used with the 337-nm (nitrogen laser) or with the 354-nm (neodymium : yttrium/aluminum/garnet, Nd : YAG laser) lights. Their properties, in brief:

(1) Gentisic acids (2,5-dihydroxybenzoic acid, M_r 154 Da) is a useful matrix for a

wide variety of peptides and proteins. It is often adopted as first choice for analysing peptide mixtures and it appears to be ideal for reflectron MALDI-MS.

(2) Sinapic (or sinapinic) acid (*trans*-3,5-dimethoxy-4-hydroxycinnamic acid, M_r 224 Da) has a strong affinity for proteins of all types and gives good results with mixtures of polypeptides.

(3) *Trans*-3-indoleacrylic acid (M_r 203 Da) has not found broad applications for proteins and peptides, but is used rather for analysing industrial polymers.

(4) 4-HCCA (α-cyano-4-hydroxycinnamic acid, M_r 189 Da) produces strong signals from peptides and proteins. This matrix produces intense, multiply charged ions in the positive-ion spectra of proteins; however, such protein ions frequently undergo metastable decay in the mass spectrometer.

Some important aspects of the solutions in which the protein is dissolved should be borne in mind. First of all, strong ionic detergents (such as SDS) should be thoroughly removed, whereas non-ionic surfactants, such as Triton X or octylglucoside, are tolerated. Sodium azide in buffers also should be avoided, since its presence suppresses protein ion formation in the mass spectrometer. As also mentioned in Chapter 13 (see Section 13.4.5 and Figs. 13.7 and 13.8) exposure to formic acid solutions should be avoided, since this compound reacts with amino groups (especially Lys residues) as well as with the –OH group of Ser and Thr (see also Section 13.4.5) resulting in a formyl derivative of the protein. Also concentrated trifluoroacetic acid can react with free amino groups. Finally, it should be remembered that, although the best stain, in view of MS analysis, is Coomassie Blue, even the latter is not immune from artefacts. Some proteins bind tenaciously to Coomassie Blue, this resulting in multiple peaks in MALDI-TOF analysis (see Fig. 14.18A). This, per se, is not dangerous, since such extra peaks can be clearly identified, due to the regular mass increments per each adduct, corresponding to the mass of Coomassie. More disturbing, though, is the fact that, during digestion with trypsin, some Coomassie residues might cover the potential hydrolytic sites, this resulting in anomalous fragmentation (it should be remembered that, since the Coomassie family contains two sulphate groups, they tend to bind to positively charged sites on proteins, such as Lys and Arg residues, i.e. to the typical sites of attack by trypsin). Another interesting example is given in Fig. 14.18B, where the native lactoglobulin peak is seen to be followed by a series of peaks, representing three different covalent adducts with lactose and another four complexes with Coomassie. Although the few examples here recalled are accidental protein modifications (except for the above lactose adducts!), resulting from the various manipulation steps, there are several post-synthetic modifications, driven by in vivo processes, which result in permanent alteration of the mass of a protein. A large number of them is listed in Table 14.3. Some of these modifications would result in a huge variation of the mass of a peptide, to the point that it might not any longer be identifiable in databases. However most databases allow for such protein modifications, thus permitting, in most cases, proper recognition of a given polypeptide chain.

It is here recalled that peptide-mass fingerprinting (PMF) compares an experimental profile of peptide masses against a theoretical profile calculated from the known sequences in a non-redundant database. The effectiveness of a PMF search is strongly influenced by the accuracy with which the peptide masses are measured. Since MALDI

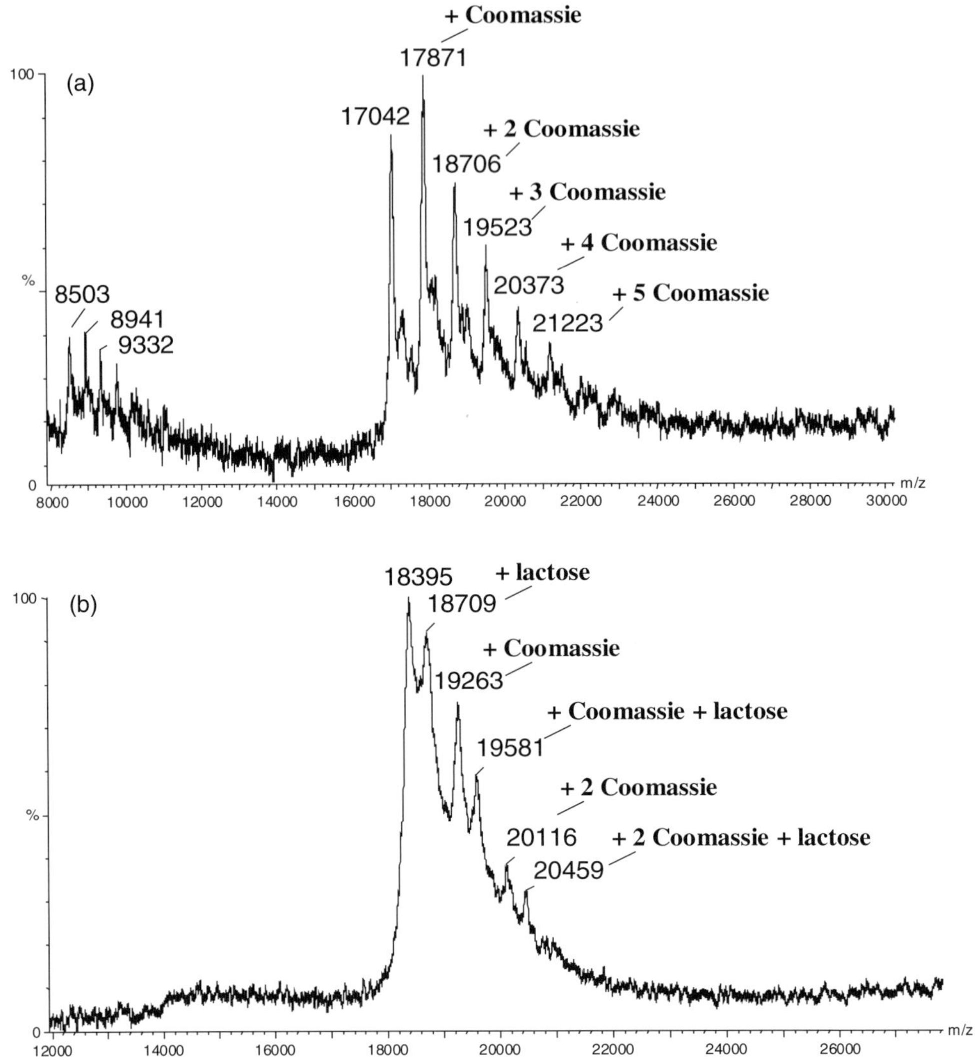

Fig. 14.18. (A) formation of protein adducts with Coomassie Blue, as detected by MALDI-TOF MS. In this particular case, up to five adducts can be seen, where each adjacent pair of peaks is spaced by *m/z* values which coincide with that of Coomassie Blue. The protein is a polypeptide detected by 2-D maps in UHT milk, still unidentified, having p*I* ca. 8 and M_r of 17,042. The peaks starting at 8503 *m/z* refer to the doubly charged species, followed by the Coomassie Blue adducts. (B) In this case, lactoglobulin (variants A and B) was fractionated in a narrow range pH 4–5 IPGs and subjected to MALDI-TOF MS after staining with Coomassie Blue. One can notice the intact protein, together with a series of adducts with both lactose and Coomassie Blue (the data were kindly provided by Drs. M. Hamdan and M. Galvani, GlaxoSmithKline Medicines Research Centre, Verona).

is today performed on TOF machines, and such instruments today incorporate both delayed extraction and reflectrons, both resolution and mass accuracy are dramatically improved. Modern electronics has increased the sampling rate, which is now as high as 4

TABLE 14.3

MASS SHIFTS DUE TO SOME POST-TRANSLATIONAL MODIFICATIONS IN PROTEINS (IN ORDER OF INCREASING MASS VALUES) (FROM [186], WITH PERMISSION)

Modification	Mass change
Disulphide bond formation	−2.0
Desamidation of Asn or Gln	+1.0
Methylation	+14.0
Hydroxylation	+16.0
Oxidation of Met	+16.0
Formylation	+28.0
Acetylation	+42.0
Phosphorylation	+80.0
Sulphation	+80.0
Cysteinylation	+119.1
Pentoses (Ara, Rib, Xyl)	+132.1
Deoxyoses (Fru, Rha)	+146.1
Hexosamines (GalN, GlcN)	+161.2
Hexoses (Fru, Gal, Glc, Man)	162.1
Lipoic acid (amide bond to Lys)	+188.3
N-acetylhexose amines (GalNac, GlcNAc)	+203.2
Farnesylation	+204.4
Myristoylation	+210.4
Biotinylation (amide bond to Lys-ε-NH_2)	+226.3
Pyridoxal phosphate (Schiff base with Lys-ε-NH_2)	+231.1
Palmitoylation	+238.4
Stearoylation	+266.5
Geranylgeranylation	+272.5
N-acetylneuraminic acid	+291.3
Addition of glutathione	+305.3
N-glycolylneuraminic acid	+307.3
5′-Adenosylation	+329.2
4′-Phosphopantetheine	+339.3
ADP-ribosylation	+541.3

GHz. Such improvements and good calibration routines allow databases to be searched with peptide masses typically measured to an accuracy of 10 ppm. The confidence in such searches is directly correlated with mass accuracy. At least four or five peptides must be matched and 20% sequence coverage is required for proper identifications.

What are the strengths and limitations of MALDI-TOF MS? The main advantages that render this approach a cornerstone of proteomics are: (a) the instrumentation is robust, relatively inexpensive and can be automated; (b) the analysis time per sample is just a few seconds, thus offering a potential for a very high throughput. There are some limitations, though:

(1) The ionisation is selective and is not quantitative. In an equimolar pool of peptides from the digest of a protein, some peptides will not be seen and there will be wide variations in signal intensity from the remainder.

(2) As the amount of protein in the gel decreases, so does the number of peptides and,

at very low levels of protein, only a few peptides may be observed. Thus, identifying a protein with confidence from just a few peptide masses is impossible and yet this situation often arises with very small amounts of protein.

(3) MALDI-TOF MS has only a limited ability to deal with mixtures. In proteomics, mixtures are common to the point at which, even in the case of apparently 'clean' spots eluted from 2-D gels, a few principal components and several minor ones might be present.

Given the above caveats, why does MALDI-TOF continue to be so popular? One of the reasons must certainly be its very high throughput. The other reason could be the possibility of adding partial peptide sequence to the mass information. In general, just one mass, accurately measured, plus partial sequence, is sufficient for identifying a protein with confidence. It just so happens that, with a judicious use of MALDI-TOF MS post-source decay (PSD), some sequence information can be obtained, although the fragmentation pattern could be difficult to interpret. MALDI-PSD (as well as ESI, see below) have now started to replace classical analytical approaches such as Edman degradation, in certain cases. MALDI-PSD, e.g. offers higher sensitivity over chemical processing (Edman) and has the possibility to analyse multicomponent samples. In addition, it has a higher tolerance to sample impurities, thus it is ideally suited for modern approaches of 2-D separation and 2-D sample handling and storage. PSD analysis is a method that allows to mass analyse fragment ions that are formed in the field-free region following the ion source of a TOF mass spectrometer. These fragment ions are produced from the decay of MALDI-generated ions. Mass determination of these PSD ions is based on using an electrostatic ion reflector as an energy analyser (see Fig. 14.17). In fact, during the linear ion path (the first part of the flight) all ions leaving the source have the same nominal kinetic energy. Most of them are still unfragmented precursor molecular ions; however, during the subsequent flight time (tens of micro-seconds), through the field-free drift region, a certain number of analyte ions undergo post-source decay into product ions. Due to their lower mass, such PSD ions have a considerably lower kinetic energy than their precursor ions. The ion reflector in PSD instruments is thus used as an energy filter and therefore as a mass analyser for PSD ions. Due to their mass-dependent kinetic energies, PSD ions are reflected at different positions within the reflector, thus impinging at different zones within the area of the reflector detector, this in turn leading to mass-dependent total flight times through the instrument. The majority of applications employing PSD-MALDI MS have concentrated on peptide characterisation. The sensitivity of the method is in the range of 30 to 100 fmol of peptide. Mass resolving power is in the range of 6000 to 10,000 for precursor ions and in the range of 1500 to 3000 for PSD ions. An example of this partial sequence determination by PSD is given in Fig. 14.19. This is an interesting case, which deserves further comments. When digesting bovine β-lactoglobulin B, Bordini et al. [192] obtained a peptide of 2113.6 Da, which could not be found in the Swiss-Prot database. On the assumption that this could be a wrong cut obtained by trypsin digestion, an attempt was made at obtaining a partial amino acid sequence by the PSD method. PSD analysis in fact resulted in the product ion spectrum of Fig. 14.19. This spectrum was obtained by operating the instrument in the reflectron mode, during which appropriate ion-gating pulses were applied allowing the selection of the parent ion (i.e. the 2113.6 Da peptide) and the monitoring of its

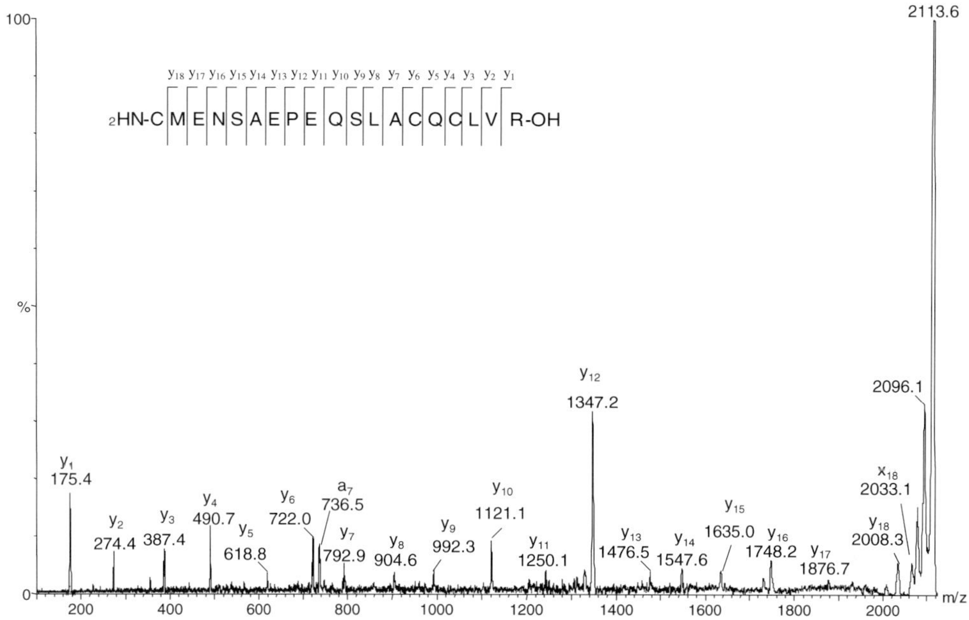

Fig. 14.19. Example of sequence determination by PSD of a peptide of m/z 2113.6 obtained by tryptic digestion of β-lactoglobulin B on the basis of the PSD spectrum and the sequence determined, it was possible to assign this fragment to the amino acid region (106–124) of the protein chain. For a detailed explanation, see text. (From [143], with permission.)

unimolecular and background collision-induced fragments. The spectrum shows the complete series of y_n-fragment ions (according to the nomenclature of Roepstorff and Fohlman [193]) which allowed the verification of the sequence given in the same figure. Although there are computer programs that, once fed the size all the fragments obtained, will reconstruct such a sequence, it is nice to see how one could do it by himself. The search begins with the nearest fragment having a size corresponding to the loss of one amino acid residue. This can be found in the fragment labelled as y_{18} (M_r 2008), which is seen to correspond to the loss of one Cys residue (103 Da). The second nearest fragment is y_{17} (1876.7), which corresponds to the loss of a dipeptide, Cys–Met (total mass loss 234 Da). The third nearest fragment (y_{16}) exhibits a total mass of 1748.2, which corresponds to the loss of the tripeptide Cys–Met–Glu (total mass of 363 Da). By continuing like that, one finds at the end a single fragment, labelled y_1, corresponding to the mass of the free amino acid Arg, positioned at the carboxyl-terminus (note that in this last case the mass of the –OH group should be added, since this last residue would not be engaged into a peptide bond). By this procedure, which resembles an Edman degradation, except that this degradation occurs in the mass spectrometer rather than being chemically induced via phenylisothiocyanate attack, one could construct this entire sequence: $_2$HN–Cys–Met–Glu–Asn–Ser–Ala–Glu–Pro–Glu–Gln–Ser–Leu–Ala–Cys–Gln–Cys–Leu–Val–Arg–OH, i.e. a sequence of 19 amino acids, found to correspond to the tryptic fragments [106–124] of β-lactoglobulin B. This is an extraordinary sequence, both due its accuracy and its length. Although it is rare to obtain such a long sequence (typically restricted, under normal operating conditions) to 8–10 amino acids, this figure proves that, when obtaining high-quality spectra, such sequencing process can give much improved results and longer readings.

Another way of controlling fragmentation would be to insert the MALDI source onto a tandem mass spectrometer. A recent approach has been to use a tandem TOF-MS, although this method has been shown to work only with standard peptides and not yet with protein digests [194]. A second way is to use MALDI with a hybrid quadrupole TOF mass spectrometer, in order to generate the needed sequence information. Two recent papers [195,196] have described this approach, by using a MALDI rather than an ESI ion source.

14.3.2. ESI mass spectrometry

Electrospray ionisation (ESI) MS is the second pillar for proteome analysis. ESI interfaced with MS has evolved into a powerful tool for the analysis of proteins, peptides, nucleic acids, carbohydrates, glycoproteins, drug metabolites and other biologically active species. The success of ESI in this area is largely due to its ability to extract these fragile chemical species from solution intact, ionise them and transfer them into the gas phase, where they may be subjected to mass analysis. A unique and powerful characteristic of the ESI source is its additional ability to generate ions carrying many charges, whereas most ionisation methods generate ions with a single charge. The majority of mass spectrometers are limited to analysing ions with mass-to-charge ratios (m/z) of only a few thousand, but by dramatically increasing the number of charges (z) through ES ionisation, compounds with masses up to a million Da can be analysed by MS [197].

ESI-MS hardware can be configured to operate with continuous sample flow introduction, discrete sample injection (as in flow-injection analysis, FIA) or on-line sample preparation systems, such as liquid chromatography (LC) and capillary zone electrophoresis (CZE). Enzymes can be used to cleave a protein selectively, and the resulting peptide mixture can be separated by either LC or CZE, which are then interfaced on-line to ESI-MS. ES can produce sample ions from solution flow rates ranging from 25 nl/min to more than 1 ml/min. In addition to producing intact parent molecule M_r information, ESI-MS can selectively provide fragment or daughter ion information from a collision-induced dissociation (CID) process, which can occur either on the ES source itself [198] or with the use of multiple mass spectrometer stages (MS/MS) [199].

The term electrospray (ES) has evolved to describe collectively a basic set of processes encompassing: (a) the formation of electrically charged micron-sized liquid droplets created from a flowing liquid sample; (b) the extraction of gas-phase ions from these same droplets under a high electric field; and (c) the subsequent transport of these ions into a vacuum suitable for MS analysis. Fig. 14.20 is a drawing of an ES ion source as interfaced to a quadrupole MS machine. Although the ES source has been successfully interfaced to magnetic sector [200], Fourier transform [201,202], ion trap [203] and TOF [204] mass spectrometers, the quadrupole is currently the most common

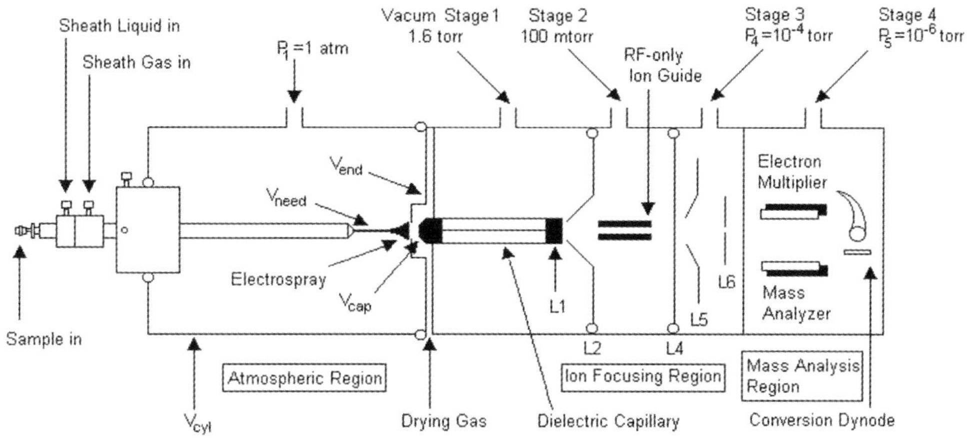

Fig. 14.20. ESI-MS system with a quadrupole mass analyser. For other details, see text. (From [181], with permission.)

instrument used with ES. As shown in Fig. 14.20, sample liquid is introduced through a tube and exits at a conductive needle tip (V_{need}) which is maintained typically at 2 to 7 kV relative to the surrounding electrodes: the cylinder (V_{cyl}), end plate (V_{end}) and capillary entrance (V_{cap}). By setting appropriate voltages on V_{cyl}, V_{cap} and V_{end}, a Taylor cone is drawn out from the tip of the capillary, from which a thin liquid filament extends (see Fig. 14.21). The filament then breaks up into charged droplets primarily due to mechanical Rayleigh fluid instability propagation. The electrosprayed droplets shrink in size until the Rayleigh limit is reached, at which point smaller droplet

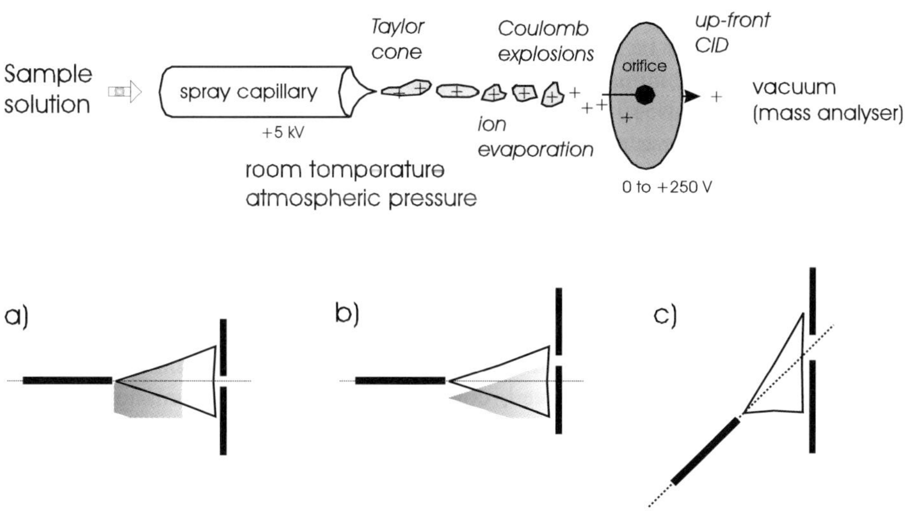

Fig. 14.21. Electrospray ion source and different sprayer arrangements relative to the ion sampling orifice: (a) on-axis; (b) off-axis; and (c) diagonal. (From [186], with permission.)

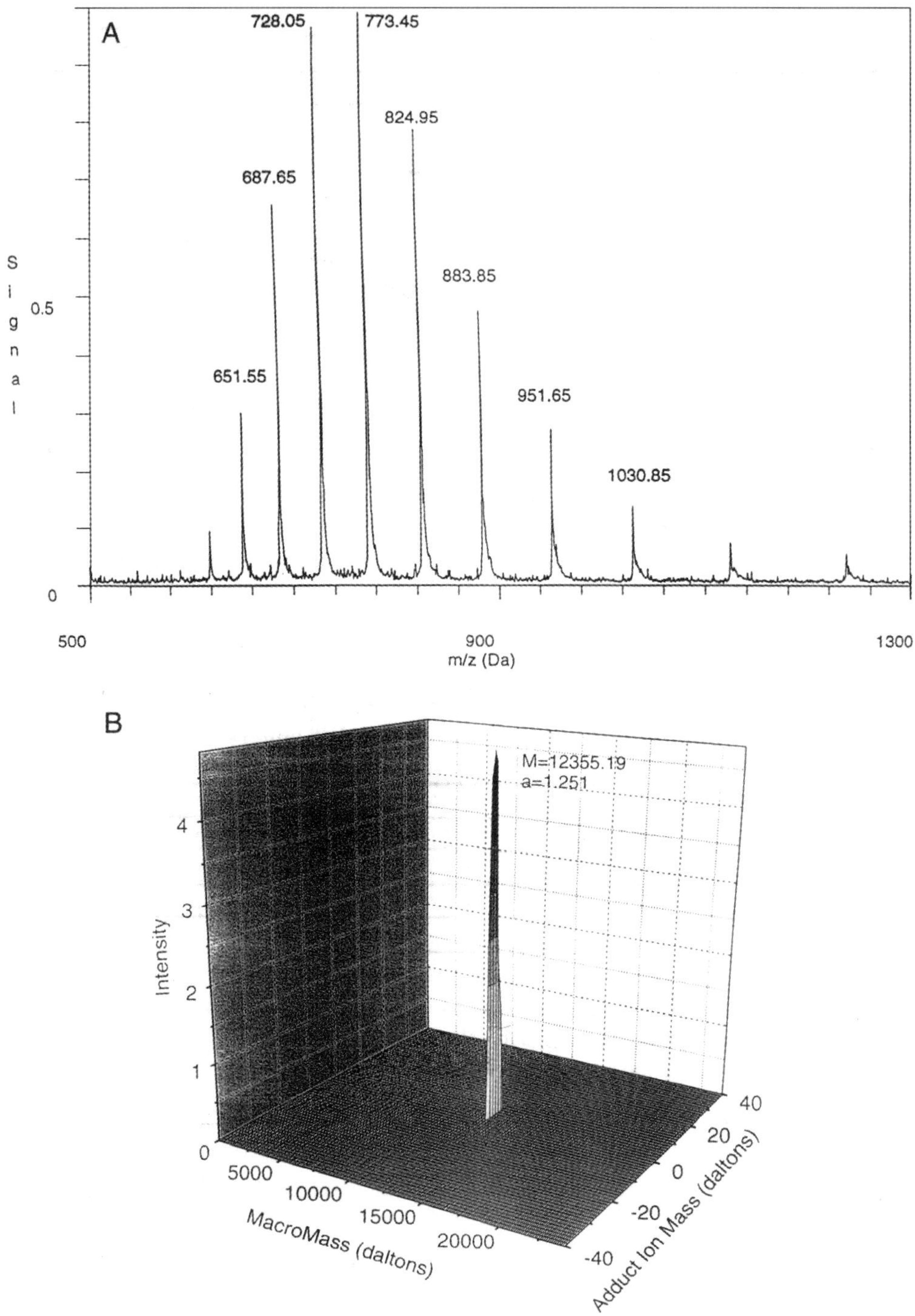
A
728.05
773.45
824.95
687.65
883.85
651.55
951.65
1030.85
Signal
0.5
0
500
900
1300
m/z (Da)
B
M=12355.19
a=1.251
Intensity
4
3
2
1
0
5000
10000
15000
20000
MacroMass (daltons)
40
20
0
-20
-40
Adduct Ion Mass (daltons)

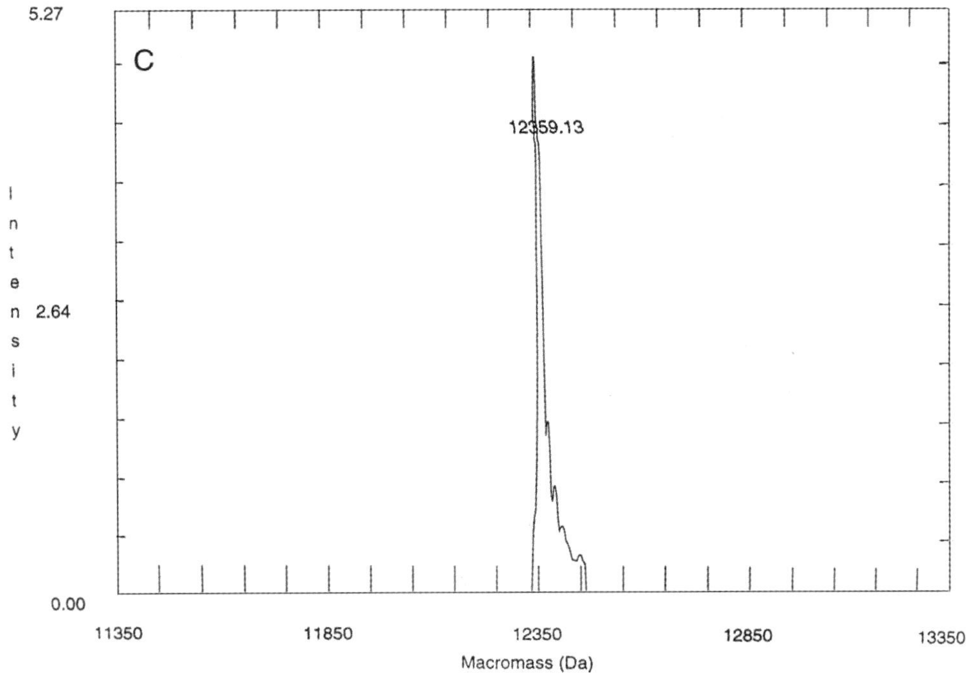

Fig. 14.22. Acquisition and deconvolution of a cytochrome *c* mass spectrum by ESI-MS. (A) ESI-MS spectrum of this protein in water and 1.0% acetic acid. (B) Deconvolution of the cytochrome *c* spectrum, by solving simultaneously for charge state, adduct ion mass and M_r as in Eq. (14.2), (C) Cross-section of the deconvoluted 3-D surface in (B), taken along the adduct ion mass 1.008 (H^+). (From [181], with permission.)

As a final comment, it might be of interest to see how scientists evaluate the two main approaches to proteome analysis, MALDI-MS vs. ESI (or nESI)-MS. We will offer here the comments of Blackstock [176]. It is now common perception that MALDI-MS is the route to high-throughput proteomics, with ESI being reserved for difficult analyses or for post-translational modification studies. However, Blackstock challenges this view. If throughput is measured solely in terms of samples analysed per hour, then MALDI-MS might be superior, but a better measure is the number of definitive protein assignments per hour. According to this criterion, the above author believes that on-line ESI is at present superior to MALDI, especially with human samples.

14.3.4. Mass spectrometry for quantitative proteomics

One of the main problems with current 2-D map protocols, including the most sophisticated ones, is the lack of proper quantitation of the resolved and detected polypeptide spots. For example, specific classes of proteins have long been known to be excluded or under-represented in 2-D gel patterns. These include very acidic or basic proteins, excessively large or small polypeptides and membrane proteins. In addition, it is now clear that 2-D maps cannot detect low-abundance proteins without a pre-enrichment step. Thus, quantitative proteomics still has to come of age. Yet it is thoroughly needed, since the ability of accurately detecting and quantifying potential protein changes induced in an organism by a specific perturbation is an essential part of the study of dynamic biological processes. It might be argued that, in principle, quantitative proteomics might be assessed by measuring the levels (and its changes) of specific mRNAs, as present in an organism at rest or as induced by given stimuli. Yet, this approach might be fallacious given the facts that: (a) in the few studies that measured the copies of mRNA and protein levels in the same system, the correlation of these two levels was very poor and not strong enough to enable prediction of one value from the other [209,210]; (b) protein activation or inactivation by post-translational mechanisms cannot be detected by looking at the mRNA levels.

One elegant way out of this impasse has been recently proposed by Gygi and Aebersold [178]: quantifying proteins by using stable-isotope dilution. This 'venerable' technique involves the addition to the sample of a chemically identical form of the analyte(s) containing stable heavy-isotopes (such as ^{2}H, ^{13}C, ^{15}N) as internal standards. The most suitable internal standard for a candidate peptide is the same peptide labelled with stable isotopes. Therefore, proteins can be profiled in a quantitative manner if two protein mixtures are compared, with one serving as the reference sample, by containing the same proteins as the other sample but at different abundance and labelled with heavy stable isotopes. In theory, then, all peptides from the combined samples exist as analyte pairs of identical sequences but with different masses. Thus, when analysed by MS techniques, the ratio between the intensities of the lower and upper mass components of these pairs of peaks provides an accurate measure of the relative abundance of such peptides (or proteins) in the original cell lysates. There are two strategies for achieving this goal: labelling either before or after extraction.

14.3.4.1. Labelling before extraction

This approach has been followed by Oda et al. [211] and Pasa-Tolic et al. [212]. The first authors grew one yeast culture on a medium containing the natural nitrogen isotope distribution (^{14}N 99.6%; ^{15}N 0.4%) and another one on the same medium enriched in ^{15}N (>96%). After a suitable growth period, the cell pools were combined and the proteins of interest extracted and separated by RP-HPLC and then by SDS-PAGE. After extraction of the proteins of interest and digestion, the peptides were subjected to MS analysis. Each incorporated ^{15}N atom shifted the mass of any given peptide upwards by one mass unit, leading to a pair of peaks from each peptide. Pasa-Tolic et al. [212], on the other hand, used stable isotope media for giving to proteins a specific isotope signature. They compared the cadmium stress response in *E. coli* grown in normal and rare-isotope-depleted media (lacking ^{13}C, ^{15}N and ^{2}H). Intact-protein mass measurements were carried out by Fourier-transform ion-cyclotron-resonance (FT-ICR) MS, a technique of growing importance for ultra-high resolution MS analysis of biopolymers [213,214]. Although no protein could be positively identified, the expression ratio for 200 different proteins could be compared.

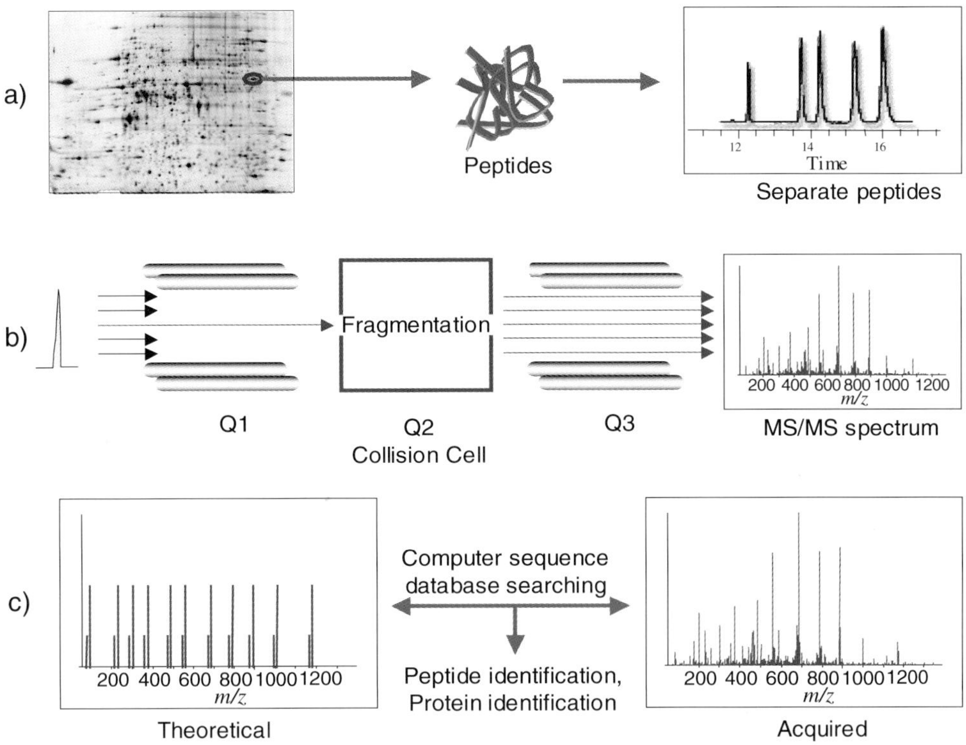

Fig. 14.23. Standard 2-D maps coupled to MS analysis. The method consists of three stages, tightly integrated. In (a) proteins are separated by 2-D PAGE and stained spots are excised and digested with trypsin. The resulting peptides are then separated by on-line HPLC. In (b), an eluted peptide is ionised by ESI-MS, enters the mass spectrometer passing through quadrupole 1 (Q1) and is fragmented in a collision cell (Q2). Finally, the fragmented-ion tandem MS spectrum is collected in Q3; this contains sequence-specific information. In (c) the tandem MS spectrum from the selected ionised peptide is seen to contain enough information to unable identification of the protein from which it originated. (From [178], with permission.)

14.3.4.2. Labelling after extraction

This is the approach taken by Gygi and Aebersold [178]. It is based on so-called isotope-coded affinity tags (ICAT). In order to understand the protocol described below, one should compare the standard method, based on 2-D PAGE and conventional MS analysis (as displayed in Fig. 14.23), with the ICAT protocol, as described in Fig. 14.24. In this novel procedure, the stable isotopes are separately incorporated by the selective alkylation of Cys residues with either a ‘heavy’ or ‘light’ reagent, after which the two protein pools to be compared are mixed. The ICAT reagent (whose general structure is depicted in Fig. 14.24a) is composed of three parts: a biotin portion, used as an affinity tag; a linker, which can incorporate either the heavy or light isotopes; and a third terminal group, which contains a reactive iodine atom able to alkylate specifically thiol groups (Cys residues). The ‘heavy’ ICAT contains eight deuterium atoms, which in the ‘light’ one are replaced by standard hydrogen atoms. Proteins from two different

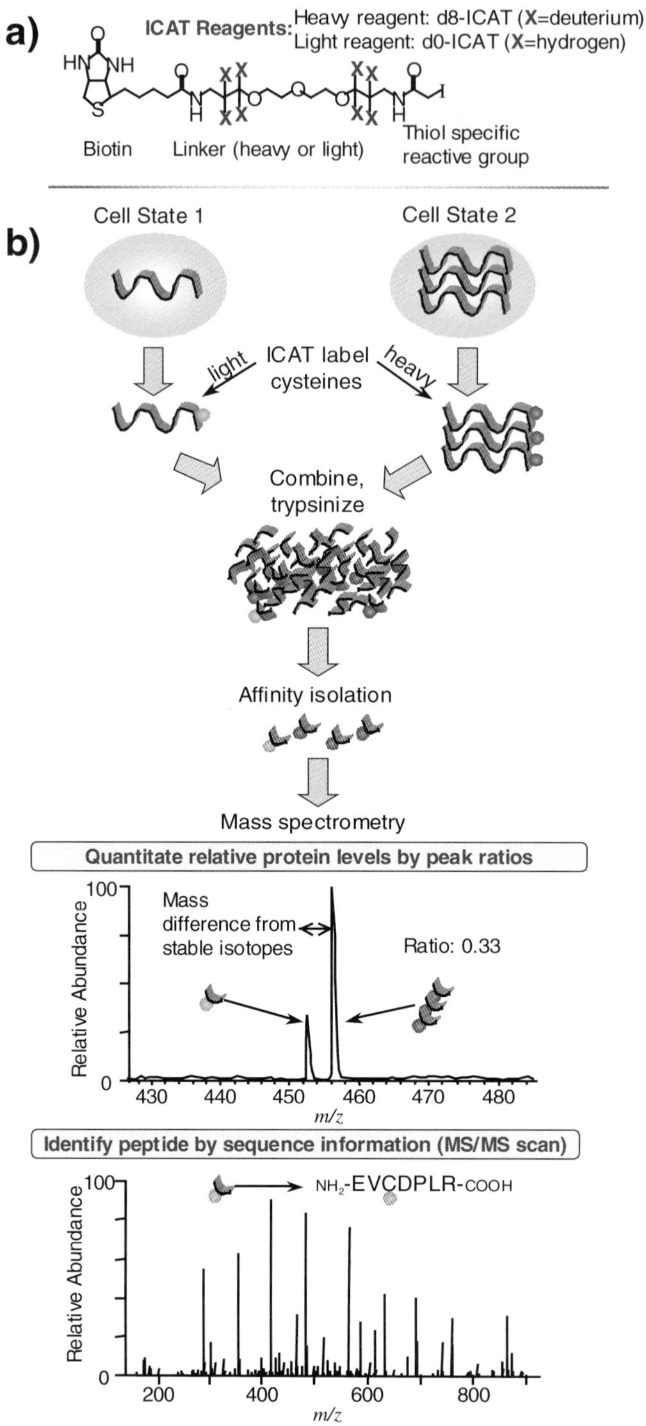
a)
ICAT Reagents:
Heavy reagent: d8-ICAT (X=deuterium)
Light reagent: d0-ICAT (X=hydrogen)
Biotin
Linker (heavy or light)
Thiol specific reactive group
b)
Cell State 1
Cell State 2
ICAT label cysteines
light
heavy
Combine, trypsinize
Affinity isolation
Mass spectrometry
Quantitate relative protein levels by peak ratios
Relative Abundance
Mass difference from stable isotopes
Ratio: 0.33
m/z
Identify peptide by sequence information (MS/MS scan)
NH2-EVCDPLR-COOH

cell states are harvested, denatured, reduced and labelled at Cys residues with either light or heavy ICAT reagent. The samples are then combined and digested with trypsin. ICAT-labelled peptides can be further isolated by biotin-affinity chromatography and then analysed by on-line HPLC coupled to tandem MS. The ratio of the ion intensities for any ICAT-labelled pair quantifies the relative abundance of its parent protein in the original cell state. In addition, the tandem MS approach produces the sequence of the peptide, and thus can unambiguously identify the protein of interest. This strategy, ultimately, results in the quantification and identification of all protein components in a mixture and, in principle, could be applied to protein mixtures as complex as the entire genome. The ICAT strategy appears to be endowed of the following advantages.

First, the method is compatible with any amount of protein harvested from body fluids, cells or tissues under any growth conditions.

Secondly, the alkylation reaction is highly specific and occurs in the presence of salts, detergents and a large body of solubilisers.

Thirdly, the complexity of the peptide mixture to be analysed is reduced by isolating only the Cys-tagged peptides.

As a last bonus, the ICAT strategy permits almost any type of biochemical, immunological or physical fractionation, which render it compatible with the analysis of low-abundance proteins.

Some drawbacks appear to be due to the fact that the ICAT label is quite large (ca. 500 Da), a fact that can complicate database searching for small peptides. Another problem could be due to the fact that small proteins, lacking Cys residues, will not be labelled. Finally, it remains to be seen if the reaction is quantitative.

14.3.5. Multidimensional chromatography coupled to mass spectrometry

In spite of the contributions of 2-D PAGE to proteomics, there are shortcomings to this technology. High-throughput analysis of proteomes is challenging because each spot from 2-D PAGE must be individually extracted, digested and analysed, a time-consuming process. In addition, owing to the limited loading capacity of 2-D PAGE gels and the detection limit of staining methods, 2-D PAGE presently has an insufficient dynamic range for complete proteome analysis. These shortcomings have encouraged development of alternative methods, which have been recently reviewed by Washburn and Yates [215]. Such methodologies could be 1-D and 2-D chromatographic/electrophoretic

---

Fig. 14.24. Scheme of the isotope-coded affinity tag (ICAT) strategy for quantification of protein expression. (a) Structure of the ICAT reagent. It consists of 3 segments: an affinity tag (biotin); a linker, which can incorporate either deuterium or hydrogen; and a reactive tail specific for thiol groups. (b) ICAT strategy. Protein from two different cell states is harvested, denatured, reduced and labelled at Cys with the light or heavy ICAT reagents. The samples are then combined and digested with trypsin. After affinity isolation of the ICAT-labelled peptides, they are analysed in a tandem mass spectrometer. The ratio of the ion intensities for each ICAT-labelled pair quantifies the relative abundance of its parent proteins in the original state. Finally, the MS spectrum enables sequencing and protein identification. (From [178], with permission.)

methods, such as high-performance liquid chromatography (HPLC), capillary isoelectric focussing (cIEF), capillary zone electrophoresis (CZE) or micro-capillary chromatography. Just as with 2-D PAGE, MS would be the method of choice for identifying proteins resolved by liquid separation protocols. This would have the advantage that the coupling with MS would be on-line, thus eliminating the lengthy steps for transferring proteins to MS equipment in 2-D PAGE.

Among the 1-D steps, in the early nineties a flurry of papers appeared hyphenating CZE to MS instrumentation. Wahl et al. [216,217], for instance, have claimed attomole level sensitivity by coupling CZE to ESI-MS in narrow-bore (barely 5 μm I.D.) capillaries. Moseley et al. [218] and Deterding et al. [219], by coupling CZE to fast-atom bombardment (FAB)-MS, reported sensitivities of the order of low femtomoles, under high separation efficiencies in the CZE step (410,000 theoretical plates) for bioactive peptides. Weinmann et al. [220,221] described coupling of CZE to ^{252}Cf plasma desorption mass spectrometry, both off- and on-line, with sensitivities in the sub-picomolar range. These authors were able to analyse not only hydrophilic, but also hydrophobic peptides, the latter dissolved into aqueous acetic acid buffers containing up to 20% 2-propanol or 25% acetonitrile. In another approach, Thompson et al. [222] devised a CZE/ESI-MS system exploiting on-column, transient isotachophoretic sample pre-concentration, able to offer two orders of magnitude improvement in sensitivity. A number of other reports have appeared, in general coupling the eluate of a CZE run to an ESI-MS system [223–225]. In a recent review [226] many of these developments have been listed and evaluated; in addition, the possibility of coupling microfabricated devices (CZE on a chip) to ESI-MS has been discussed.

Nevertheless, these early reports were not concerned at all with proteome analysis, but simply with development of such hyphenated techniques; thus, very simple samples were analysed and only a handful of peptides separated. Only in recent times CZE was re-assessed as a potential tool in the proteome field. Among the different CZE variants, cIEF appeared promising, since it would combine a high-resolution step, based on surface charge (p*I*) with an MS step, the latter providing a highly accurate M_r value. Tang et al. [227] coupled cIEF to ESI-MS, whereas Yang et al. [228] and Jensen et al. [229] hyphenated cIEF to FTICR-MS. With both approaches, however, very few proteins were electrophoretically resolved. When an *E. coli* lysate was analysed by cIEF coupled to FTCIR-MS, a few hundred proteins were resolved, but only a few of them were identified, a situation not so adequate in proteomics studies. An alternative way to the process of separation would be to digest the entire protein mixture into a peptide mixture and use tandem MS (MS-MS) for generating amino-acid-sequence-specific data. It is in fact known that the knowledge of the number of peptides generated by proteolysis, coupled to the assessment of their relative masses, allows superior protein identification. The two best methods for achieving that, today, are MALDI-TOF MS and MS-MS. To this aim, Eng et al. [230] have developed a popular algorithm, called SEQUEST, for matching the observed tandem mass spectrum of a peptide to theoretical mass spectra existing in protein and DNA databases. By using this search algorithm, McCormack et al. [231] identified proteins from complex mixtures by first digesting them into peptides and subsequently loading the generated peptides onto a reverse-phase (RP) column. Upon elution, the isolated peptides were fed into an ESI-MS instrument.

However, in order to match the unique resolving power of 2-D PAGE, 1-D chromatography is not the best approach; multidimensional chromatography would definitely be a better choice. Davidsson et al. [232] and Nilsson et al. [233] have devised such an approach, consisting in preparative 2-D liquid-phase electrophoresis (LPE) coupled to MALDI-TOF MS for final identification of the separated analytes. The samples under analysis were cerebrospinal fluid, in the first case, and human pleural eluate in the case of the last authors. Since 2-D LPE has a high loading capacity, low-abundance proteins could be detected and identified. Another interesting approach consists in using a variety of 2-D HPLC approaches, coupled to MS. In one instance, Raida et al. [234] coupled two chromatographic columns, first a cation-exchanger followed by RP-HPLC. The effluent of this last column was fed into an ESI MS machine and the peptide masses assessed. Opiteck et al. [235,236] described two different chromatographic approaches for separation of complex protein mixtures. In one approach, an *E. coli* lysate was injected onto a cation-exchange column and the eluates fed stepwise into a RP HPLC column. The eluate after this last step was sprayed into an ESI mass spectrometer. In a second system, size-exclusion chromatography was coupled to a RP HPLC column, again for analysis of *E. coli* cells. In both instances, however, too few proteins could be identified. In yet another approach, Link et al. [237] devised a discontinuous 2-D methodology, utilising strong-anion exchange HPLC for a 1st dimension separation. Portions of the eluate were digested with trypsin and analysed by RP-microcolumn HPLC. The eluted peptides were finally characterised by tandem mass spectrometry. Whereas this is an off-line method, Link et al. [238] subsequently devised an on-line approach, by which a peptide mixture from a digested *S. cerevisiae* was fed into a biphasic 2-D capillary column packed with a strong cation exchanger (SCX) juxtaposed to RP beads. Peptides were displaced iteratively by salt from the SCX resin into the RP pearls; this was followed by a classic eluant for RP columns, feeding the eluted peptides into an ESI MS-MS. After re-equilibrating the RP beads, raising the salt concentration displaced more peptides from the SCX resin onto the RP pearls, thus reiterating the process. In a single experiment, this method could resolve and identify 189 unique proteins from a *S. cerevisiae* whole-cell lysate via SEQUEST interrogation.

It might be of interest to see how the above techniques would score in proteome analysis, as compared to the well-ingrained 2-D PAGE mapping methodology. This has been assessed by Washburn and Yates [215]. According to these authors, none of the above alternative methods has yet attained the proficiency of 2-D PAGE, but they believe that they will in the near future. In their opinion, additionally, the most promising system appears to be the one developed by Link et al. [237,238], i.e. the one hyphenating a binary SCX-RP column to MS-MS for on-line protein identification, as depicted in Fig. 14.25.

14.4. INFORMATICS AND PROTEOME: INTERROGATING DATABASES

As stated in the introduction to this chapter, proteome analysis can be roughly divided into the following main areas: (1) 30% of it belongs to the fine art of electrokinetic methodologies, combining with proper skills an isoelectric focussing step

On-line

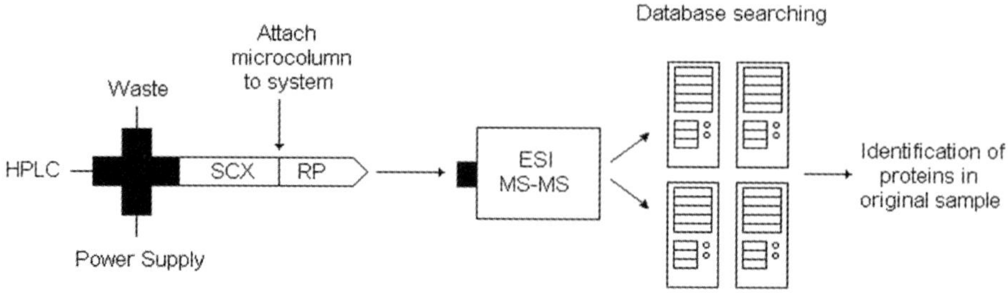

Fig. 14.25. Scheme of the on-line SCX-RP biphasic column hyphenated to ESI MS-MS. A whole cell lysate is first digested to peptides and subsequently loaded on a strong cation-exchanger (SCX) column. The eluate of this resin is directly fed into reverse-phase beads. In turn, the final eluate is sprayed into an ESI interface and the resulting ions analysed by tandem mass spectrometry (MS-MS). The process is reiterated many times, by eluting the first SCX resin with progressively higher salt gradients. The mass spectra generated are correlated with theoretical ones to be found in protein and DNA databases by the SEQUEST algorithm (modified from [215], with permission.)

to an orthogonal SDS-PAGE; (2) 30% of it is proper use and knowledge of mass spectrometry, especially in the variant MALDI-TOF, for a proper assessment of the precise M_r value of polypeptide chains and their fragments; (3) the remaining 30% is computer science coupled to good skills in interrogating databases and extracting pertinent data; and (4) finally, perhaps the remaining 10% is fantasy and intuition, sprinkled in the above cocktail for added flavour and depth of field.

The field of informatics in proteome analysis is growing at an impressive rate and it would be impossible here to summarise the present state of the art, since this summary would be rapidly outdated. It is however important, for the readers, to understand the basics of it, in order to be able to screen properly the databases and derive the relevant information. To this aim, one should consult at least three major chapters in the book of Wilkins et al. [1], which give a unique reconstruction and provide excellent guidelines for database browsing. They are: the chapter by Bairoch [239] on proteome databases; the one by Appel [240], on interfacing and integrating databases; and the third one by Peitsch and Guex [241] on large-scale comparative protein modelling. In addition, for understanding how these databases started and evolved over the years, it is of great interest to read some reviews highlighting the evolution of the informatisation of this area. These are: an article by Sanchez et al. [242] offering a tour inside Swiss 2-D PAGE database, and its update by Hoogland et al. [243]; a review by Bairoch and Apweiler [244] on the SWISS-PROT sequence database and its supplement TrEMBL; an article by Crawford et al. [245] on databases and knowledge resources for proteomics research; and, finally, an article by Beavis and Fenyö [177] on database searching with mass-spectrometric information. While this list is certainly non-exhaustive, it will provide the readers with at least the basic tools for a good start.

Identifying protein components from physical evidence, such as mobility on gels or the masses of the intact protein, is often the starting point. After this comes the full protein sequence, usually predicted from the known DNA sequence, and then structural

TABLE 14.4

PRIMARY AND SECONDARY ATTRIBUTES FOR PROTEIN RECOGNITION

Protein attribute	Origin of analytical data
Primary	
Species of origin	Starting biological material
Isoelectric point (p*I*)	From 1st dimension IPG
Apparent mass	From 2nd dimension SDS-PAGE
Real mass	From mass spectrometry, e.g. MALDI-TOF of intact protein
Protein specific sequence at N- and C-termini	From chemical sequencing of proteins immobilised onto membranes
Extended N-terminus sequence	From chemical sequencing of proteins immobilised onto membranes
Secondary	
Fingerprinting of peptide masses	MALDI-TOF or ESI-MS of digested peptides
Peptide fragmentation and de novo sequencing via MS	Fragmentation of peptides via MS-MS or PSD MALDI-TOF
Amino acid composition	Multiple radio-labelling analysis; chromatography of total protein hydrolysates

and functional predictions. These initial steps are summarised in Table 14.4. In model organisms such as yeast, more than half of the proteins have been already functionally analysed and, of the proteins coded for by the human genome, roughly 10% have already been studied in one or more laboratories. The big revolution in proteomics, though, has come from MS analysis. Highly accurate masses can be obtained for peptide fragments isolated from proteolytic digests of gel-separated proteins, and these are sufficient for protein identification given effective databases. Many resources today exist for identifying proteins by peptide masses, and these are listed in Table 14.5. Perhaps the most commonly used databases are SWISS-PROT, TrEMBL and the non redundant (nr) collection of protein sequences at the US National Centre for Biotechnology Information (NCBI). SWISS-PROT is an annotated collection of protein sequences on the Expasy server; TrEMBL is a large collection of predicted protein sequences given automatic annotation until they can be fully annotated and entered into SWISS-PROT; the NCBI nr database contains translated protein sequences from the entire collection of DNA sequences kept at GenBank, and also protein sequences in the PDB, SWISS-PROT and PIR databases. Such databases also offer additional information, including brief functional descriptions (if known), an annotation of sequence features (e.g. modification signals), secondary and tertiary structure predictions, key references and links to other databases.

14.4.1. An example of navigation on 2-D map sites

At this point, we feel a practical example is due. Although everybody knows the site SWISS-PROT, we have chosen to perform an excursion onto a site, perhaps not so well known, but quite interesting. By selecting the location: http//linux.farma.unimi.it/

TABLE 14.5

URLs FOR PROTEOMIC RESOURCES

Name of resource	Web location
Protein identification	
SWISS-2DPAGE	http//www.expasy.ch/ch2d/
PeptIdent	http//www.expasy.ch/tools/peptident.html
PROWL	http://prowl.rockefeller.edu/
SEQUEST	http://thompson.mbt.washington.edu/sequest/
MS-FIT/TAG	http://prospector.ucsf.edu/
Sequence databases	
NCBI/BLAST tools	http//www.ncbi.nlm.nih.gov(BLAST/
SWISS-PROT	http//www.expasy.ch/sprot/
3-D structure	
SCOP/PDBL-ISL	http://scop.mrc-Imb.cam.ac.uk/scop/
3Dcrunch	http//www.expasy.ch/swissmod/SM_3DCrunch.html
SWISS-MODEL	http//www.expasy.ch/swissmod/SWISS-MODEL.html
MODBASE	http://guitar.rockefeller.edu/modbase/
VAST	http//www..ncbi.nlm.nih.gov(Structure/VAT/vast.html
PDB	http//www.rcsb.org.pdb/
Domain structure	
PROSITE	http//www.expasy.ch/tools/scnpsit1.html
PRINTS	http//www.biochem.ucl.ac.uk/bsm/dbbrowser/PRINTS/PRINTS.html
BLOCKS	http//www.blocks.fhcrc.org/
Pfam	http://pfam.wustl.edu/
ProDOM	http://protein.toulouse.inra.fr/prodom.html
SMART	http://smart.embl-heldelberg.de/index.shtml
InterPro	http//www.ebi.ac.uk/interpro/
Subcelluar-location prediction	
PSORT-II	http://psort.nibb.ac.jp/
Model organism resources	
SGD	http://genome-www-stanford.edu/Sccharomyces/
WormBase/AceDB	http//www.wormbase.org/
FlyBase	Http//www.fruitfly.org/
Literature-based databases	
YPD/WormPD/PombePD/	http//www.proteome.com/dabases/

homeframed.html, one enters a site called 'rat serum protein study group', run jointly by the University of Milano, Faculty of Pharmacy (Dr. E. Gianazza), the University of Wien, Veterinary School (Dr. I. Miller) and the University of Washington (Protein Laboratory of Dr. R. Aebersold) (see Fig. 14.26a). By clicking on the right upper side (rat serum protein identification in 2-DE) one gets a page giving the aims and scopes of the project, as well as the methods adopted in the study. Major relevant publications are also given (lower left side) as well as two images of miniaturised 2-D maps of normal and inflamed rat sera (Fig. 14.26b). By double-clicking on the left one, one obtains a full view of the reference map of rat serum (Fig. 14.26c). All the proteins so far identified in this map are marked by a red cross. By clicking onto any of these crosses, one can

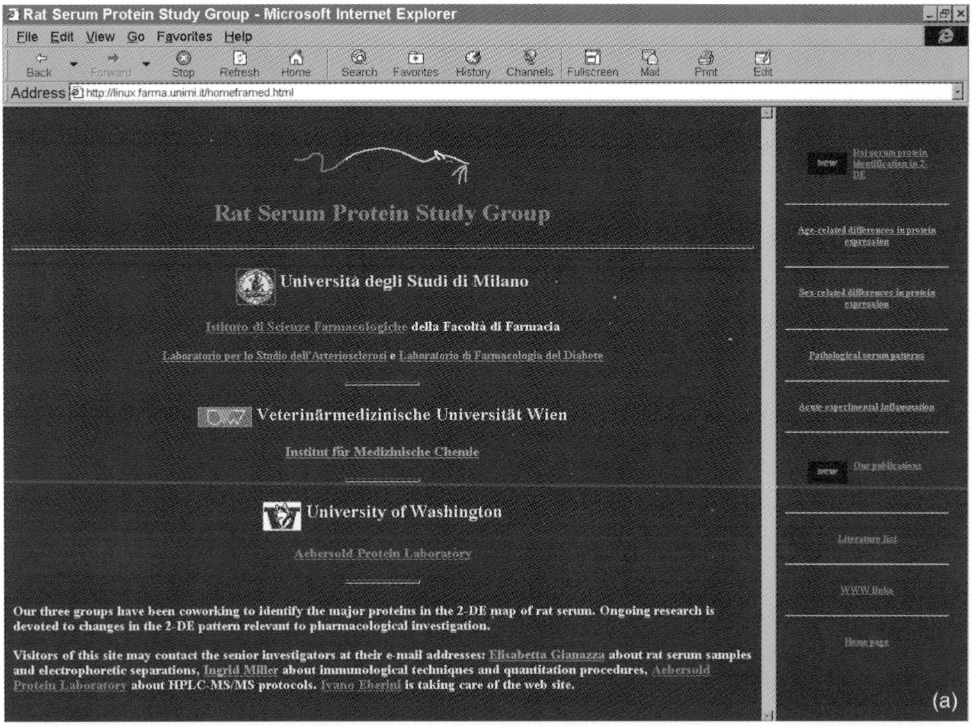
Rat Serum Protein Study Group - Microsoft Internet Explorer
File Edit View Go Favorites Help
Back Forward Stop Refresh Home Search Favorites History Channels Fullscreen Mail Print Edit
Address http://linux.farma.unimi.it/homeframed.html
Rat Serum Protein Study Group
Università degli Studi di Milano
Istituto di Scienze Farmacologiche della Facoltà di Farmacia
Laboratorio per lo Studio dell'Arteriosclerosi e Laboratorio di Farmacologia del Diabete
Veterinärmedizinische Universität Wien
Institut für Medizinische Chemie
University of Washington
Aebersold Protein Laboratory
Our three groups have been coworking to identify the major proteins in the 2-DE map of rat serum. Ongoing research is devoted to changes in the 2-DE pattern relevant to pharmacological investigation.
Visitors of this site may contact the senior investigators at their e-mail addresses: Elisabetta Gianazza about rat serum samples and electrophoretic separations, Ingrid Miller about immunological techniques and quantitation procedures, Aebersold Protein Laboratory about HPLC-MS/MS protocols. Ivano Eberini is taking care of the web site.
Rat serum protein identification in 2-DE
Age-related differences in protein expression
Sex-related differences in protein expression
Pathological serum patterns
Acute experimental inflammation
Our publications
Literature list
WWW links
Home page
(a)

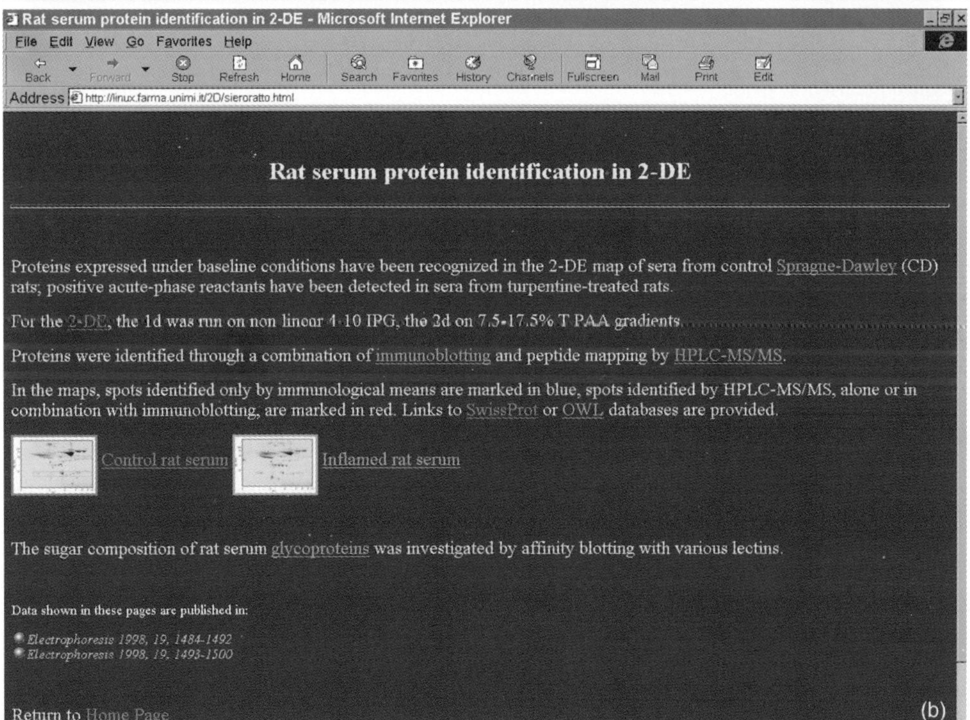
Rat serum protein identification in 2-DE - Microsoft Internet Explorer
File Edit View Go Favorites Help
Back Forward Stop Refresh Home Search Favorites History Channels Fullscreen Mail Print Edit
Address http://linux.farma.unimi.it/2D/sieroratto.html
Rat serum protein identification in 2-DE
Proteins expressed under baseline conditions have been recognized in the 2-DE map of sera from control Sprague-Dawley (CD) rats; positive acute-phase reactants have been detected in sera from turpentine-treated rats.
For the 2-DE, the 1d was run on non linear 4-10 IPG, the 2d on 7.5-17.5% T PAA gradients.
Proteins were identified through a combination of immunoblotting and peptide mapping by HPLC-MS/MS.
In the maps, spots identified only by immunological means are marked in blue, spots identified by HPLC-MS/MS, alone or in combination with immunoblotting, are marked in red. Links to SwissProt or OWL databases are provided.
Control rat serum
Inflamed rat serum
The sugar composition of rat serum glycoproteins was investigated by affinity blotting with various lectins.
Data shown in these pages are published in:
Electrophoresis 1998, 19, 1484-1492
Electrophoresis 1998, 19, 1493-1500
Return to Home Page
(b)

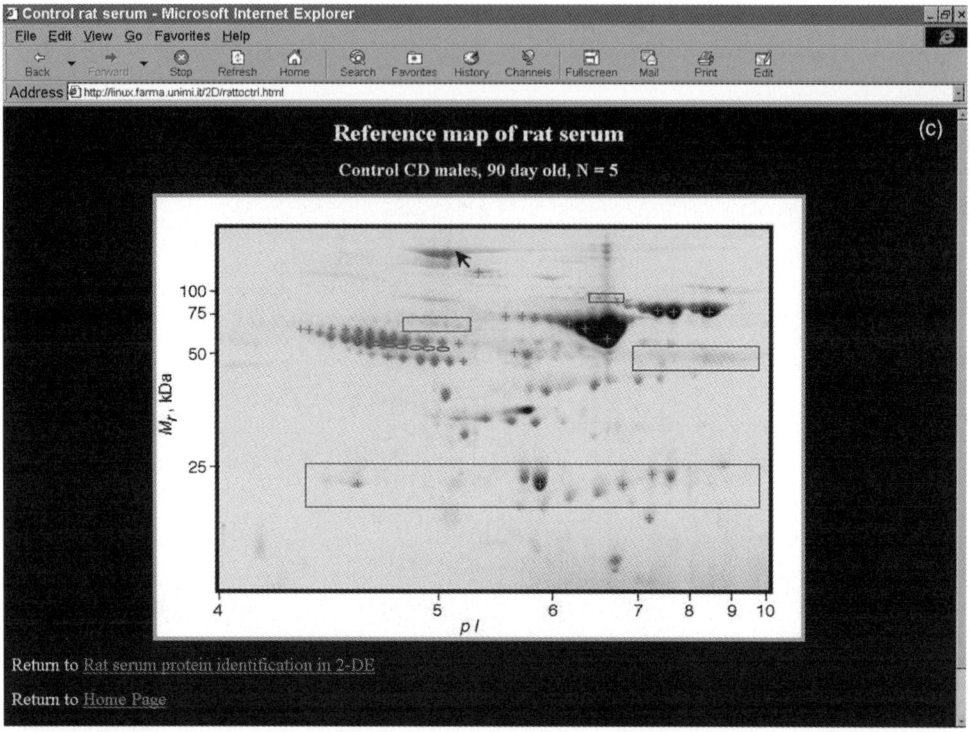

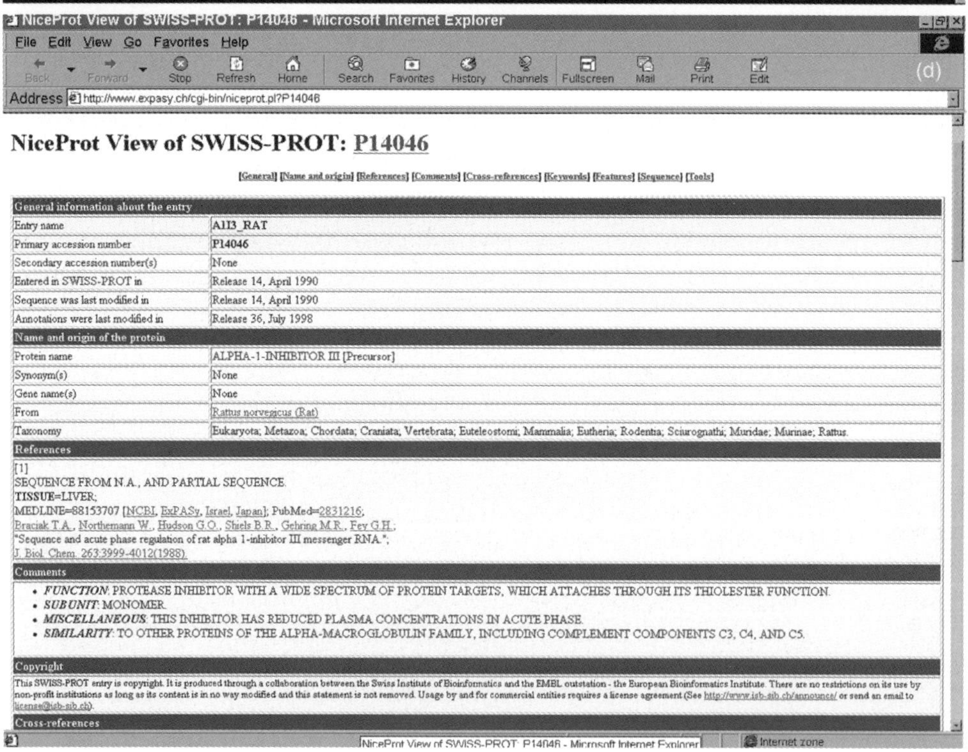

NiceProt View of SWISS-PROT: P14046

[General] [Name and origin] [References] [Comments] [Cross-references] [Keywords] [Features] [Sequence] [Tools]

General information about the entry	
Entry name	**A1I3_RAT**
Primary accession number	**P14046**
Secondary accession number(s)	None
Entered in SWISS-PROT in	Release 14, April 1990
Sequence was last modified in	Release 14, April 1990
Annotations were last modified in	Release 36, July 1998
Name and origin of the protein	
Protein name	ALPHA-1-INHIBITOR III [Precursor]
Synonym(s)	None
Gene name(s)	None
From	Rattus norvegicus (Rat)
Taxonomy	Eukaryota; Metazoa; Chordata; Craniata; Vertebrata; Euteleostomi; Mammalia; Eutheria; Rodentia; Sciurognathi; Muridae; Murinae; Rattus.

References

[1]
SEQUENCE FROM N.A., AND PARTIAL SEQUENCE.
TISSUE=LIVER;
MEDLINE=88153707 [NCBI, ExPASy, Israel, Japan]; PubMed=2831216;
Braciak T.A., Northemann W., Hudson G.O., Shiels B.R., Gehring M.R., Fey G.H.;
"Sequence and acute phase regulation of rat alpha 1-inhibitor III messenger RNA.";
J. Biol. Chem. 263:3999-4012(1988).

Comments

- ***FUNCTION***: PROTEASE INHIBITOR WITH A WIDE SPECTRUM OF PROTEIN TARGETS, WHICH ATTACHES THROUGH ITS THIOLESTER FUNCTION.
- ***SUBUNIT***: MONOMER.
- ***MISCELLANEOUS***: THIS INHIBITOR HAS REDUCED PLASMA CONCENTRATIONS IN ACUTE PHASE.
- ***SIMILARITY***: TO OTHER PROTEINS OF THE ALPHA-MACROGLOBULIN FAMILY, INCLUDING COMPLEMENT COMPONENTS C3, C4, AND C5.

Copyright

This SWISS-PROT entry is copyright. It is produced through a collaboration between the Swiss Institute of Bioinformatics and the EMBL outstation - the European Bioinformatics Institute. There are no restrictions on its use by non-profit institutions as long as its content is in no way modified and this statement is not removed. Usage by and for commercial entities requires a license agreement (See http://www.isb-sib.ch/announce/ or send an email to license@isb-sib.ch).

Cross-references

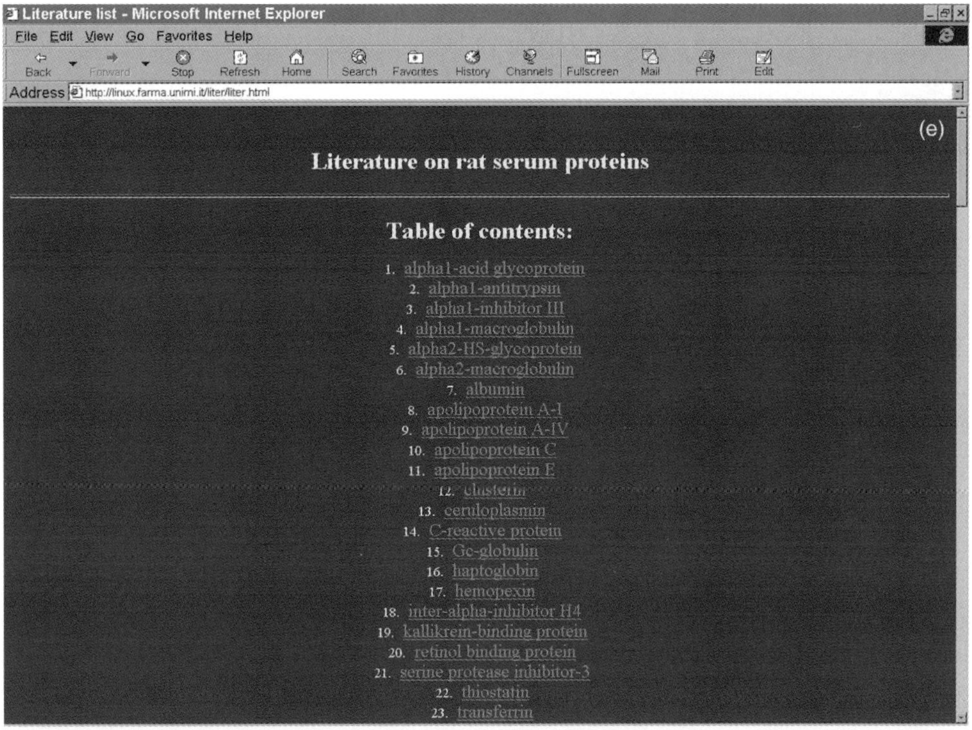
Literature list - Microsoft Internet Explorer
File Edit View Go Favorites Help
Back Forward Stop Refresh Home Search Favorites History Channels Fullscreen Mail Print Edit
Address http://linux.farma.unimi.it/liter/liter.html
(e)
Literature on rat serum proteins
Table of contents:
1. alpha1-acid glycoprotein
2. alpha1-antitrypsin
3. alpha1-inhibitor III
4. alpha1-macroglobulin
5. alpha2-HS-glycoprotein
6. alpha2-macroglobulin
7. albumin
8. apolipoprotein A-I
9. apolipoprotein A-IV
10. apolipoprotein C
11. apolipoprotein E
12. clusterin
13. ceruloplasmin
14. C-reactive protein
15. Gc-globulin
16. haptoglobin
17. hemopexin
18. inter-alpha-inhibitor H4
19. kallikrein-binding protein
20. retinol binding protein
21. serine protease inhibitor-3
22. thiostatin
23. transferrin

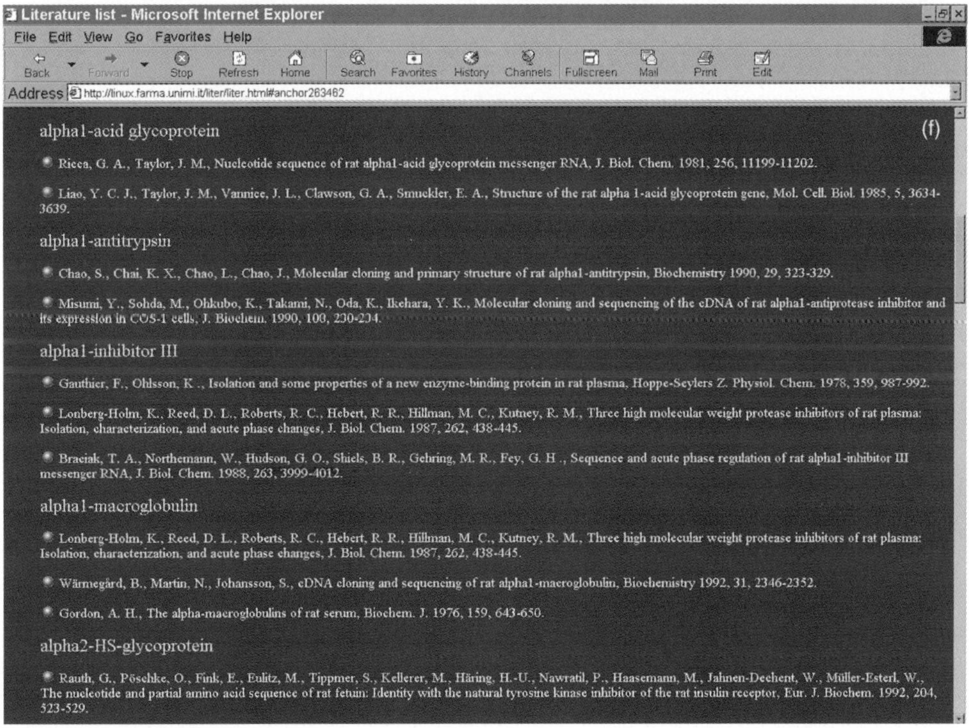
Literature list - Microsoft Internet Explorer
File Edit View Go Favorites Help
Back Forward Stop Refresh Home Search Favorites History Channels Fullscreen Mail Print Edit
Address http://linux.farma.unimi.it/liter/liter.html#anchor263462
(f)
alpha1-acid glycoprotein
Ricca, G. A., Taylor, J. M., Nucleotide sequence of rat alpha1-acid glycoprotein messenger RNA, J. Biol. Chem. 1981, 256, 11199-11202.
Liao, Y. C. J., Taylor, J. M., Vannice, J. L., Clawson, G. A., Smuckler, E. A., Structure of the rat alpha 1-acid glycoprotein gene, Mol. Cell. Biol. 1985, 5, 3634-3639.
alpha1-antitrypsin
Chao, S., Chai, K. X., Chao, L., Chao, J., Molecular cloning and primary structure of rat alpha1-antitrypsin, Biochemistry 1990, 29, 323-329.
Misumi, Y., Sohda, M., Ohkubo, K., Takami, N., Oda, K., Ikehara, Y. K., Molecular cloning and sequencing of the cDNA of rat alpha1-antiprotease inhibitor and its expression in COS-1 cells, J. Biochem. 1990, 108, 230-234.
alpha1-inhibitor III
Gauthier, F., Ohlsson, K ., Isolation and some properties of a new enzyme-binding protein in rat plasma, Hoppe-Seylers Z. Physiol. Chem. 1978, 359, 987-992.
Lonberg-Holm, K., Reed, D. L., Roberts, R. C., Hebert, R. R., Hillman, M. C., Kutney, R. M., Three high molecular weight protease inhibitors of rat plasma: Isolation, characterization, and acute phase changes, J. Biol. Chem. 1987, 262, 438-445.
Braciak, T. A., Northemann, W., Hudson, G. O., Shiels, B. R., Gehring, M. R., Fey, G. H ., Sequence and acute phase regulation of rat alpha1-inhibitor III messenger RNA, J. Biol. Chem. 1988, 263, 3999-4012.
alpha1-macroglobulin
Lonberg-Holm, K., Reed, D. L., Roberts, R. C., Hebert, R. R., Hillman, M. C., Kutney, R. M., Three high molecular weight protease inhibitors of rat plasma: Isolation, characterization, and acute phase changes, J. Biol. Chem. 1987, 262, 438-445.
Wärmegård, B., Martin, N., Johansson, S., cDNA cloning and sequencing of rat alpha1-macroglobulin, Biochemistry 1992, 31, 2346-2352.
Gordon, A. H., The alpha-macroglobulins of rat serum, Biochem. J. 1976, 159, 643-650.
alpha2-HS-glycoprotein
Rauth, G., Pöschke, O., Fink, E., Eulitz, M., Tippmer, S., Kellerer, M., Häring, H.-U., Nawratil, P., Haasemann, M., Jahnen-Dechent, W., Müller-Esterl, W., The nucleotide and partial amino acid sequence of rat fetuin: Identity with the natural tyrosine kinase inhibitor of the rat insulin receptor, Eur. J. Biochem. 1992, 204, 523-529.

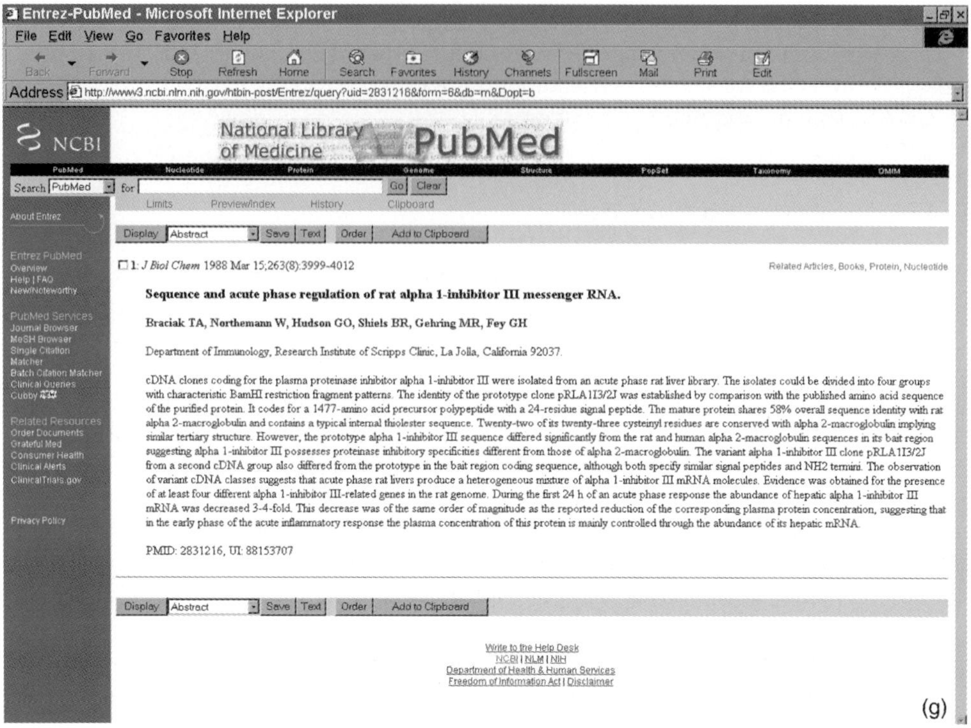

Fig. 14.26. Series of panels (a–g) giving an example of navigation in the site: ‘Rat serum protein study group’, jointly run by the Universities of Milano, Wien and Washington. The address of the site is: http//linux.farma.unimi.it/homeframed.html. For additional details, see text.

automatically obtain a page giving the proper identification of the selected spot. For example, one might be intrigued by the large spot marked here by an arrow, indicating a rather large M_r protein. Clicking on this spot allows you to navigate into the outer world, since this operation opens a link, sure enough, with SWISS-PROT. As shown in Fig. 14.26d, it turns out that this protein (entry name AII3_RAT; primary accession number PI4046) is indeed a protease inhibitor (α1-inhibitor III), which has similarities with other proteins of the α-macroglobulin family, including complement components C3, C4 and C5. If one now goes back to the home page and hits the text ‘literature list’ (lower right side) one obtains the list of Fig. 14.26e, which offers a catalogue of the 24 proteins (the 24th, transthyretin, missing from the list) so far characterised in rat sera. The spot we had previously selected in Fig. 14.26c (α1-inhibitor III) is listed here as No. 3. Below this catalogue of proteins, one can find a long list of relevant publications about all 24 of them, of which we have selected, in Fig. 14.26f, a number pertaining to α1-inhibitor III plus some contiguous species. To the left of each reference, there is a yellow button which, when pressed, will transfer us to the National Library of Medicine (PubMed). In our case, we have selected the last paper on the protein of interest (α1-inhibitor III) and this paper appears now in Fig. 14.26g, which gives the full reference and abstract of the relevant paper. It is thus seen that, while comfortably sitting at our desk computer, it is possible to make a nice journey around the world. Most databases on proteome are structured in a similar fashion, and most of them carry links to a few other databases, particularly to SWISS-PROT, which appears to be a kind of 'Central Dogma' in proteome analysis.

14.4.2. The SWISS-PROT database

As stated above, SWISS-PROT is perhaps the central pillar in proteome and 2-D maps analysis. SWISS-PROT is an annotated protein sequence database, which was created at the Department of Medical Biochemistry of the University of Geneva and has been a collaborative effort with the European Molecular Biology Laboratory (EMBL) since 1987 [244]. SWISS-PROT is available at: http//www.expasy.ch/ch/sprot/ and http//www.ebi.ac.uk/swissprot/. The SWISS-PROT database is set apart from other types of databases by three distinct criteria: (a) annotations; (b) minimal redundancy; and (c) integration with other databases.

In SWISS-PROT, two classes of data can be distinguished: core and annotation data. For each sequence entry, the core data consist of: (i) sequence data; (ii) citation (bibliographic references); and (iii) taxonomic data (description of the biological source of the protein). Conversely, the annotation comprises the following items.

(1) Function(s) of the protein.

(2) Post-translational modifications (if any), such as glycosylation, phoshorylation, acetylation etc. (see Table 14.4).

(3) Domains and sites. As an example, ATP-binding sites, zinc fingers, homeoboxes, Ca-binding regions and the like.

(4) Secondary structures, such as α-helix, β-sheets etc.

(5) Quaternary structure, such as subunit composition and number, homodimers, hetero-oligomers etc.

(6) Similarities (if any) to other proteins.

(7) Potential disease(s) associated with deficiency in the protein.

(8) Sequence conflicts, variants, etc.

In terms of potential redundancy, it is known that many databases contain, for a given protein sequence, separate entries which correspond to different literature reports. In SWISS-PROT, all these data are merged, so as to minimise the redundancy of the database. In terms of integration with other databases, SWISS-PROT provides links with the three main types of sequence-related databases (nucleic acid sequences, protein sequences and protein tertiary structure) as well as with specialised data collection. For example, any sample sequence contains, among others, DR (Databank references) lines that point to EMBL, PDB, OMIM, Pfam and PROSITE. Thus, for a given protein, when available, it is possible to retrieve the nucleic acid sequence that codes for that protein (EMBL), the description of any potential disease associated with the lack (or mutation) of that protein (OMIM), the 3D structure (PDB) and the information specific to the protein family to which it belongs (PROSITE and Pfam). As a recent development, a large number of organisms, whose genome has been recently sequenced, have been added, rendering SWISS-PROT perhaps one of the largest databases encompassing eukariotes as well as prokariotes. Recently added links are those directed to the Zebrafish Information network (ZFIN) database (http://zfish.uoregon.edu/ZFIN/) and to the CarbBank Complex Carbohydrate Structure Database (CCSD) (http://128.192.9.29/carbbank/). As of October 1999, SWISS-PROT contained ca. 81,000 sequence entries, comprising 30 million amino acids abstracted from ca. 65,000 references. The most efficient and friendly way to browse interactively in SWISS-PROT or TrEMBL is to use the WWW molecular biology server ExPASy, at the following address: http//www.expasy.ch/, or at its mirror sites in Australia (http://expasy.proteome.org.au/), Canada (http://expasy.cbr.nrc.ca/) and Taiwan (http://expasy.nhri.org.tw/).

14.4.3. TrEMBL: a supplement to SWISS-PROT

TrEMBL is a computer-annotated supplement to SWISS-PROT, with entries derived from the translation of all coding sequences (CDSs) in the EMBL database (except from the CDSs already existing in the SWISS-PROT). TrEMBL (Translation of EMBL nucleotide sequence database) was introduced in 1996, due to the ever increasing data flow from genome projects, rendering problematic the database annotations in SWISS-PROT. The release 11 of TrEMBL was produced in July 1999 and was based on the translation of all 379,000 CDSs in the EMBL Nucleotide Sequence Database release No. 58. Around 119,000 of these CDSs were already as sequence reports in SWISS-PROT and thus were excluded from TrEMBL. The remaining 260,000 sequence entries were automatically merged so as to reduce the redundancy in TrEMBL. This step has led to 245,761 entries in TrEMBL. For TrEMBL to act as a computer-annotated supplement to SWISS-PROT, new procedures have been introduced for removing redundancy [246] and for automatically adding highly reliable annotation [247].

14.4.4. The SWISS-2DPAGE database

The SWISS-2DPAGE database collects data on proteins identified on various 2-D maps [248,243]. The identification of proteins on such 2-D maps was obtained by a number of techniques, including gel comparison, microsequencing, immunoblotting, amino acid composition analysis, peptide mass fingerprinting via mass spectrometry, or a combination of some of these approaches. The core of this database consists therefore in the description of the proteins identified, including mapping procedures, physiological and pathological relevance of each protein, experimental data (isoelectric point, molecular mass, amino acid composition, peptide masses) and bibliographic references, in addition to the 2-D gel images showing the protein location on the map. Cross-references are provided to MEDLINE, to other 2-D gel databases (such as ECO2DBASE, HSC-2DPAGE, YPD) and to two newly built SIENA-2DPAGE and PHCI-2DPAGE sites, as well as to SWISS-PROT, which in turn provides many links to other molecular databases (EMBL, GenBank, PROSITE, MIM, and the like). The release 11 of SWISS-2DPAGE (October 1999) contains 24 reference maps. Fourteen of them are from human cells or tissues (kidney, liver, lymphoma, platelet cells, red blood cells, colorectal epithelial cells), body fluids (cerebrospinal fluid, plasma), culture cells (colorectal adenocarcinoma cells, erythroleukemia cells, hepatoblastoma carcinoma-derived cells, hepatoblastoma carcinoma-derived cell line secreted proteins, macrophage like cell line, promyelocytic leukemia derived cells); four are from mouse cells or tissues (liver, gastrocnemious muscle, pancreatic islet cells, epididymal fat pad); the other maps are from *S. cerevisiae*, *E. coli* (obtained from four pH ranges: 3.5–10, 4–5, 4.5–5.5 and 5–6) and *Dictyostelium discoideum* origin. On these maps a total of 2824 spots have been identified by a number of techniques, including gel matching (51%), immunoblotting (24%), co-migration (3%), microsequencing (17%), amino acid composition (6%) and mass spectrometry (15%). The reason why the total exceeds 100% is because some spots have been identified by more than one method. These identifications correspond, at present, to only 614 different protein entries from human, mouse, yeast, *E. coli* and *D. discoideum* origin. One might ask why, out of such a large number of spots seen in all these 2-D maps, there are only so few different proteins. One reason for that could be the tremendous increase in spot numbers due to post-translation modifications (a long list of which can be found in Table 14.3). In principle, the product of a single gene could appear into as many as 10 (or more) different spots (each representing an individual peptide chain) in a 2-D map. Often, such spots form 'trains' or 'ladders', as clearly visible in human plasma protein patterns. In the case of such plasma proteins, Sariouglu et al. [249] have recently reported a subtle and often undetected cause for such heterogeneity: deamidation of asparagine residues. This process can hardly be detected even with the most sophisticated MALDI-TOF-MS techniques, since each deamidation event of Asn to Asp produces only a mass difference of exactly 1 Da, well within the experimental error of MS analysis of intact proteins. Thus, in order to detect such minute mass differences, these authors had to digest spots eluted from a 2-D maps of human serum, separate the relevant peptides by reversed-phase HPLC and then analyse them by mass spectrometry, in the ESI-FT-ICR (electrospray, Fourier-transform, ion-cyclotron resonance) mode. According to these authors, deamidation of Asn is a widespread phenomenon, strongly relevant at least in 2-D maps of human blood plasma proteins.

In order to understand the complexity and the architecture of these databases in general, the reader is referred to a comprehensive chapter by Bairoch [239], which should be mandatory reading for those navigating into the domain of 2-D maps. Just as an example, let us try a short cruise into the realm of 2-D maps available on Internet. By entering the SWISS-2DPAGE site, one can select, e.g. a map of human serum. In this map (see Fig. 14.27a) all the spots marked by a red cross represent polypeptides which have been identified by various means. By clicking on any of those crosses (e.g. on the one marked by an arrow and a circle on the lower left side) one obtains the response shown in Fig. 14.27b: all the personal data of the protein are recorded, including some basic references to it. In addition, the table shows not only when the spot was first entered in the site, but also when it was last modified. In addition, by clicking on the line 'View entry in original SWISS-2DPAGE format', one is offered the table shown in Fig. 14.27c, which gives a series of headings offering some basic information on the spot. The significance of the various lines in this figure is given below.

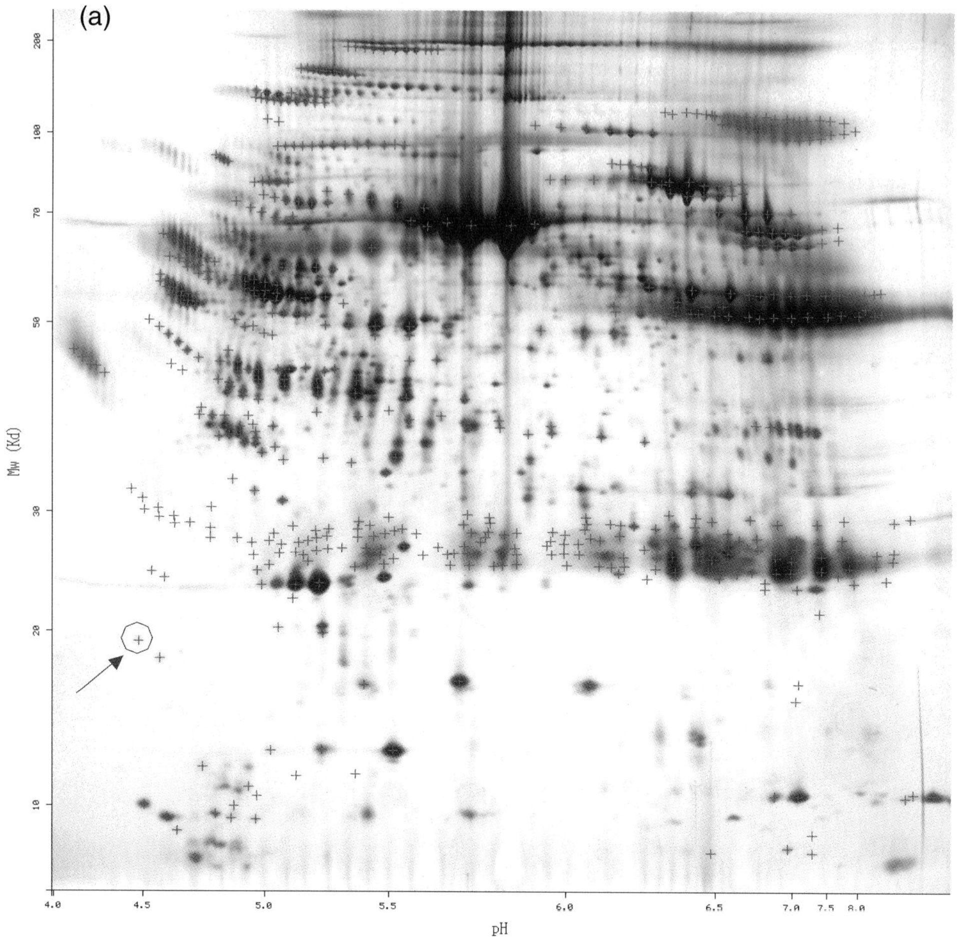

Fig. 14.27. Looking for the identity of protein spots on a 2-D map in SWISS-2DPAGE. (a) Published 2-D map of human serum. The red crosses represent identified polypeptide chains. By clicking on the spot marked by a circle and an arrow, one obtains the table (b), in which the post is identified as hemopexin (β-1B-glycoprotein). By clicking on the line 'view new entry in original SWISS-2DPAGE format', one obtains the table (c). The significance of the various lines is explained in the text.

ID: it gives the identification of such a spot and it contains the entry name, useful for retrieving this particular entry from the database.

AC: accession number. It is a stable identifier, since the entry name could change over time. If more than one number is listed, the first one is 'the primary accession number', which is the one typically cited in most databases.

DT: date (in general three lines). The first line tells when the sequence was first entered in SWISS-PROT (or 2DPAGE); the second when this sequence was last modified and the third when the most recent information was added to the entry.

DE: description line, which gives the name(s) of the protein.

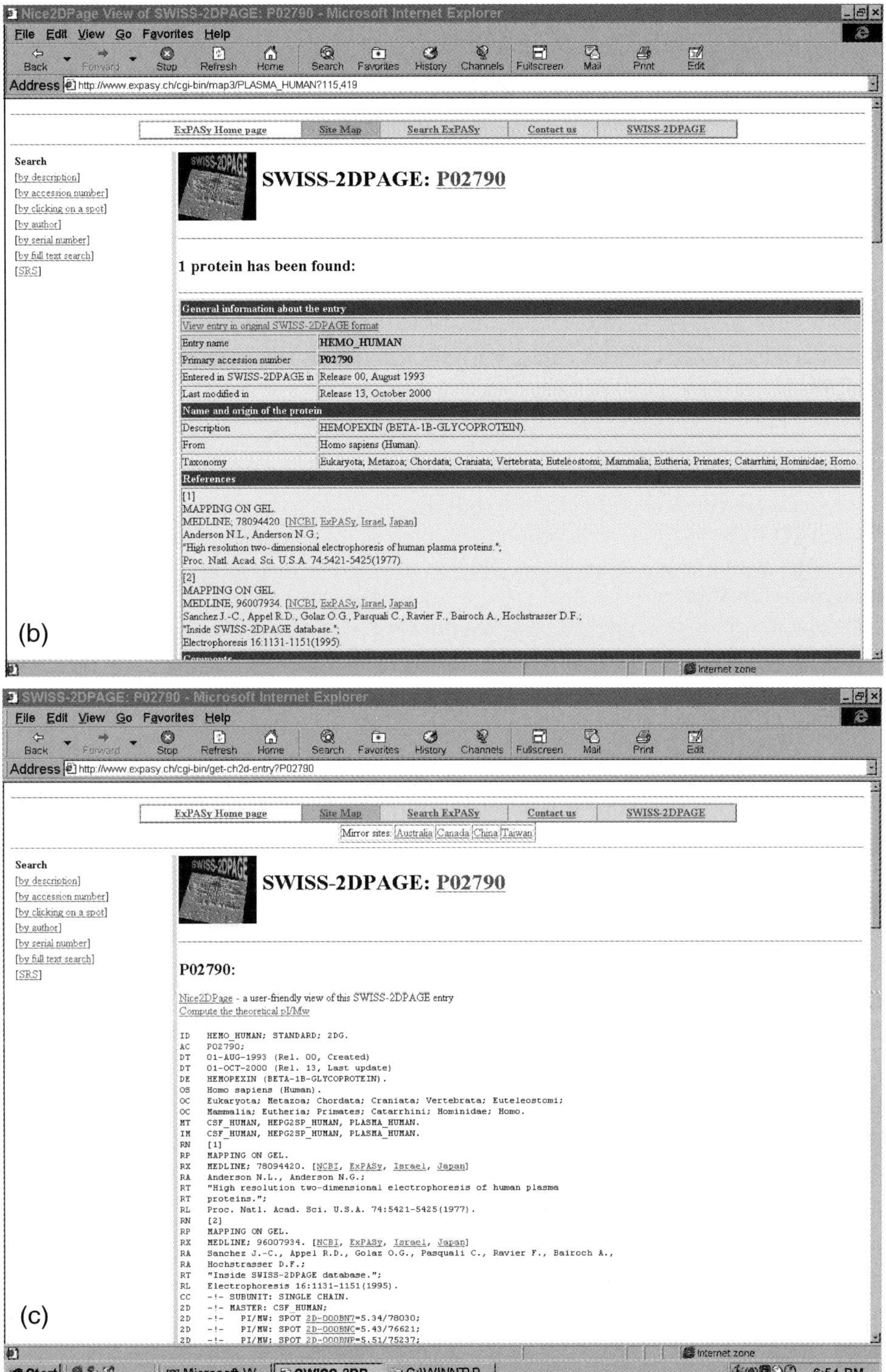
Nice2DPage View of SWISS-2DPAGE: P02790 - Microsoft Internet Explorer
File Edit View Go Favorites Help
Back Forward Stop Refresh Home Search Favorites History Channels Fullscreen Mail Print Edit
Address http://www.expasy.ch/cgi-bin/map3/PLASMA_HUMAN?115,419
ExPASy Home page | Site Map | Search ExPASy | Contact us | SWISS-2DPAGE
Search
[by description]
[by accession number]
[by clicking on a spot]
[by author]
[by serial number]
[by full text search]
[SRS]
SWISS-2DPAGE: P02790
1 protein has been found:
General information about the entry
View entry in original SWISS-2DPAGE format
Entry name HEMO_HUMAN
Primary accession number P02790
Entered in SWISS-2DPAGE in Release 00, August 1993
Last modified in Release 13, October 2000
Name and origin of the protein
Description HEMOPEXIN (BETA-1B-GLYCOPROTEIN).
From Homo sapiens (Human).
Taxonomy Eukaryota; Metazoa; Chordata; Craniata; Vertebrata; Euteleostomi; Mammalia; Eutheria; Primates; Catarrhini; Hominidae; Homo.
References
[1]
MAPPING ON GEL.
MEDLINE; 78094420. [NCBI, ExPASy, Israel, Japan]
Anderson N.L., Anderson N.G.;
"High resolution two-dimensional electrophoresis of human plasma proteins.";
Proc. Natl. Acad. Sci. U.S.A. 74:5421-5425(1977).
[2]
MAPPING ON GEL.
MEDLINE; 96007934. [NCBI, ExPASy, Israel, Japan]
Sanchez J.-C., Appel R.D., Golaz O.G., Pasquali C., Ravier F., Bairoch A., Hochstrasser D.F.;
"Inside SWISS-2DPAGE database.";
Electrophoresis 16:1131-1151(1995).
(b)
Internet zone
SWISS-2DPAGE: P02790 - Microsoft Internet Explorer
File Edit View Go Favorites Help
Back Forward Stop Refresh Home Search Favorites History Channels Fullscreen Mail Print Edit
Address http://www.expasy.ch/cgi-bin/get-ch2d-entry?P02790
ExPASy Home page | Site Map | Search ExPASy | Contact us | SWISS-2DPAGE
Mirror sites: Australia Canada China Taiwan
Search
[by description]
[by accession number]
[by clicking on a spot]
[by author]
[by serial number]
[by full text search]
[SRS]
SWISS-2DPAGE: P02790
P02790:
Nice2DPage - a user-friendly view of this SWISS-2DPAGE entry
Compute the theoretical pI/Mw
ID HEMO_HUMAN; STANDARD; 2DG.
AC P02790;
DT 01-AUG-1993 (Rel. 00, Created)
DT 01-OCT-2000 (Rel. 13, Last update)
DE HEMOPEXIN (BETA-1B-GLYCOPROTEIN).
OS Homo sapiens (Human).
OC Eukaryota; Metazoa; Chordata; Craniata; Vertebrata; Euteleostomi;
OC Mammalia; Eutheria; Primates; Catarrhini; Hominidae; Homo.
MT CSF_HUMAN, HEPG2SP_HUMAN, PLASMA_HUMAN.
IM CSF_HUMAN, HEPG2SP_HUMAN, PLASMA_HUMAN.
RN [1]
RP MAPPING ON GEL.
RX MEDLINE; 78094420. [NCBI, ExPASy, Israel, Japan]
RA Anderson N.L., Anderson N.G.;
RT "High resolution two-dimensional electrophoresis of human plasma
RT proteins.";
RL Proc. Natl. Acad. Sci. U.S.A. 74:5421-5425(1977).
RN [2]
RP MAPPING ON GEL.
RX MEDLINE; 96007934. [NCBI, ExPASy, Israel, Japan]
RA Sanchez J.-C., Appel R.D., Golaz O.G., Pasquali C., Ravier F., Bairoch A.,
RA Hochstrasser D.F.;
RT "Inside SWISS-2DPAGE database.";
RL Electrophoresis 16:1131-1151(1995).
CC -!- SUBUNIT: SINGLE CHAIN.
2D -!- MASTER: CSF_HUMAN;
2D -!- PI/MW: SPOT 2D-000BN7=5.34/78030;
2D -!- PI/MW: SPOT 2D-000BNC=5.43/76621;
2D -!- PI/MW: SPOT 2D-000BNP=5.51/75237;
(c)
Internet zone
Start Microsoft W... SWISS-2DP... C:\WINNT\P...
6:54 PM

GN: gene name. It lists the name(s) of the protein's gene, but it could be absent if no gene name had been assigned.

OS: organism species. It lists the organism name in Latin (first the genus, followed by the species to which it belongs).

OC: organism classification. It gives the taxonomic tree; this could be followed by an OG (organelle) line, if applicable.

MT: master line. It gives the reference map where the entry has been identified.

IM: image line. It catalogues the 2-D images available for the entry.

RN: reference number.

RP: reference on the protein. It outlines the type of work carried out on such a protein.

RC: comments to the reference (e.g. for indicating the tissue or strain from which the protein was extracted).

RX: for cross-reference (e.g. it assigns a specific reference in a bibliographic database, such as Medline).

RA: authors in the given reference.

RL: location of reference. It gives the standard journal citation.

2D: the 2D lines integrate different topics, such as mapping procedure, spot co-ordinates, protein amino acid composition, peptide mass from MS data, protein expression levels and modification.

DR: database cross-reference. It provides links from SWISS-PROT to other biomolecular databases.

It should be here recalled that, for standardisation purposes, the format of SWISS-2DPAGE entries is similar to that of the SWISS-PROT database, so that what we have illustrated above is valid for both sites. Since most of the above databases and sites have been given as acronyms, we have listed, in Table 14.6, the signification, when available, of such terms, since we feel that this will help the readers to better remember such sites.

14.4.5. Database searching via mass-spectrometric information

MS combined with database searching is becoming perhaps the most popular method for identifying proteins during proteome projects. In a typical proteomics experiment, the protein of interest is first enriched (when possible) and then separated by 2-D electrophoresis. The spots thus separated are digested with an enzyme (typically trypsin) and then the proteolytic peptides are analysed by mass spectrometry. Thus, the starting point for MS analysis is in general peptide mapping, followed by a search in appropriate databases. The simplest scoring method for peptide mapping is to count the number of measured peptide masses that match calculated peptide masses within the accuracy of the measurement. This scoring method is adequate for high-quality experimental data, but has the weakness that it usually gives higher scores to larger proteins, for which more possible peptides can be calculated and thus provide better random matching. Table 14.7 gives the URLs for the most popular database-search programs today available, whereas Table 14.8 offers a comparison of the different algorithm characteristics for MS-related protein identification.

TABLE 14.6

SIGNIFICANCE OF A NUMBER OF ACRONYMS RELATED TO DATABASES

Acronym	Signification
AMSdb	AntiMicrobial Sequences of proteins
BLOCKS	Database of blocks of multiple sequence aligments containing no gaps
DDBJ	DNA DataBase of Japan
DSSP	Dictionary of Secondary Structure of Proteins
EcoCyc	*Escherichia coli* metabolic pathways
EMP/WIT	Enzyme and Metabolic Pathways database; What Is There
EST	Expressed Sequence Tags
FSSP	Families of Structurally Similar Proteins
GCRdb	G-protein Coupled Receptors database
HSSP	Homology-derived Secondary Structure of Proteins
KEGG	Kyoto Encyclopedia of Genes and Genomes
MIM	Mendelian Inheritance in Man
NCBI	National Center for Biotechnology Information
O-GLYBASE	O-linked glycosylation sites database
OMIM	On-line MIM
PDB	Protein Data Bank
Pfam	Protein Families (multiple alignments of protein domains)
PRINTS	Protein fingerprints (series of conserved sequence alignments)
ProDom	Protein Domain database
PROSITE	Protein families and domains site
PTM	Post-Translational Modifications
TrEMBL	Translation of European Molecular Biology Laboratory sequence database
YPD	Yeast Protein Database

TABLE 14.7

URLs FOR THE PRIMARY SITES ASSOCIATED WITH DATABASE-SEARCH ALGORITHMS

Name	Address
ProFound	http//www.proteometrics.com/prowl-cgi/ProFound.exe
Mascot	http//www.matrixscience.com/cgi/search_form.pl?SEARCH=PMF
PepSea	http://pepsea.protana.com/PA_PepSeaForm.html
	http://pepsea.protana.com/PA_PeptidePatternForm.html
MS-Fit	http://prospector.ucsf.edu/ucsfhtml3.2/msfit.htm
MOWSE	http://srs.hgmp.mrc.ac.uk/cgi-bin/mowse
PepIdent	http//www.expasy.ch/tools/peptident.html
Multident	http//www.expasy.ch/tools/multident/
SEQUEST	http://thomposn.mbt.washignton.edu/sequest/
Mascot	http//www.matrixscience.com/cgi/search_form.pl?SEARCH=MIS
PepFrag	http//www.proteometrics.com/prowl/PepFragch.html
MS-Tag	http://prospector.ucsf.edu/ucsfhtml3.2/mstagfd.htm

It is of interest to illustrate here a typical search of databases starting with experimental data for a spot eluted from a 2-D map. The search here shown was made with PeptIdent. Fig. 14.28a shows a portion of a 2-D map of human serum, blotted on an Immobilon membrane and stained with colloidal Coomassie Blue. The white holes with

TABLE 14.8

COMPARISON OF DATABASE SEARCH ALGORITHM CHARACTERISTICS FOR MS-BASED PROTEIN IDENTIFICATION

Name	MS type	Taxonomy data	Enzymes	Sequence modifications	Protein properties	Masses	Other inputs
ProFound	MS and MS-MS	Yes	8 + user defined	User defined	Mass + p*I*	Mo and A	Aa
Mascot	MS and MS-MS	Yes	10	Predefined	Mass	Mo or A	–
PepSea	MS and MS-MS	No	8	Cys blocking + Met oxidation	Mass	Mo or A	Sequence tags
MS-Fit	MS	Yes	11 + 12 mixtures	Predefined	Mass + p*I*	Mo or A	AA
MOWSE	MS	Yes	8	None	Mass	Mo and A	AA + sequence tags
PepIdent	MS	Yes	1	Cys blocking + Met oxidation	Mass + p*I*	Mo or A	–
Multident	MS	Yes	9	Cys blocking + Met oxidation	Mass + p*I*	Mo or A	AA + sequence tags
SEQUEST	MS-MS	Yes	User defined	User defined	Mass + p*I*	Mo or A	–
PepFrag	MS-MS	Yes	5	Cys blocking + phosphorylation	Mass + p*I*	Mo or A	AA
MS-Tag	MS-MS	Yes	11 + 12 mixtures	Predefined	Mass + p*I*	Mo or A	AA

Abbreviations: A, average chemical mass; AA, amino acid composition; Mo, monoisotopic mass; MS, mass spectrometry; MS-MS, tandem MS. (Modified from [177], with permission.)

numbers show the location of the spots excised for protein identification. We will follow here the fate of spots Nos. 4 and 5. After digestion with trypsin and analysis of the derived peptides by MALDI-TOF MS, the data thus obtained are listed under an Excel sheet, as shown in Fig. 14.28b. In this figure, out of the first 19 lines only the first one is filled with the spot number; the remaining will be automatically filled at the end of the search (if successful). At the end of these empty lines we see four columns, which list the peptide M_r values (as obtained by MALDI-TOF) with the relative peak intensity for the two spots under analysis. Although this is only a partial list, spot 4 gave 57 readable peptides, whereas spot 5 produced as many as 67 peptides. Fig. 14.28c shows the initial image when entering PeptIdent at the Expasy server. At the top left side (name of unknown protein) one enters the spot No. (in this case No. 5). The database for the search is given typically as a combined SWISS-PROT TrEMBL. The species to be searched is here *Homo sapiens*. On the right side, one has to enter the p*I* (here given as 8) and the mass value (here given as 60, 000 ± 50%, both data pertaining to the intact polypeptide. At this point, all the list of peptide masses obtained for spot 5 (67 fragments) are entered in the appropriate window. Below this, some chemical information has to be given, such as the type of mass obtained (typically the mono-protonated mass) and whether it is monoisotopic or average. Other informations about the state of the peptide are how

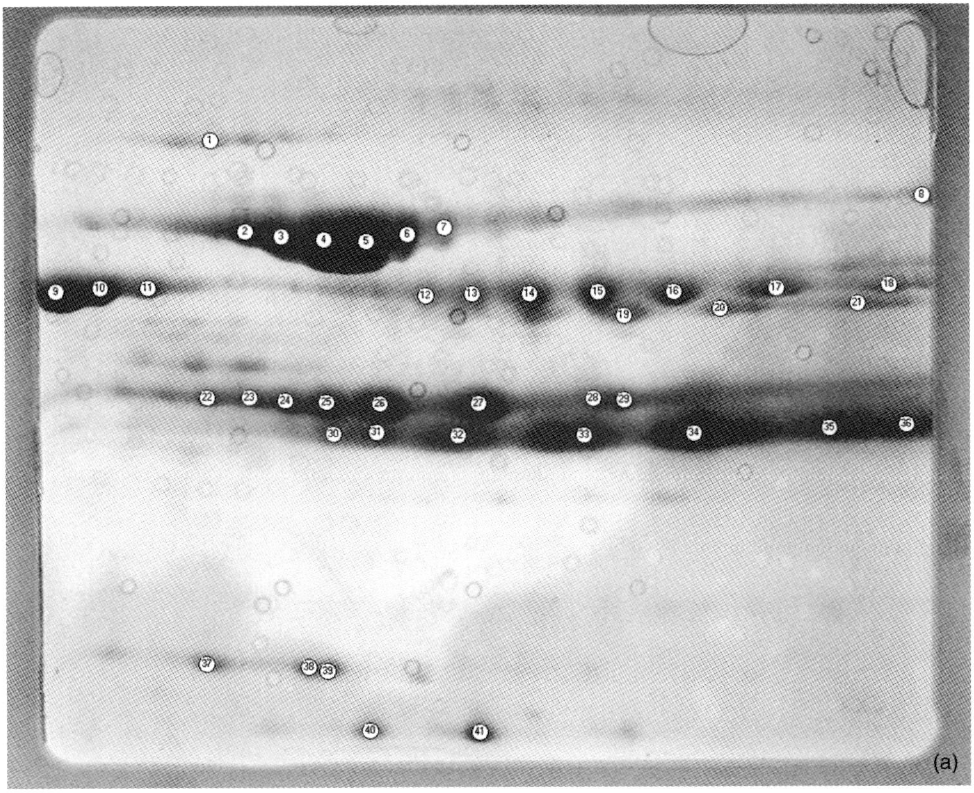

Fig. 14.28. Series of panels (a–g) giving an example of identification of a protein eluted from a 2-D map of human serum, via the PeptIdent algorithm. The address of the site is: http//www.expasy.ch/tools/peptident.html. For additional details, see text. (From A. Castagna, B. Herbert and P.G. Righetti, unpublished.)

the Cys residues are blocked (by default with iodoacetamide) and if Met residues are oxidised (in general, one allows for such oxidation event). The mass tolerance is of course minute, barely ±0.1 Da, as it should when analysing peptides. On the right side, another important information to be added is the type of enzyme used for cleavage (in general trypsin) and if there have been any missed cleavages. Fig. 14.28d summarises all the input data, as described above. At this point the search begins. Fig. 14.28e gives the results of such a search: a long list of potential candidates is given, but it is seen that the one with the highest peptide matches (a total of 31) and the highest possible score is human transferrin, having a p*I* of 6.7 and an M_r value of 75,181 Da (the same search for spot No. 4 gave the same identity, in fact all the train of bands 2–7 were found to be isotransferrins). It can be additionally noticed that, except for the second protein listed, which could have given some uncertainties in identification, due to the rather close number of peptide matches (27 vs. 31) and a rather high score (0.40 vs. 0.46) all the other proteins listed could have been easily excluded from the search, given the very low number of peptide matches and the quite low score. Fig. 14.28f,g lists the 31

Spot number	a4		a5	
ID confidence				
final ID name				
final ID accession number				
Final number of hits				
Spot quality				
Spectrum quality				
Search result				
Name of protein				
Accession/contig number				
2D location				
Theoretical pI				
Theoretical MW				
Subcellular location				
Function				
Coverage				
Within gel spot matches				
Number of hits				
Added hits				
Final number of hits				
	Peptide Mr	Intensity of signal	Peptide Mr	Intensity of signal
Peaklists	828.373	152.77	827.407	619.76
	876.423	242.17	849.407	154.63
	997.523	430.78	864.507	301.3
	1000.52	257.21	874.457	268.93
	1138.57	175.64	876.438	371.39
	1167.62	145.16	887.457	141.26
	1178.57	673.4	909.457	176.36
	1180.57	252.56	934.457	203.05
	1195.62	5550.73	964.557	563.86
	1211.57	296.97	978.507	122.46
	1239.62	214.9	997.507	1883.07
	1249.67	274.65	1000.56	489.7
	1258.67	192.05	1029.61	169.09
	1265.62	805.61	1125.61	291.14
	1273.72	364.51	1138.56	260.73
	1276.67	1531.98	1167.61	228.25
	1283.62	7757.03	1178.56	1346.47
	1317.62	401.71	1180.56	234.25
	1323.67	213.8	1195.61	7371.44
	1354.67	1042.26	1211.56	555.27
	1438.52	281.77	1217.56	309.68
	1448.67	1133.72	1249.66	450.8
	1462.72	204.89	1258.66	331.22
	1474.67	296.03	1265.61	1275.93
	1478.77	5074.1	1273.71	595.26
	1492.77	280.61	1276.66	2021.64
	1494.72	857.89	1283.61	10322.58
	1496.72	192.62	1305.61	328.66
	1514.67	185.99	1317.61	467.53
	1521.77	333.39	1323.66	291.94
	1531.67	2739.64	1354.66	1536.11
	1539.72	2324.2	1368.66	296.42
	1553.72	188.15	1448.66	1268.86
	1577.67	1776.31	1462.71	329.95
	1586.77	3561.18	1474.66	271.9
	1591.67	290.28	1478.76	6553.35

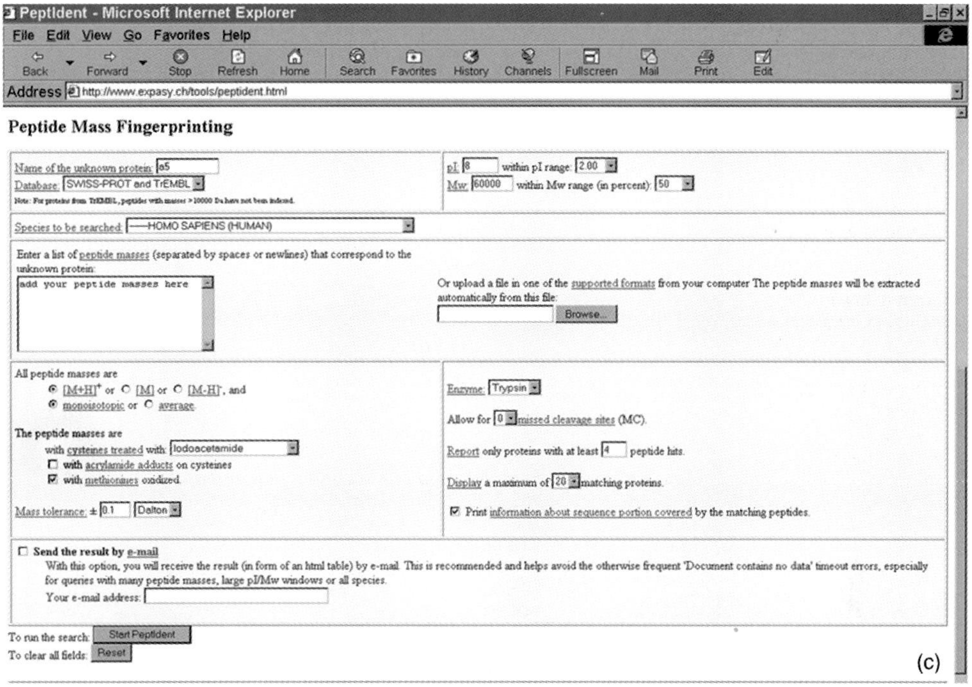
PeptIdent - Microsoft Internet Explorer
File Edit View Go Favorites Help
Back Forward Stop Refresh Home Search Favorites History Channels Fullscreen Mail Print Edit
Address http://www.expasy.ch/tools/peptident.html
Peptide Mass Fingerprinting
Name of the unknown protein: a5
Database: SWISS-PROT and TrEMBL
pI: 8 within pI range: 2.00
Mw: 60000 within Mw range (in percent): 50
Species to be searched: ----HOMO SAPIENS (HUMAN)
Enter a list of peptide masses (separated by spaces or newlines) that correspond to the unknown protein:
add your peptide masses here
Or upload a file in one of the supported formats from your computer The peptide masses will be extracted automatically from this file:
Browse...
All peptide masses are
[M+H]+ or [M] or [M-H]-, and
monoisotopic or average.
The peptide masses are
with cysteines treated with: Iodoacetamide
with acrylamide adducts on cysteines
with methionines oxidized.
Mass tolerance: ± 0.1 Dalton
Enzyme: Trypsin
Allow for 0 missed cleavage sites (MC).
Report only proteins with at least 4 peptide hits.
Display a maximum of 20 matching proteins.
Print information about sequence portion covered by the matching peptides.
Send the result by e-mail
With this option, you will receive the result (in form of an html table) by e-mail. This is recommended and helps avoid the otherwise frequent 'Document contains no data' timeout errors, especially for queries with many peptide masses, large pI/Mw windows or all species.
Your e-mail address:
To run the search: Start PeptIdent
To clear all fields: Reset
(c)

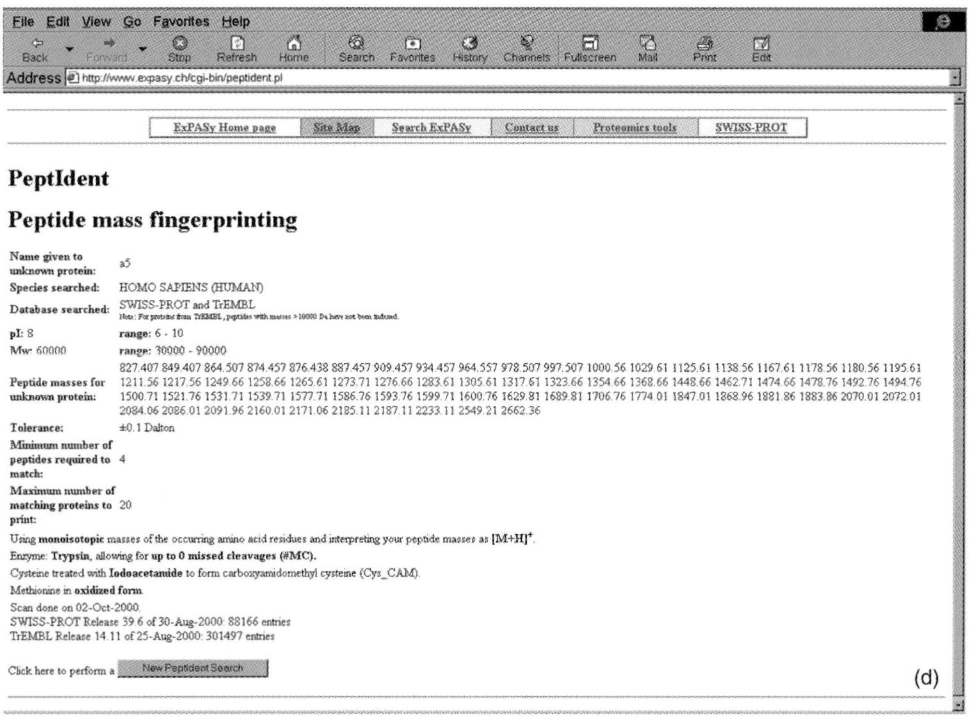
File Edit View Go Favorites Help
Back Forward Stop Refresh Home Search Favorites History Channels Fullscreen Mail Print Edit
Address http://www.expasy.ch/cgi-bin/peptident.pl
ExPASy Home page Site Map Search ExPASy Contact us Proteomics tools SWISS-PROT
PeptIdent
Peptide mass fingerprinting
Name given to unknown protein: a5
Species searched: HOMO SAPIENS (HUMAN)
Database searched: SWISS-PROT and TrEMBL
pI: 8 range: 6 - 10
Mw: 60000 range: 30000 - 90000
Peptide masses for unknown protein: 827.407 849.407 864.507 874.457 876.438 887.457 909.457 934.457 964.557 978.507 997.507 1000.56 1029.61 1125.61 1138.56 1167.61 1178.56 1180.56 1195.61 1211.56 1217.56 1249.66 1258.66 1265.61 1273.71 1276.66 1283.61 1305.61 1317.61 1323.66 1354.66 1368.66 1448.66 1462.71 1474.66 1478.76 1492.76 1494.76 1500.71 1521.76 1531.71 1539.71 1577.71 1586.76 1593.76 1599.71 1600.76 1629.81 1689.81 1706.76 1774.01 1847.01 1868.96 1881.86 1883.86 2070.01 2072.01 2084.06 2086.01 2091.96 2160.01 2171.06 2185.11 2187.11 2233.11 2549.21 2662.36
Tolerance: ±0.1 Dalton
Minimum number of peptides required to match: 4
Maximum number of matching proteins to print: 20
Using monoisotopic masses of the occurring amino acid residues and interpreting your peptide masses as [M+H]+.
Enzyme: Trypsin, allowing for up to 0 missed cleavages (#MC).
Cysteine treated with Iodoacetamide to form carboxyamidomethyl cysteine (Cys_CAM).
Methionine in oxidized form.
Scan done on 02-Oct-2000.
SWISS-PROT Release 39.6 of 30-Aug-2000: 88166 entries
TrEMBL Release 14.11 of 25-Aug-2000: 301497 entries
Click here to perform a New PeptIdent Search
(d)

PeptIdent - Microsoft Internet Explorer

Address http://www.expasy.ch/cgi-bin/peptident.pl

Displaying the first 50 matches.

(e)

Score	# peptide matches	AC	ID	Description	pI	Mw
0.46	31	P02787	TRFE_HUMAN_1	CHAIN 1: SEROTRANSFERRIN. - Homo sapiens (Human).	6.70	75181.44
0.40	27	Q9UHV0	Q9UHV0	PRO1400. - Homo sapiens (Human).	6.95	63407.09
0.22	15	Q9UHU8	Q9UHU8	PRO1557. - Homo sapiens (Human).	7.42	37542.69
0.12	8	Q9UJV9	ABS_HUMAN	DEAD-BOX PROTEIN ABSTRAKT HOMOLOG. - Homo sapiens (Human).	8.07	69737.92
0.12	8	Q92771	Q92771	HELICASE (FRAGMENT). - Homo sapiens (Human).	undefined	undefined
0.12	8	O00471	O00471	BRAIN SECRETORY PROTEIN HSEC10P. - Homo sapiens (Human).	6.27	81852.87
0.12	8	Q15597	Q15597	TRANSLATION INITIATIONFACTOR EIF-4GAMMA (FRAGMENT). - Homo sapiens (Human).	undefined	undefined
0.10	7	Q9UBB3	Q9UBB3	ZXDA, ZXDB (ZINC FINGER X-LINKED PROTEIN). - Homo sapiens (Human).	6.46	84792.33
0.10	7	Q99759	M3K3_HUMAN	MITOGEN-ACTIVATED PROTEIN KINASE KINASE KINASE 3 (EC 2.7.1.-) (MAPK/ERK KINASE KINASE 3) (MEK KINASE 3) (MEKK 3). - Homo sapiens (Human).	8.91	70969.75
0.10	7	P40939	ECHA_HUMAN_1	CHAIN 1: MITOCHONDRIAL TRIFUNCTIONAL ENZYME ALPHA - Homo sapiens (Human).	8.98	78969.76
0.10	7	Q16679	Q16679	GASTRIN-BINDING PROTEIN. - Homo sapiens (Human).	9.13	83041.73
0.09	6	Q13476	Q13476	KINASE SUPPRESSOR OF RAS-1 (KSR1) (FRAGMENT). - Homo sapiens (Human).	undefined	undefined
0.09	6	Q9UMS6	Q9UMS6	MYOPODIN PROTEIN (FRAGMENT). - Homo sapiens (Human).	undefined	undefined
0.09	6	Q9UNA4	Q9UNA4	RAD30B. - Homo sapiens (Human).	6.37	80346.29
0.09	6	Q16249	Q16249	MYOSIN HEAVY CHAIN 12 (FRAGMENT). - Homo sapiens (Human).	undefined	undefined
0.09	6	P01833	PIGR_HUMAN_2	CHAIN 2: SECRETORY COMPONENT. - Homo sapiens (Human).	6.34	64287.44
0.09	6	P13497	BMP1_HUMAN_1	CHAIN 1: BONE MORPHOGENETIC PROTEIN 1. - Homo sapiens (Human).	8.70	69700.68
0.09	6	Q06187	BTK_HUMAN	TYROSINE-PROTEIN KINASE BTK (EC 2.7.1.112) (BRUTON'S TYROSINE KINASE) (AGAMMAGLOBULINAEMIA TYROSINE KINASE) (ATK) (B CELL PROGENITOR KINASE) (BPK). - Homo sapiens (Human).	7.83	76281.24
0.09	6	O94845	O94845	KIAA0744 PROTEIN. - Homo sapiens (Human).	8.86	65886.59
0.09	6	Q92498	Q92498	CHL1 PROTEIN. - Homo sapiens (Human).	8.24	75728.03
0.09	6	Q93000	Q93000	CHL1 PROTEIN (FRAGMENT). - Homo sapiens (Human).	undefined	undefined
0.09	6	Q15052	Q15052	KIAA0006 PROTEIN (FRAGMENT). - Homo sapiens (Human).	undefined	undefined
0.09	6	Q99423	Q99423	BONE MORPHOGENETIC PROTEIN BMP1-6. - Homo sapiens (Human).	8.40	81082.48
0.07	5	O14992	O14992	HS24/P52. - Homo sapiens (Human).	8.51	52367.48
0.07	5	O60592	O60592	ARG/ABL-INTERACTING PROTEIN ARGBP2A. - Homo sapiens (Human).	8.99	74804.05
0.07	5	O15297	O15297	WIP1. - Homo sapiens (Human).	9.14	66675.10
0.07	5	P23276	KELL_HUMAN	KELL BLOOD GROUP GLYCOPROTEIN (EC 3.4.24.-). - Homo sapiens (Human).	8.09	82823.86
0.07	5	Q05655	KPCD_HUMAN	PROTEIN KINASE C, DELTA TYPE (EC 2.7.1.-) (NPKC-DELTA). - Homo sapiens (Human).	7.93	77477.00
0.07	5	Q00537	KPT2_HUMAN	SERINE/THREONINE-PROTEIN KINASE PCTAIRE-2 (EC 2.7.1.-). - Homo sapiens (Human).	9.10	59664.37
0.07	5	O60640	O60640	HYPOTHETICAL 76.7 KDA PROTEIN. - Homo sapiens (Human).	6.63	76729.39

PeptIdent - Microsoft Internet Explorer

Address http://www.expasy.ch/cgi-bin/peptident.pl

Score: 0.46, 31 matching peptides: P02787 (TRFE_HUMAN) pI: 6.70, Mw: 75181.44
CHAIN 1: SEROTRANSFERRIN. - Homo sapiens (Human).

(f)

user mass	matching mass	Δmass (Dalton)	#MC	modification	position	peptide	links
827.407	827.405	-0.0023	0		565-571	NPDPWAK	
864.507	864.413	-0.0938	0		652-659	DDTVCLAK	
874.457	874.442	-0.0152	0		316-323	DSAHGFLK	
887.457	887.415	-0.0417	0	Cys_CAM: 137	136-143	SCHTGLGR	
887.457	887.415	-0.0417	0	Cys_CAM: 469	468-475	SCHTAVGR	
964.557	964.532	-0.0246	0		601-609	APNHAVVTR	
978.507	978.489	-0.0179	0		216-225	DGAGDVAFVK	
997.507	997.477	-0.0298	0	Cys_CAM: 67	62-69	ASYLDCIR	
1000.56	1000.499	-0.0613	0		669-676	YLGEEYVK	
1138.56	1138.521	-0.0388	0		363-371	WCALSHHER	
1195.61	1195.543	-0.0674	0	Cys_CAM: 364	363-371	WCALSHHER	
1195.61	1195.552	-0.0575	0		123-132	DSGFQMNQLR	
1211.56	1211.547	-0.0126	0	MSO: 128	123-132	DSGFQMNQLR	
1249.66	1249.606	-0.054	0		454-464	SASDLTWDNLK	
1273.71	1273.654	-0.0564	0		226-236	HSTIFENLANK	FindMod
1276.66	1276.632	-0.0279	0		300-310	EFQLFSSPHGK	
1283.61	1283.569	-0.0408	0		531-541	EGYYGYTGAFR	PeptideMass
1317.61	1317.589	-0.0207	0	Cys_CAM: 28	27-37	WCAVSEHEATK	
1354.66	1354.631	-0.0292	0	Cys_CAM: 582	577-587	DYELLCLDGTR	BioGraph
1474.66	1474.666	0.0065	0	1xCys_CAM	684-696	CSTSSLLEACTFR	
1478.76	1478.735	-0.0251	0		332-343	MYLGYEYVTAIR	
1494.76	1494.73	-0.0302	0	MSO: 332	332-343	MYLGYEYVTAIR	
1531.71	1531.688	-0.022	0	2xCys_CAM	684-696	CSTSSLLEACTFR	
1539.71	1539.711	0.0008	0	Cys_CAM: 246	240-251	DQYELLCLDNTR	
1577.71	1577.658	-0.0523	0	Cys_CAM: 503	495-508	FDEFFSEGCAPGSK	
1586.76	1586.774	0.0144	0	Cys_CAM: 596	588-600	KPVEEYANCHLAR	
1593.76	1593.809	0.0493	0	MSO: 483	476-489	TAGWNIPMGLLYNK	
1629.81	1629.816	0.0059	0		108-121	EDPQTFYYAVAVVK	
1689.81	1689.849	0.0389	0	Cys_CAM: 260	259-273	DCHLAQVPSHTVVAR	
1706.76	1706.766	0.0059	0	2xCys_CAM	516-530	LCMGSGLNLCEPNNK	

Internet zone

Start | PeptIdent - ... | E:\Annalisa\... | E:\Annalisa\... | Microsoft W... | a5.txt - Note... | 6:03 PM

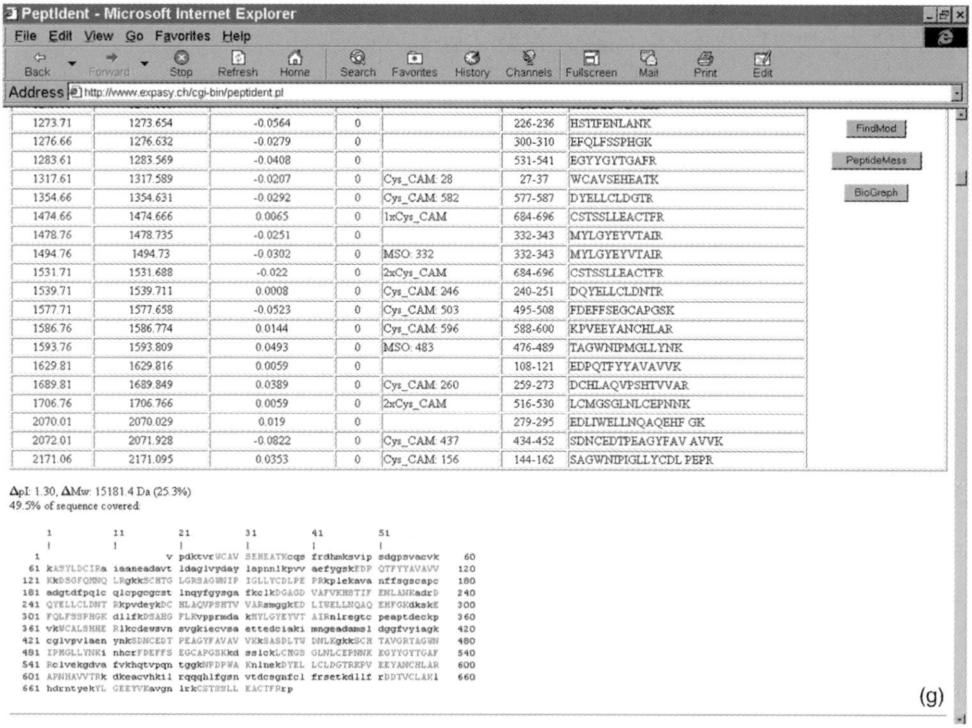

1273.71	1273.654	-0.0564	0		226-236	HSTIFENLANK
1276.66	1276.632	-0.0279	0		300-310	EFQLFSSPHGK
1283.61	1283.569	-0.0408	0		531-541	EGYYGYTGAFR
1317.61	1317.589	-0.0207	0	Cys_CAM: 28	27-37	WCAVSEHEATK
1354.66	1354.631	-0.0292	0	Cys_CAM: 582	577-587	DYELLCLDGTR
1474.66	1474.666	0.0065	0	1xCys_CAM	684-696	CSTSSLLEACTFR
1478.76	1478.735	-0.0251	0		332-343	MYLGYEYVTAIR
1494.76	1494.73	-0.0302	0	MSO: 332	332-343	MYLGYEYVTAIR
1531.71	1531.688	-0.022	0	2xCys_CAM	684-696	CSTSSLLEACTFR
1539.71	1539.711	0.0008	0	Cys_CAM: 246	240-251	DQYELLCLDNTR
1577.71	1577.658	-0.0523	0	Cys_CAM: 503	495-508	FDEFFSEGCAPGSK
1586.76	1586.774	0.0144	0	Cys_CAM: 596	588-600	KPVEEYANCHLAR
1593.76	1593.809	0.0493	0	MSO: 483	476-489	TAGWNIPMGLLYNK
1629.81	1629.816	0.0059	0		108-121	EDPQTFYYAVAVVK
1689.81	1689.849	0.0389	0	Cys_CAM: 260	259-273	DCHLAQVPSHTVVAR
1706.76	1706.766	0.0059	0	2xCys_CAM	516-530	LCMGSGLNLCEPNNK
2070.01	2070.029	0.019	0		279-295	EDLIWELLNQAQEHF GK
2072.01	2071.928	-0.0822	0	Cys_CAM: 437	434-452	SDNCEDTPEAGYFAV AVVK
2171.06	2171.095	0.0353	0	Cys_CAM: 156	144-162	SAGWNIPIGLLYCDL PEPR

ΔpI: 1.30, ΔMw: 15181.4 Da (25.3%)
49.5% of sequence covered:

```
      1          11         21         31         41         51
      |          |          |          |          |          |
  1                       v pdktvrWCAV SEHEATKcqs frdhmksvip sdgpsvacvk    60
 61 kASYLDCIRa iaaneadavt ldaglvyday lapnnlkpvv aefygskEDP QTFYYAVAVV   120
121 KkDSGFQMNQ LRgkkSCHTG LGRSAGWNIP IGLLYCDLPE PRkplekava nffsgscapc   180
181 adgtdfpqlc qlcpgcgcst lnqyfgysga fkclkDGAGD VAFVKHSTIF ENLANKadrD   240
241 QYELLCLDNT RkpvdeykDC HLAQVPSHTV VARsmggkED LIWELLNQAQ EHFGKdkskE   300
301 FQLFSSPHGK dllfkDSAHG FLKvpprmda kMYLGYEYVT AIRnlregtc peaptdeckp   360
361 vkWCALSHHE Rlkcdewsvn svgkiecvsa ettedciaki mngeadamsl dggfvyiagk   420
421 cglvpvlaen ynkSDNCEDT PEAGYFAVAV VKkSASDLTW DNLKgkkSCH TAVGRTAGWN   480
481 IPMGLLYNKi nhcrFDEFFS EGCAPGSKkd ssickLCMGS GLNLCEPNNK EGYYGYTGAF   540
541 Rclvekgdva fvkhqtvpqn tggkNPDPWA KnlnekDYEL LCLDGTRKPV EEYANCHLAR   600
601 APNHAVVTRk dkeacvhkil rqqqhlfgsn vtdcsgnfcl frsetkdllf rDDTVCLAKl   660
661 hdrntyekYL GEEYVKavgn lrkCSTSSLL EACTFRrp
```

(g)

matching peptides, with their position in the native transferrin chain and their sequence. At the end of the panel in Fig. 14.28g, the entire sequence of transferrin is given, in red colour the sequences of the matching peptides. It is seen that 49.5% of the entire sequence has been covered and that there are no uncertainties in assigning this spot to transferrin.

Some comments to the above protocol, largely followed in proteomics, are due. Is this a correct procedure or not? In general, this is what everybody does, and it seems to be working properly in most cases. However, for difficult identifications, this might not be the best procedure. It is a fact that most users start the search with not quite accurate values of the basic physico-chemical parameters of the intact polypeptide chain, namely its p*I* and M_r values. In terms of p*I*, this is only an approximate value when using conventional IEF in soluble carrier ampholytes, but certainly more precise p*I* values (correct to the second decimal digit) can be given when using IPGs in the first dimension. In terms of M_r, here too only approximate values can be given, since most people assess them via electrophoretic mobilities in SDS-PAGE against a set of M_r markers. Yet, if the protocol of Fig. 14.29 were to be followed, namely measuring both, the accurate M_r value of the intact protein by MS, followed by MS assessment of all peptides derived by the protein digestion, one would have a very narrow search window to start with, thus dramatically increasing the probability of a correct answer at the end of the search. This procedure, suggested to us by Dr. M. Hamdan (GlaxoSmithKline,

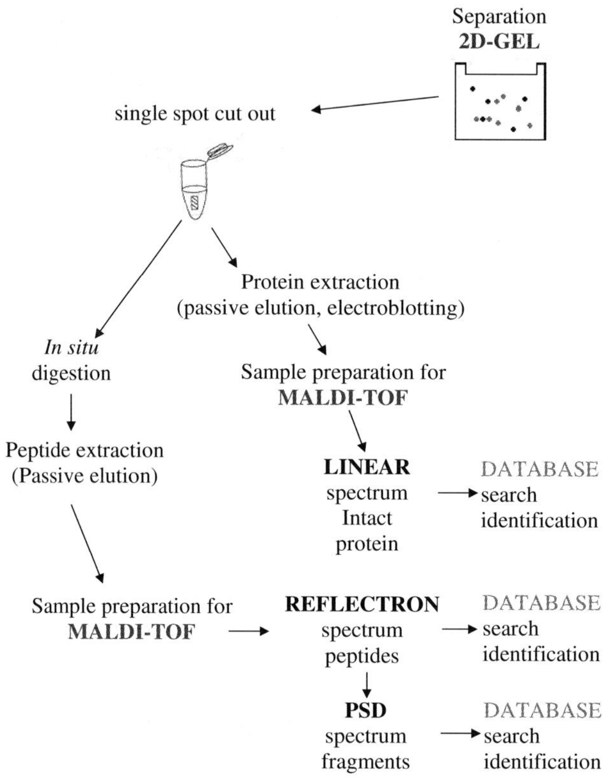

Fig. 14.29. Scheme of the series of steps to be taken for identifying a polypeptide chain eluted from SDS-PAGE or from a 2-D map. Note that, in addition to enzyme digestion and searching of databases via peptide masses (and partial peptide sequences obtained via PSD spectra), it is here suggested that an aliquot of the intact protein should be taken for assessment of native M_r (courtesy of Dr. M. Hamdan, GlaxoSmithKlein, Verona.)

Verona) is certainly to be followed in difficult cases because, at the small price of an added burden of making an extra M_r measurement (which can be done by simply harvesting an aliquot of the eluted protein, prior to enzymatic digestion), it greatly enhances the chances of success in protein identification. As seen in Fig. 14.28c, in fact, the default errors are ± 2 pH units in the value of p*I* inserted and 50% of the M_r range, terribly crude estimates, indeed, unjustified at the light of the highly sophisticated and accurate techniques today available. It might also be asked why, being the two spots Nos. 4 and 5 both transferrins, in one case one obtains 57 and in the other 67 peptides. Of course the number of peptides generated by trypsin digestion should be the same (unless point mutants lacking either Lys or Arg were present), but the number of identifiable peptides turned out to be quite different, most likely due to extraction conditions for each spots. The total amount of protein being different, the weaker spot might have generated a number of peptides in a too low concentration, so that the MS signal was rejected by the standard identification protocols. It should also be remembered that, when eluting a spot directly from a gel piece, it is customary to adsorb the liquid onto a Zip-Tip, i.e. a tiny plastic cone of automatic pipettors filled with a C_{18}-resin, for purification purposes. Upon elution of the adsorbed polypeptide, some of it could be irreversibly bound to the hydrophobic beads. Finally, it should be borne in mind that the data thus generated are subjected to a number of filtering procedures, so as to eliminate spurious data, such as those which could correspond to contaminants. As an example, all mass values which could be attributed to trypsin, Coomassie Blue adducts, cheratins, are automatically rejected. In addition, each peptide list is compared with an analogous list obtained by analysing a control spot, i.e. the eluate from a small gel segment which does not appear to contain any detectable polypeptide spot: in this way, all peaks belonging to contaminants are rejected.

Some general information on the various search tools listed in Table 14.7 is here due. All the search programs work by comparing the measured masses with masses calculated using protein sequences in a database. Software tools that rank the proteins in the database according to the number of matching peptides include PepSea, MS-Fit, MS-Tag, PepFrag and PeptIdent/Multident. According to Beavis and Fenyö [177], the MOWSE algorithm has a higher selectivity and sensitivity than these algorithms, which simply count the number of peptide matches. MOWSE takes into account the relative abundances of the peptides of a given mass in the database and also compensates for protein size. MASCOT, on the other hand, is based on the MOWSE algorithm but also calculates an approximate probability that the observed match between experimental data and a protein sequence is a random event. MASCOT, additionally, can use information from both peptide maps and tandem mass spectra for identifying proteins. What could the future development of these search engines be? Probably the following:

(1) Improvements in the signal-processing algorithms used for generating mass spectra.

(2) The automation of protein-identification searches and the rational storage of results in large databases.

(3) The development of data-dependent search engines that can guide the data-gathering process in real time.

(4) Post-processing adopting additional statistical data from compiled databases of theoretical and experimentally determined sequence properties.

14.5. PRE-FRACTIONATION TOOLS IN PROTEOME ANALYSIS

As stated above, although both the first (IEF/IPG) and second (SDS-PAGE) separation dimensions appear to be performing at their best, there remains the fact that, due the vast body of polypeptide chains present in a tissue lysate, especially at acidic pH values (pH 4–6), resolution is still not enough. Some recent papers suggest that this problem could be overcome by running a series of narrow-range IPG strips (covering no more than 1 pH unit) on large size gels (18 cm or longer in the first dimension, large-format slabs in the second dimension), which would dramatically increase the resolution [250]. The entire, wide-range 2-D map would then be electronically reconstructed by stitching together the narrow-range maps. This might turn out to be a fata morgana, though:

there remains the fact that, even when using very narrow IPG strips, they have to be loaded with the entire cell lysate, containing proteins focussing all along the pH scale. Massive precipitation will then ensue, due to aggregation among unlike proteins, with the additional drawback that the proteins which should focus in the chosen narrow-range IPG interval will be strongly under-represented, since they will be only a small fraction of the entire sample loaded. That this is a serious problem has been debated in a recent work by Gygi et al. [251]. These authors analysed a yeast lysate by loading 0.5 mg total protein on a narrow-range (nr) IPG pH 4.9–5.7. Although they could visualise by silvering ca. 1500 spots, they lamented that a large number of polypeptides simply did not appear in such a 2-D map. In particular, proteins from genes with codon bias values of <0.1 (low abundance proteins) were not found, even though fully one half of all yeast genes fall into that range. The codon bias value for a gene is its propensity to use only one of several codons to incorporate a specific amino acid into the polypeptide chain. It is known that highly expressed proteins have large codon bias values (>0.2). Thus, these authors conclude that, in reality, when analysing protein spots from 2-D maps by mass spectrometry, only generally abundant proteins (codon bias >0.2) can be properly identified. Thus, the number of spots on a 2-D gel is not representative of the overall number or classes of expressed genes that can be analysed. Gygi et al. [251] have calculated that, when loading only 40 μg total yeast lysate, as done in the early days of 2-D mapping, only polypeptides with an abundance of at least 51,000 copies/cell could be detected. With 0.5 mg of starting protein, proteins present at 1000 copies/cell could now be visualised by silvering, but those present at 100 and 10 copies/cell could not be detected. These authors thus concluded that the large range of protein expression levels limits the ability of the 2-D-MS approach to analyse proteins of medium to low abundance, and thus the potential of this technique for proteome analysis is likewise limited. This is a severe limitation, since it is quite likely that the portion of proteome we are presently missing is the most interesting one from the point of view of understanding cellular and regulatory proteins, since such low-abundance polypeptide chains will typically be regulatory proteins. Thus, presently available techniques of IEF/IPG-SDS-PAGE, coupled to MS or MS-MS, although being the best possible ones, do not appear to be suitable for global detection of proteins expressed by cell. As a corollary, the construction of complete, quantitative protein maps based on this approach will be very challenging, even for relatively simple, unicellular organisms.

Due to such major drawbacks, we believe that the only way out of this impasse will be pre-fractionation. At present, two approaches have been described: chromatographic and electrophoretic. They will be illustrated below.

14.5.1. Sample pre-fractionation via different chromatographic approaches

Fountoulakis' group has extensively developed this approach, as presented below. In a first procedure, they adopted affinity chromatography on heparin gels as a pre-fractionation step for enriching certain protein fractions in the bacterium *Haemophilus influenzae* [252,253]. This Gram-negative bacterium is of pharmaceutical interest and has been recently sequenced; its complete genome has been found to comprise ca. 1740 open reading frames, although not more than 100 proteins had been characterised by 2-D map analysis. Heparin is a highly sulphated glucosaminoglycan with affinity for a broad range of proteins, such as nucleic acid-binding proteins and growth and protein synthesis factors. On account of its sulphate groups, heparin also functions as a high-capacity cation exchanger. Thus, pre-fractionation on a Heparin actigel was deemed suitable for enriching low-copy number gene products. In fact, about 160 cytosolic proteins bound with different affinities to the heparin matrix and were thus highly enriched prior to 2-D PAGE separation. As a result, more than 110 new protein spots, detected in the heparin fraction, were identified, thus increasing the total identified proteins of *H. influenzae* to more than 230. In a second approach [254], the same lysate of *H. influenzae* was pre-fractionated by chromatofocussing on Polybuffer Exchanger. In the eluate, two proteins, major ferric iron-binding protein (HI0097) and 5′-nucleotidase (HI0206) were obtained in a pure form with another hypothetical protein (HI0052) purified to near homogeneity. Four other proteins, aspartate ammonia lyase (HI0534), peptidase D (HI0675), elongation factor Ts (HI0914) and 5-methyltetrahydropteroyltriglutamate methyltransferase (HI1702) were strongly enriched by the chromatofocussing process. Approximately 125 proteins were identified in the eluate collected from the column. Seventy of these were for the first time identified after chromatography on the Polybuffer Exchanger, the majority of them being low-abundance enzymes with various functions. Thus, with this additional step, a total of 300 protein could be identified in *H. influenzae* by 2-D map analysis, out of a total of ca. 600 spots visualised on such maps from the soluble fraction of this micro-organism. In yet another approach, the cytosolic soluble proteins of *H. influenzae* were pre-fractionated by Fountoulakis et al. [255] by hydrophobic interaction chromatography (HIC) on a TSK Phenyl column. The eluate was subsequently analysed by 2-D mapping, followed by spot characterisation by MALDI-TOF MS. Approximately 150 proteins, bound to the column, were identified, but only 30 for the first time. In addition, most of the proteins enriched by HIC were represented by major spots, so that enrichment of low-copy-number gene products was only modest. In total, with all the various chromatographic steps adopted, the number of proteins so far identified could be increased to 350.

The same heparin chromatography procedure was subsequently applied by Karlsson et al. [257] to the pre-fractionation of human fetal brain soluble proteins. Approximately 300 proteins were analysed, representing 70 different polypeptides, 50 of which were bound to the heparin matrix. Eighteen brain proteins were identified for the first time. The polypeptides enriched by heparin chromatography included both minor and major components of the brain extract. The enriched proteins belonged to several classes, including proteasome components, dihydropirimidinase-related proteins, T-complex protein 1 components and enzymes with various catalytic activities.

In yet another variant, Fountoulakis et al. [255,256] reported enrichment of low abundance proteins of *E. coli* by hydroxyapatite chromatography. The complete genome of *E. coli* has now been sequenced and its proteome analysed by 2-D mapping. To date, 223 unique loci have been identified and 201 protein entries were found in the SWISS-PROT 2-D PAGE on ExPASy server using the Sequence retrieval System query tool (http//www.expasy.ch/www/sitemap.html). Of the 4289 possible gene products of *E. coli*, about 1200 spots could be counted on a typical 2-D map when ca. 2-mg total protein was applied. Possibly, most of the remaining proteins were not expressed in sufficient amounts to be visualised following staining with Coomassie Blue. Thus, it was felt necessary to perform a pre-fractionation step, this time on hydroxyapatite beads. By this procedure, approximately 800 spots, corresponding to 296 different proteins, were identified in the hydroxyapatite eluate. About 130 new proteins that had not been detected in 2-D gels of the total extract were identified for the first time. This chromatographic step, though, enriched both low-abundance as well as major components of the *E. coli* extract. In particular, it enriched many low-M_r proteins, such as cold-shock proteins. After such a long excursus, it is of interest to see some of the results obtained by this research group. Fig. 14.30 offers a series of panels of partial 2-D gel images of unfractionated (A) and of pools 2 (B), 4 (C), 5 (D) and 10 (E) of fractions collected from the hydroxyapatite pre-fractionation of *E. coli* total lysate. An efficient enrichment of many low-M_r cold-shock proteins (CSP), such as CSP-G, Fig. 14.30B), CSP-A (Fig. 14.30C–E) and CSP-C and CSP-E (Fig. 14.30D) could be obtained. Fig. 14.31 shows the analysis of hydroxyapatite pools 11 (A) and 13 (B), whereas Fig. 14.32 displays analogous 2-D fractionations from pools 19 (A) and 21 (B) in the same chromatographic step. On a similar line of thinking, Harrington et al. [258] reported the use of cation-exchange chromatography for enriching DNA-binding proteins on tissue extracts, prior to 2-D analysis. Here too, since the basic domains do not seem to influence the overall p*I* of such proteins, one sees a general spread of the pre-fractionated sample across the complete p*I* range of a 2-D map.

Although the work presented by Fountoulakis' group is impressive, and truly innovative in proteome analysis, we believe that, nevertheless, there are some inherent drawbacks to this approach. We list here some of them.

(1) In general, huge amounts of salts are needed for elution from chromatographic columns, up to 2.5 M NaCl.

(2) As a consequence, loss of entire groups of proteins could ensue during the various manipulations (dialysis of large salt amounts, concentration of highly diluted pools of eluted fractions).

(3) The eluted fractions do not represent narrow p*I* cuts, but in general are constituted by proteins having p*I*s in the pH 3–10 range. Thus, one should still use wide pH gradients for their analysis.

In particular, it is known that high salt amounts could induce irreversible adsorption of proteins to the resin used for fractionation, this resulting in irreversible loss of spots

Fig. 14.30. Partial 2-D gel images of the input (A) and of the pools 2 (B), 4 (C), 5 (D) and 0 (E) of fractions collected from a hydroxyapatite pre-fractionation of a total *E. coli* cell lysate, demonstrating the enrichment of low M_r proteins. The proteins were eluted with progressively increasing salt gradients, including 50 mM $MgCl_2$ (B), 1.5 M $MgCl_2$ (C), 1.5 M NaCl (D) and 2.5 M NaCl (E). The samples were analysed on a pH 3–10 non-linear IPG strip, followed by 9–16% *T* gradient SDS-PAGE gel. The gels were stained with colloidal Coomassie Blue, destained with water and scanned in an Agfa DUOSCAN machine. Processing of the images was with Adobe Photoshop and Power Point softwares. The proteins were identified by MALDI-TOF MS. The abbreviated names or numbers next to the protein spots are those of the *E. coli* database (ftp://ncbi.nlm.nih.gov/genbank/genomes/bacteria/Ecoli/ecoli.ptt). (From [255], with permission.)

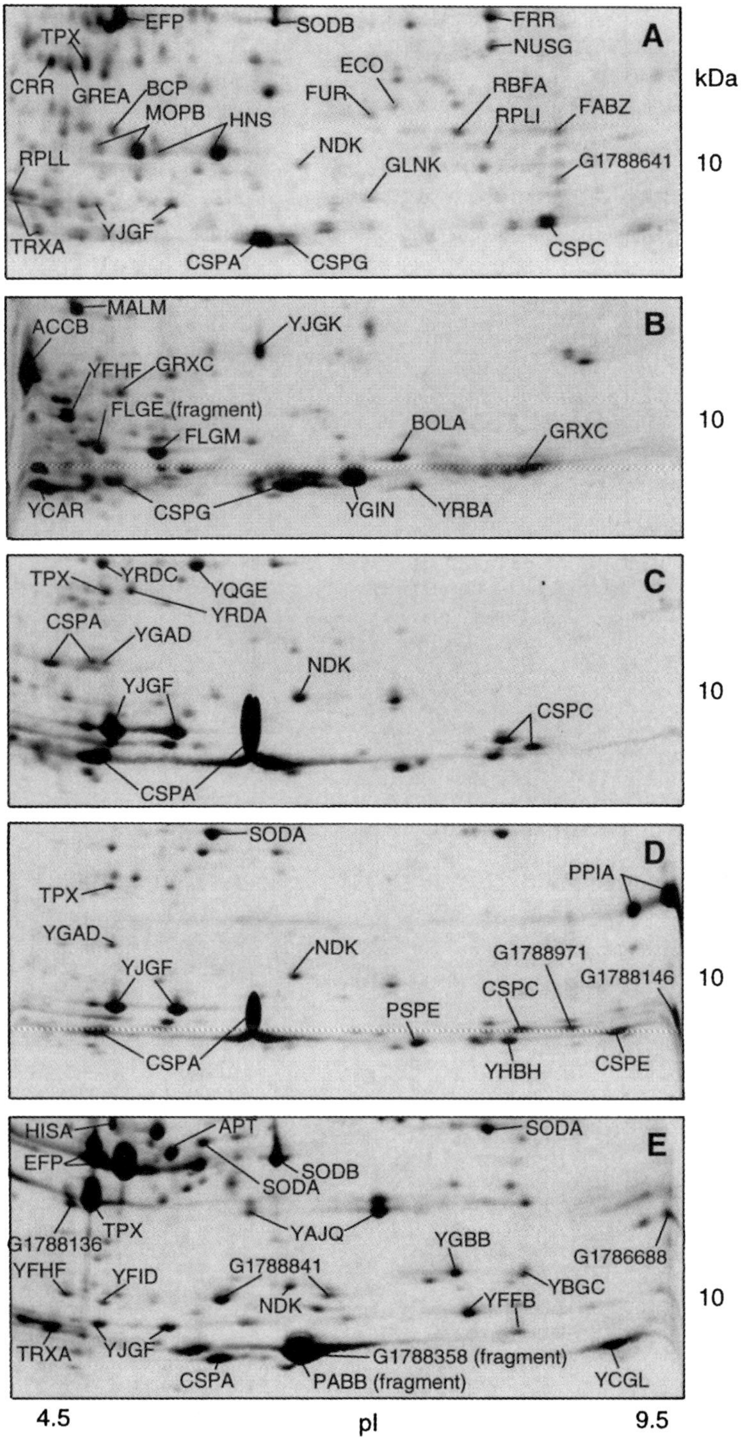
A
TPX
EFP
SODB
FRR
NUSG
CRR
GREA
BCP
ECO
FUR
RBFA
FABZ
RPLI
MOPB
HNS
NDK
RPLL
GLNK
G1788641
YJGF
TRXA
CSPA
CSPG
CSPC
B
MALM
ACCB
YJGK
YFHF
GRXC
FLGE (fragment)
FLGM
BOLA
GRXC
YCAR
CSPG
YGIN
YRBA
C
TPX
YRDC
YQGE
YRDA
CSPA
YGAD
NDK
YJGF
CSPC
CSPA
D
SODA
TPX
PPIA
YGAD
NDK
G1788971
YJGF
G1788146
CSPC
PSPE
CSPA
YHBH
CSPE
E
HISA
APT
SODA
EFP
SODB
SODA
G1788136
TPX
YAJQ
YGBB
G1786688
YFHF
YFID
G1788841
YBGC
NDK
YFFB
TRXA
YJGF
G1788358 (fragment)
CSPA
PABB (fragment)
YCGL
kDa
10
10
10
10
10
4.5
pI
9.5

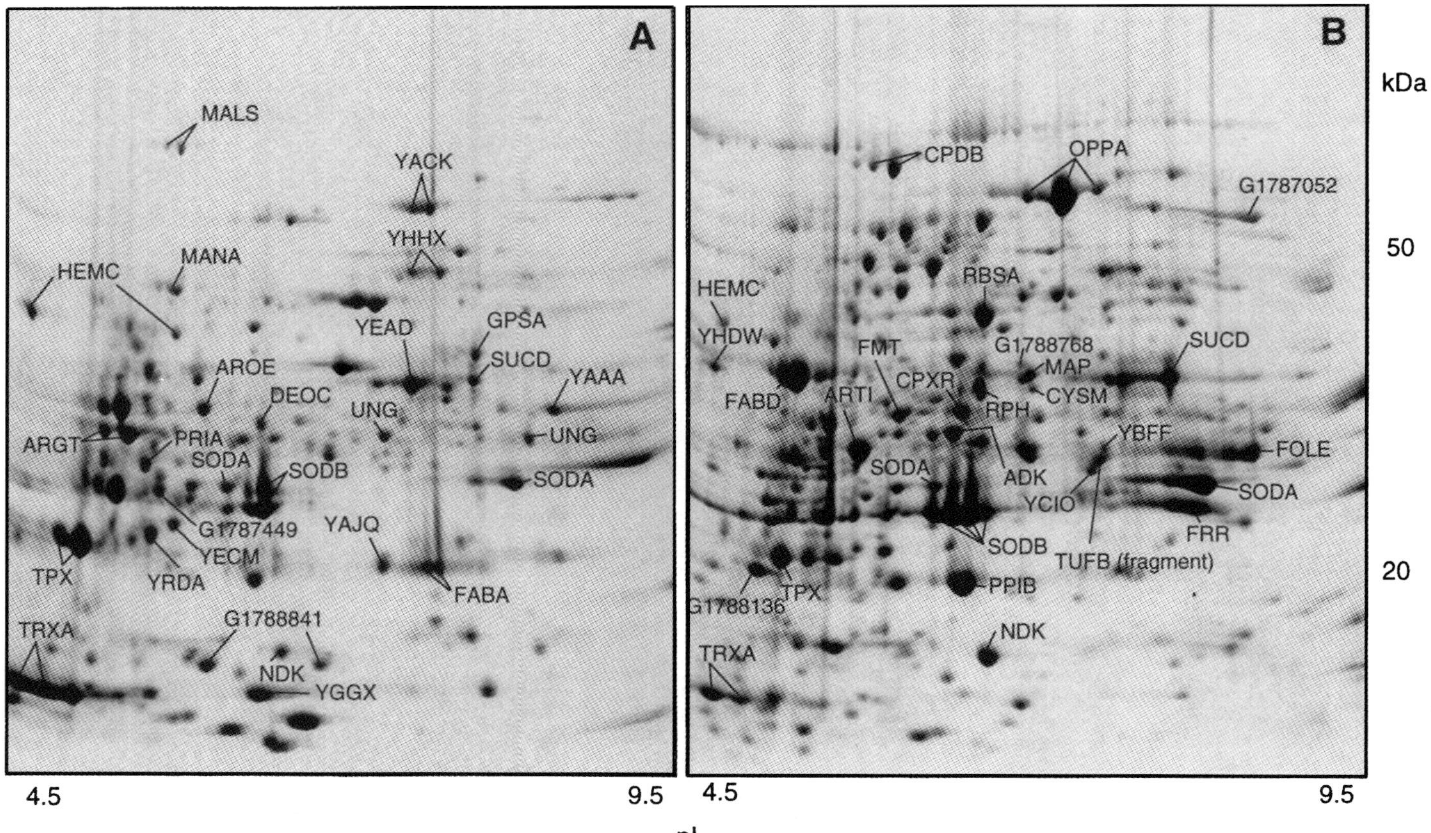

Fig. 14.31. 2-D gel analysis of pools 11 (A) and 13 (B) eluted from the hydroxyapatite column of a total *E. coli* cell lysate. The proteins were eluted with an ascending gradient of 1.5 M $MgCl_2$. All other conditions as in the legend to Fig. 14.30. (From [255], with permission.)

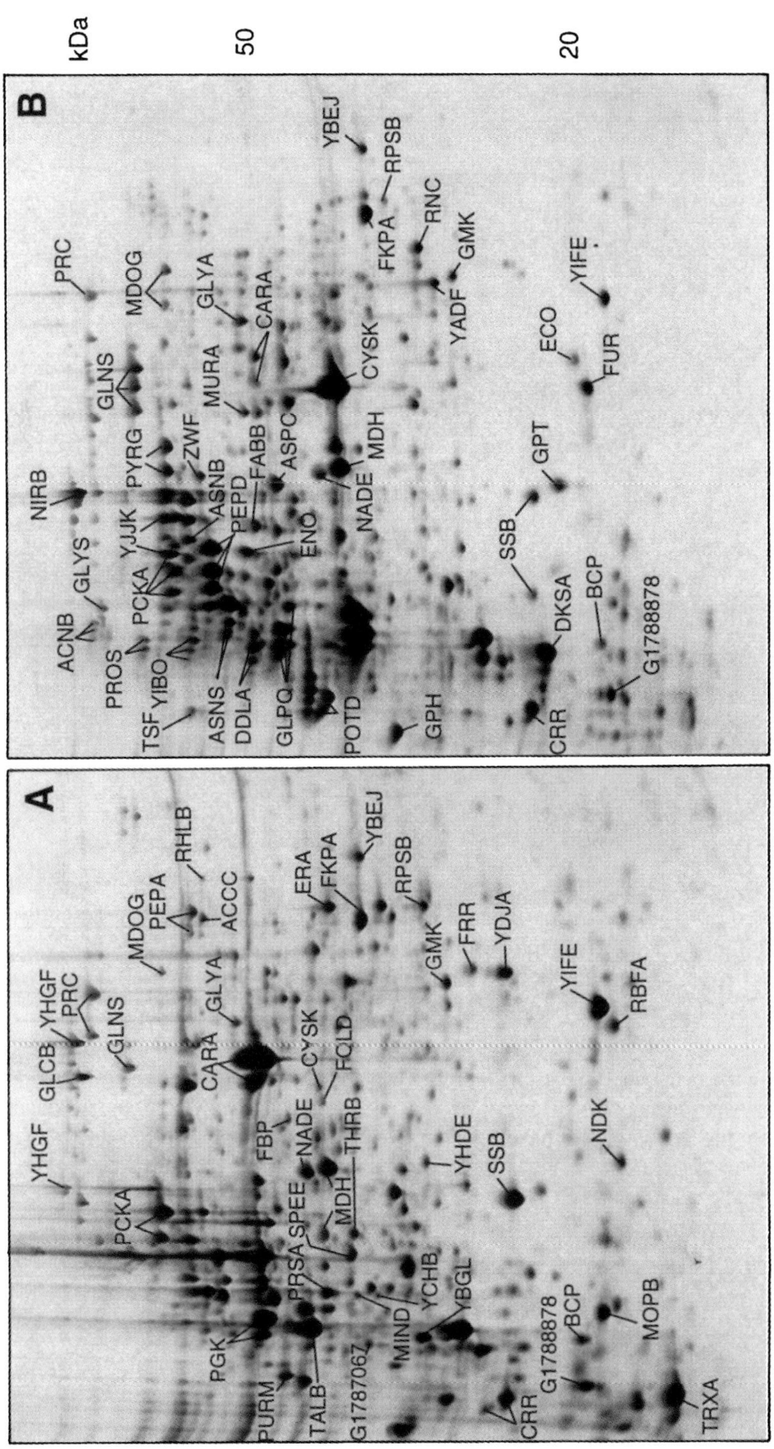

Fig. 14.32. 2-D gel analysis of pools 19 (A) and 21 (B) eluted from an hydroxyapatite pre-fractionation of a total *E. coli* cell lysate. The proteins were eluted with an ascending gradient of 5.5 M NaCl. All other conditions as in the legend to Fig. 14.30. (From [255], with permission.)

Fauel M.D. & Righetti P.G. :
ISOELECTRIC FOCUSING PROCESS AND A MEANS FOR CARRYING OUT SAID PROCESS
US Patent No. 4,971,670
Issued: November-20-1990

Righetti P.G.:
IMMOBILIZED ENZYME REACTOR
US Patent No. 5,834,272
Issued: November-10-1998

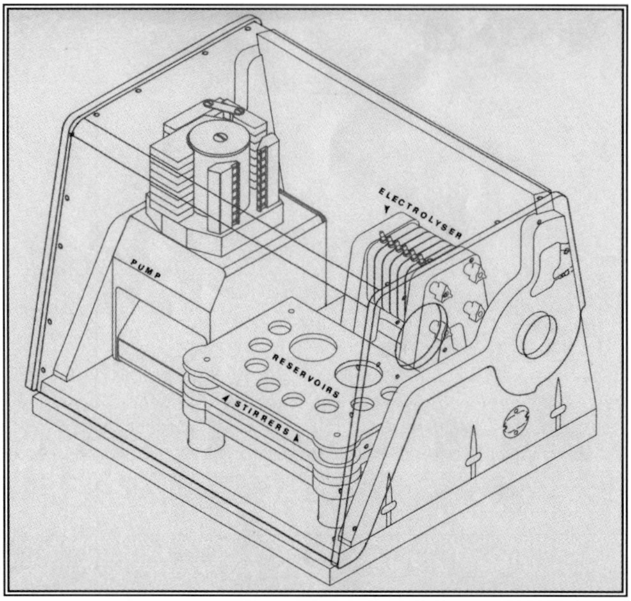

Fig. 14.33. Schematic drawing of the multicompartment electrolyser operating with isoelectric membranes. Power supply, reservoirs and connecting tubings are not shown. (P.G. Righetti, unpublished.)

during the subsequent 2-D mapping. When eluting from ion-exchange columns, moreover, not only the collected fractions will be quite dilute, necessitating a concentration step, but they will also contain huge amounts of salt, rendering them incompatible with the IEF/IPG first dimension step. Thus, although the approach by Fountoulakis' group could be quite attractive for analysis of some protein fractions in a total cell lysate, it would clearly not work for exploring the entire proteome, due to all potential losses described above. We believe that there could be a unique fractionation step, fully compatible with the subsequent 2-D mapping, and this would be the use of multicompartment electrolysers with isoelectric membranes, as illustrated below.

14.5.2. Sample pre-fractionation via multicompartment electrolysers with Immobiline membranes

This kind of equipment represents, perhaps, the most advanced evolutionary step in all preparative electrokinetic fractionation processes. The original apparatus devised by Righetti et al. [165,166,259] is shown schematically in Fig. 14.33: it consists in a stack of chambers sandwiched between an anodic and a cathodic reservoir. The apparatus is

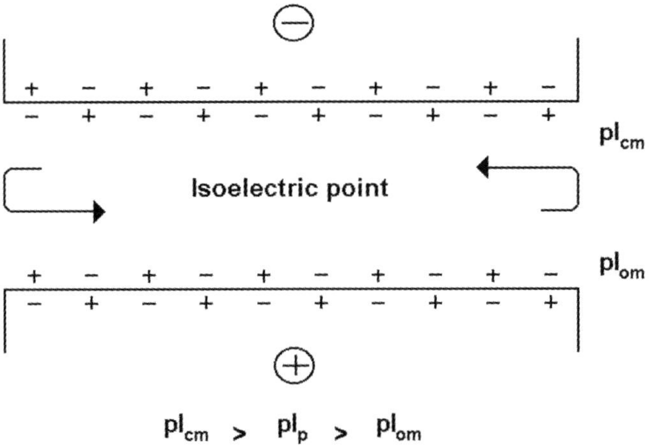

Fig. 14.34. Mechanism of the purification process in the multicompartment electrolyser. The isoelectric membranes facing each recycling chamber act by titrating the protein of interest to its isoelectric point (p*I*), thus keeping its mobility constant and equal to zero all throughout the purification process. For this to occur, it is necessary that $pI_{cm} > pI_p > pI_{am}$, where the subscripts indicate cathodic membrane, protein and anodic membrane, respectively. In addition, the two Immobiline membranes satisfy the condition of having high buffering capacity at their pI value. The curved arrows indicate protein recycling in the flow chamber. (From [142], with permission.)

modular and can accommodate up to eight flow chambers, six for sample collection and the two extreme ones as electrodic reservoirs. A multichannel peristaltic pump connects such flow chambers to as many reservoirs, positioned outside the electric field, the latter acting as heat sinks and as reservoirs for purification of larger sample volumes than could possibly be accommodated inside the flow chambers (typically restricted to a total of 5 ml). This apparatus is thus based on two orthogonal flows: a hydraulic flow, bringing all sample components in and out of the electric field, till steady-state conditions are found for all species, and an electric flow, allowing each protein in the system to reach the chamber in which it will find isoelectric conditions. The reason why this system works is shown in Fig. 14.34: two isoelectric membranes, facing each flow chamber, act by continuously titrating the protein of interest to its isoelectric point. They can be envisaged as highly selective membranes, which retain any proteins having p*I*s in between their limiting values, and which will allow transmigration of any non-amphoteric, non-isoelectric species. For that, the only condition required is that $pI_{cm} > pI_p > pI_{am}$, where the subscripts 'cm' and 'am' denote cathodic and anodic membranes, respectively, and 'p' is the protein having a given isoelectric point between the two membranes. For this mechanism to be operative, it is necessary that the two isoelectric membranes possess good buffering capacity (β), so as to be able to effectively titrate the protein present in the flow chamber to its p*I*, while ensuring good current flow through the system. Wenger et al. [260] had in fact synthesised amphoteric, isoelectric Immobiline membranes and demonstrated that indeed they are good buffers at their p*I*. These membranes are, of course, made with the same technology used to make IPG strips, i.e. they exploit the Immobiline buffers, but not for creating a pH gradient, but simply for defining any single pH value along the pH scale, such value being then the p*I* of a given membrane. The concept of isoelectric Immobiline membranes is quite revolutionary and deserves further comments. Such membranes can be envisaged as pH-dictating assemblies in an IEF separation, much like a pH-stat unit is set-up for controlling, e.g. the pH during a biochemical reaction or during in vitro tissue growth. Each species that is tangent or crosses such isoelectric membranes is titrated to the pH of the membrane (provided it does not overcome its intrinsic β value). For amphoteric compounds, this results in a drastic change in mobility, which could reach zero if the two membranes delimiting a single flow chamber have p*I*s encompassing the protein of interest to be trapped in said chamber. Any other amphoteric species with lower or higher p*I* value will be forced to exit from such a chamber either towards the anode or towards he cathode, respectively. Thus it is clear that, with a proper set of membranes, it is possible to define in a given chamber isoelectric conditions for just a single component of a protein mixture, which ultimately will be arrested as the sole isoelectric species in such a chamber. Moreover, if the protein concentration is high enough, the macroion present in the liquid stream will possess enough buffering power to control the pH, in the absence of exogenous ions migrating through the system.

Although this system was used originally for extreme purification of a single protein form, in preparation for crystallisation experiments or other biochemical assays requiring absolute purity, it was easy to envisage its application to proteome analysis. In this last case, instead of using very narrow p*I* gaps in between two membranes delimiting a chamber, one could have used large p*I* gaps, from one up to several pH units, for trapping groups of proteins in a given chamber. Such narrow-p*I* range families could then be analysed on an IPG strip encompassing just the p*I* values of such a family. The advantages of such a procedure are immediately apparent:

(1) It offers a method that is fully compatible with the subsequent first dimension separation, a focussing step based on the Immobiline technology. Such a pre-fractionation protocol is precisely based on the same concept of immobilised pH gradients and thus protein mixtures harvested from the various chambers of this apparatus can be loaded onto IPG strips without any need for further treatment, in that they are isoelectric and devoid of any non-amphoteric ionic contaminant.

(2) It permits harvesting a population of proteins having p*I* values precisely matching the pH gradient of any narrow (or wider) IPG strip.

(3) As a corollary of the above point, much reduced chances of protein precipitation will occur. In fact, when an entire cell lysate is analysed in a wide gradient, there are fewer risks of protein precipitation; on the contrary, when the same mixture is analysed in a narrow gradient, massive precipitation of all non-isoelectric proteins could occur, with a strong risk of co-precipitation of proteins that would otherwise focus in the narrow pH interval.

(4) Due to the fact that only the proteins co-focussing in the same IPG interval will be present, much higher sample loads can be operative, permitting detection of low-abundance proteins.

(5) Finally, in samples containing extreme ranges in protein concentrations (such as human serum, where a single protein, albumin, represents >60% of the total species),

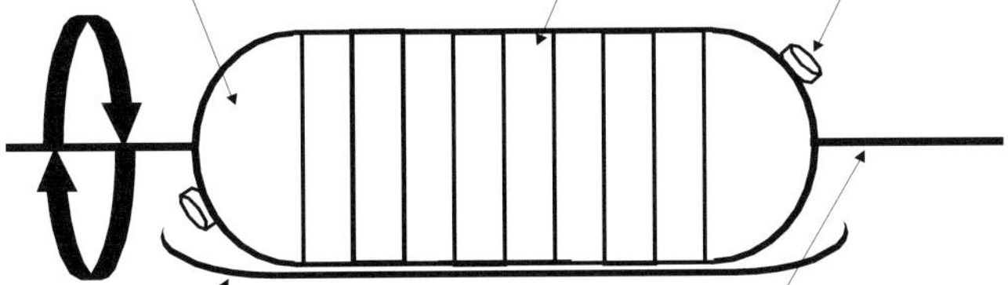

Fig. 14.35. Scheme of the rotating multichamber apparatus. The cell is seen assembled with 8 fraction collection chambers and 2 electrodic reservoirs. The entire unit is rotated onto its axis and gas escape from the electrodic chambers is secured by gas-permeable membranes. During rotation, the chamber is partly immersed in a thermostatted bath. The septa between the various chambers are isoelectric, buffering membranes, cast onto a support of glass fibres or other suitable material. Such membranes are flow-tight and ensure proper pH control. (From [261], with permission.)

one could assemble an isoelectric trap narrow enough to just eliminate the unwanted protein from the entire complex, this too permitting much higher sample loads without interference from the most abundant species.

Herbert and Righetti [261] have recently applied precisely this technology to sample pre-fractionation in preparation for 2-D map analysis. However, since the previous apparatus, as shown in Fig. 14.33, was too complex to be operated for proteome analysis, and in addition required as much as 25 ml per chamber, when operated in the recycling mode, they first had to miniaturise and render more suitable such an instrument for 2-D mapping. Fig. 14.35 shows a modified, advanced version of the previous instrument. For applications which do not require sample recycling from external reservoirs, the chambers of the multicompartment electrolyser can be sealed and the entire apparatus can be rotated onto its axis. This efficiently prevents electrodecantation and provides for proper mixing of the content of each chamber. For heat dissipation, the rotating electrolyser is partially immersed into a semi-cylindrical basin containing liquid thermostatted with Peltier elements. Ideally, such an instrument would hold in each chamber just the total liquid volume that could be adsorbed in the IPG strip for the 1st dimension run. In properly made strips, such reswelling sample volume could be as much 1 ml. However, total volumes of 3–4 ml per chamber could be a just valid alternative, since they would permit running a few IPG strips for duplicate or triplicate analysis. It then had to be demonstrated that such an instrument would perform well in proteome analysis. Fig. 14.36 displays the analysis of the p*I* 4.0–5.0 fraction collected from the p*I* 4.0–5.0 chamber of the apparatus depicted in Fig. 14.35. An IPG 3-10 strip was utilised for the first, IEF dimension, thus demonstrating that indeed, in the 2-D map, only this acidic fraction could be exhibited, all other proteins in the remainder of the pH scale having been efficiently removed. Fig. 14.37 evidences two 2-D maps, both in the pH 4–5 interval, one as obtained with such a pre-fractionation protocol, the

Fig. 14.36. Silver-stained 2-D map of the p*I* 4–5 fraction of *E. coli*, as purified in the multicompartment electrolyser of Fig. 14.35, run in a 7-cm pH 3–10 IPG strip in the first dimension. (From [261], with permission.)

other as available on Internet at the SWISS-2D-PAGE site. Although the gel on the right is stained with Coomassie Brilliant Blue it is apparent that more spots are visible, as compared to the silver-stained SWISS-2DPAGE *E. coli* map. Moreover, because the gel in the right panel is stained with Coomassie Brilliant Blue and not with silvering, almost all of the visible spots will be present in sufficient quantities for mass spectrometry. Although this is a clear indication that the pre-fractionation procedure should not lead to loss of components via trapping into the isoelectric membranes, Herbert and Righetti [261] performed a protein assay on the starting and ending products and could confirm a 95% protein recovery. Additionally, a similar assay performed on the ground and extracted isoelectric membranes failed to reveal any appreciable amount of proteinaceous material bound or adsorbed onto those surfaces. The advantages of such a pre-fractionation procedure can be appreciated from Fig. 14.38: a new 2-D map was made, with 5 times more loaded proteins as compared with Fig. 14.37 (right panel). Yet, even under such overloading conditions, no sample streaks or smears appeared, but simply more intense spots and more low-abundance proteins could be detected (note that the streaking at the gel bottom, in the low M_r region, is due to mixed ASB14-SDS micelles migrating with the salt front).

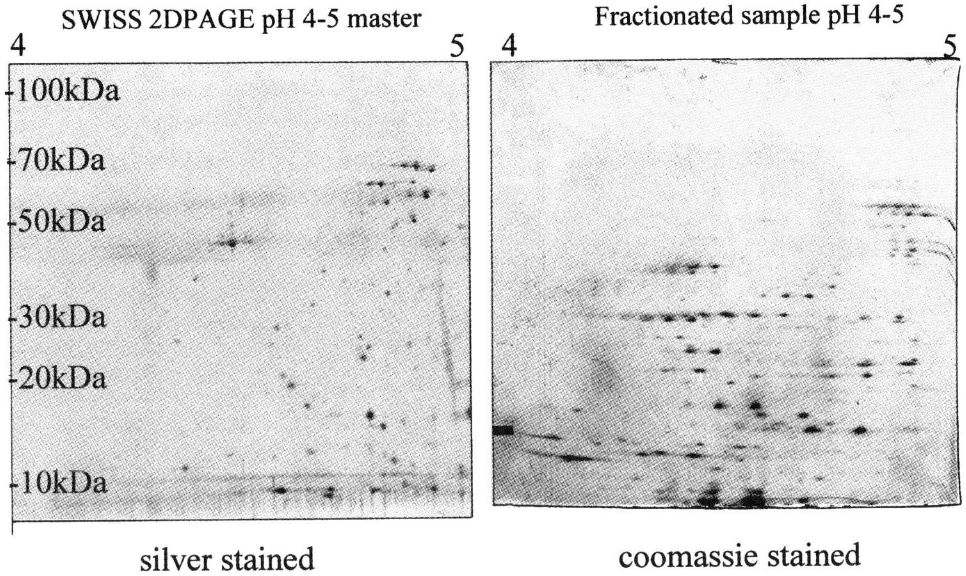

Fig. 14.37. Silver-stained 2-D map of an *E. coli* entire cell lysate (left panel) vs. the colloidal Coomassie G-250 stained gel of the sample pre-fractionated with the instrument of Fig. 14.35 (right panel), run in a 18-cm pH 4–5 gradient in the first dimension. (From [261], with permission.)

An interesting example on the power of this pre-fractionation technique comes from experiments with human serum, a most difficult sample due to the overwhelming presence of albumin. Fig. 14.39 shows three silver-stained 2-D gels representing the three sample chambers from the MCE plasma fractionation. Each gel used a pH 3–10 linear IPG in the first dimension, to clearly show the extent of the fractionation. It can be seen that both the acidic and basic chambers are free from albumin and contain acidic and basic proteins only, respectively. The albumin chamber (p*I* interval 5.6–6.1) has a very narrow p*I* spectrum of proteins, which reflects the 0.5 pH unit window of the albumin chamber membranes in the MCE. The basic chamber (p*I* interval 6.1–10.5) is highly enriched in alkaline proteins, especially in the mildly alkaline region up to approximately pH 7.5, but many proteins spots can be seen up to pH 10 and higher. Fig. 14.40 shows two silver-stained pH 3–6 IPGs with whole plasma in the left panel and fractionated plasma from the acidic chamber depicted in Fig. 14.39. The three black arrows show matching areas on the two gels. The vertical arrow on the right panel indicates pH 5.6 which was the pH of the delimiting membrane between the acidic and albumin chambers. It is clear that the fractionation has removed albumin, which is the most abundant protein on the whole plasma gel (left panel). In addition, the three abundant spots at the pH 3 end of the whole plasma gel are not present in the fractionated gel, which indicates that they may be precipitated albumin (as also indicated by the horizontal filament of unfocussed proteins uniting them to the large albumin spot on the far right side of the map). Fig. 14.41 shows two pH 3–6 IPGs with whole plasma in the left panel and fractionated plasma from the acidic chamber depicted in Fig. 14.39,

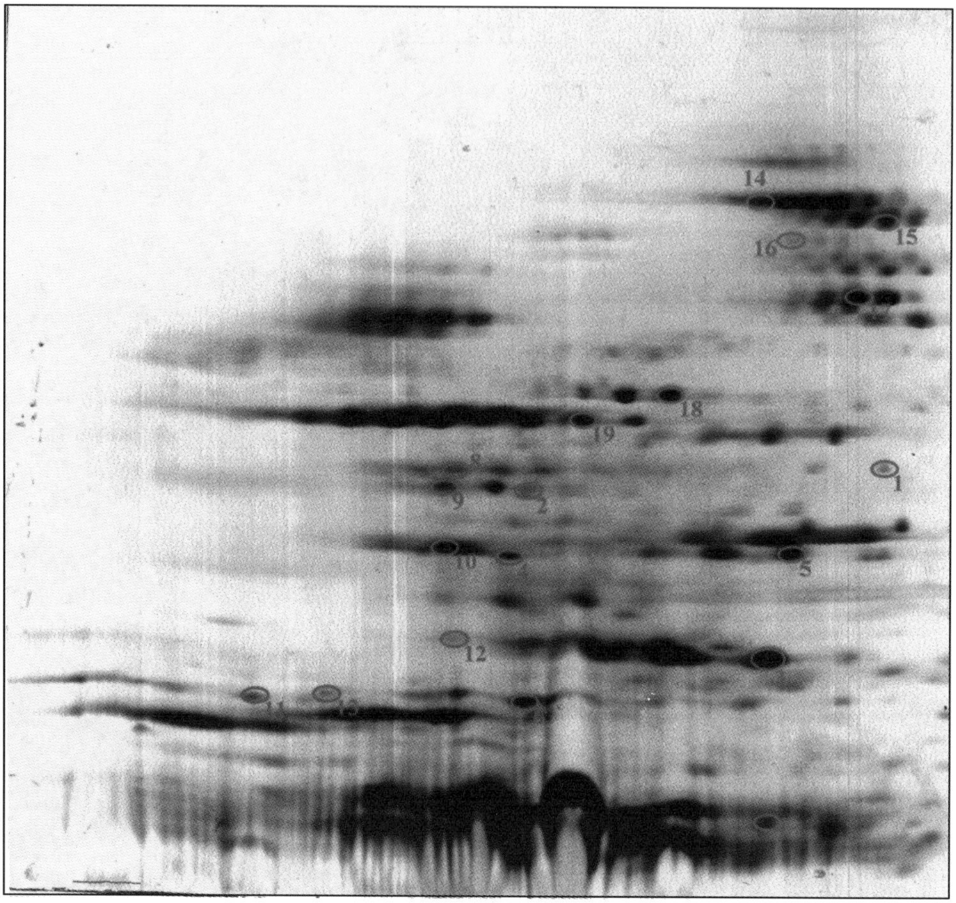

Pre-fractionated sample, pH 4-5

Fig. 14.38. 2-D map of an *E. coli* sample as in Fig. 14.37, but run with a 5-fold increased protein concentration (gel stained with colloidal Coomassie Blue G-250). The circles and numbers indicate proteins eluted and identified by MALDI-TOF MS of the recovered peptides, after trypsin digestion. (From [261], with permission.)

similar to the two panels of Fig. 14.40, except that the protein patterns are revealed with Coomassie staining. The arrowheads indicate a train of spots which is highly enriched in the pre-fractionated sample on the right side. Note additionally, in the right panel, the complete absence of albumin and the enrichment of a number of other spots.

In conclusion, the present pre-fractionation method could prove a formidable tool in proteome analysis, since not only it will provide the much needed improvement in resolution, but also the highly desirable increment in sensitivity, due to the possibility of loading a much higher sample amount in any desired narrow pH interval. The philosophy of the protocol of Herbert and Righetti [261] is to allow for such a high

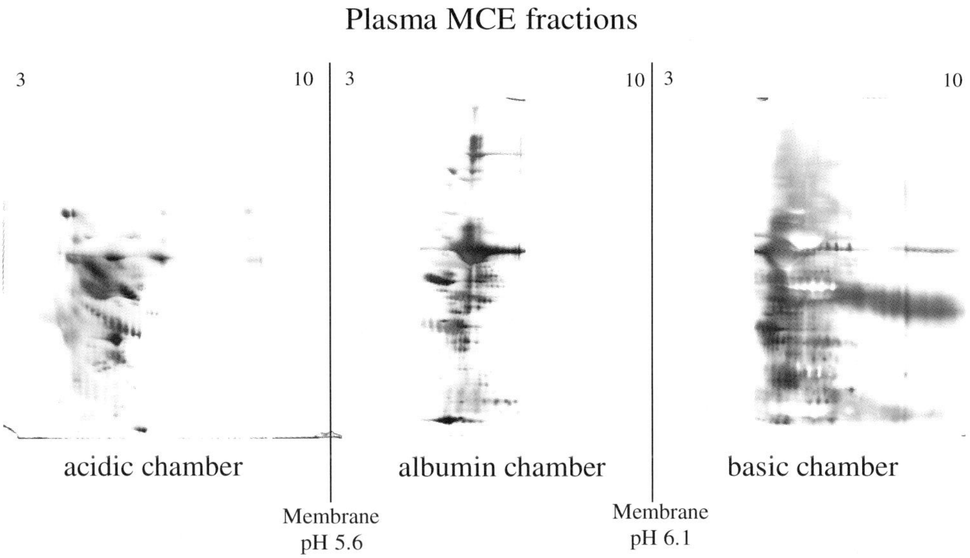

Fig. 14.39. Example of the pH 3–10 IPG 2-D analysis of plasma proteins sub-fractionated in the MCE instrument of Fig. 14.35, with membranes and conditions as given in Section 14.2.3. The three fractions are denoted as acidic chamber (left panel), albumin chamber (middle panel) and basic chamber (right panel). (From [261], with permission.)

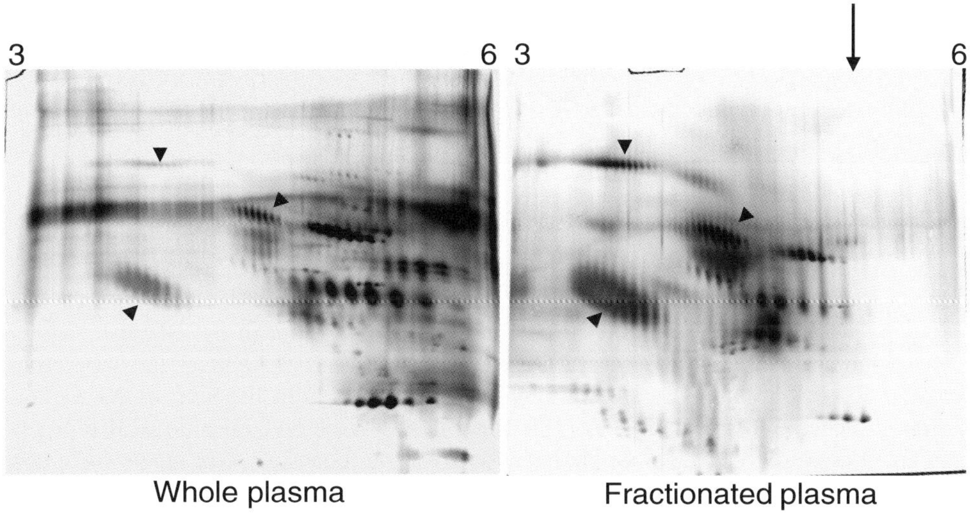

Fig. 14.40. 2-D map of a plasma sample in a narrow pH 3–6, first-dimension Immobiline gel. Left side: unfractionated plasma; right side: pre-fractionated plasma from the acidic fraction shown in Fig. 14.39. The three arrowheads denote three major trains of spots as major landmarks in the map. The vertical arrow in the right panel marks the pH position of the p*I* 5.6 membranes used in the pre-fractionation: it can be appreciated that no proteins can be found on its right side. Both maps silver-stained. (From [261], with permission.)

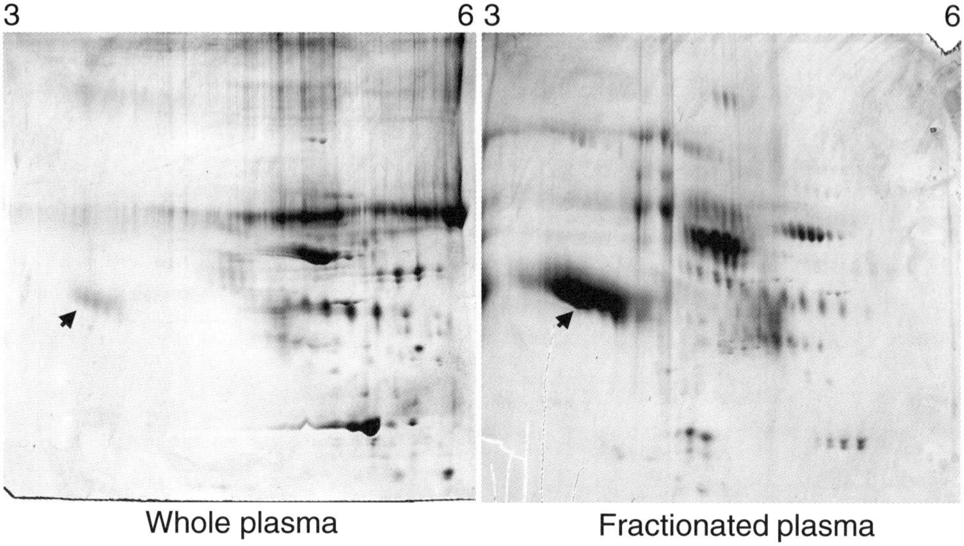

Fig. 14.41. 2-D map of a plasma sample in a narrow pH 3–6, first-dimension Immobiline gel. Left side: whole plasma; right side: pre-fractionated plasma from the acidic fraction shown in Fig. 14.39. All conditions as in Fig. 14.40 except that both maps are stained with Coomassie Blue. (From [261], with permission.)

protein loading that only Coomassie stain will be needed for detection; silver staining is in fact not very suitable for MALDI-TOF procedures, since the aldehydes used in most protocols generally cross-link proteins and render them unavailable for further analysis. In addition, the glutaraldehyde-free silver stains commonly used for MS analysis are less sensitive and are often no better than colloidal Coomassie.

Are there other examples of electrophoretic pre-fractionation protocols? One approach comes from Hochstrasser et al. [262], who reported the possibility of obtaining narrow p*I* cut by using the Rotophor. This equipment, developed by Bier's group [263], and available commercially from Bio-Rad (Hercules, CA, USA), consists of a single piece, a rotating chamber provided with a cooling finger and divided into 20 compartments by nylon nets, The mixture to be fractionated is present in the chamber, which rotates slowly on its axis to prevent electrodecantation. At the end of the separation process, all of the chambers are simultaneously emptied with the help of 20 suction needles connected to a vacuum port. This apparatus is based on conventional IEF protocols, i.e. it uses soluble carrier ampholytes for driving a focussing process. As such, it is subjected to all problems of conventional IEF, namely unreproducibility form run to run due to batch variability of the synthetic buffers, cathodic drift, and protein precipitation in overloaded chambers. In addition, since the various chambers are only divided by simple nylon nets and are not flow tight, there is no control of the content of each chamber, which could drift into neighbouring ones. In fact, although this pre-fractionation method was reported long ago by Hochstrasser's group, it does not seem to have found widespread applications. Another approach might consist of the use of the Octopus

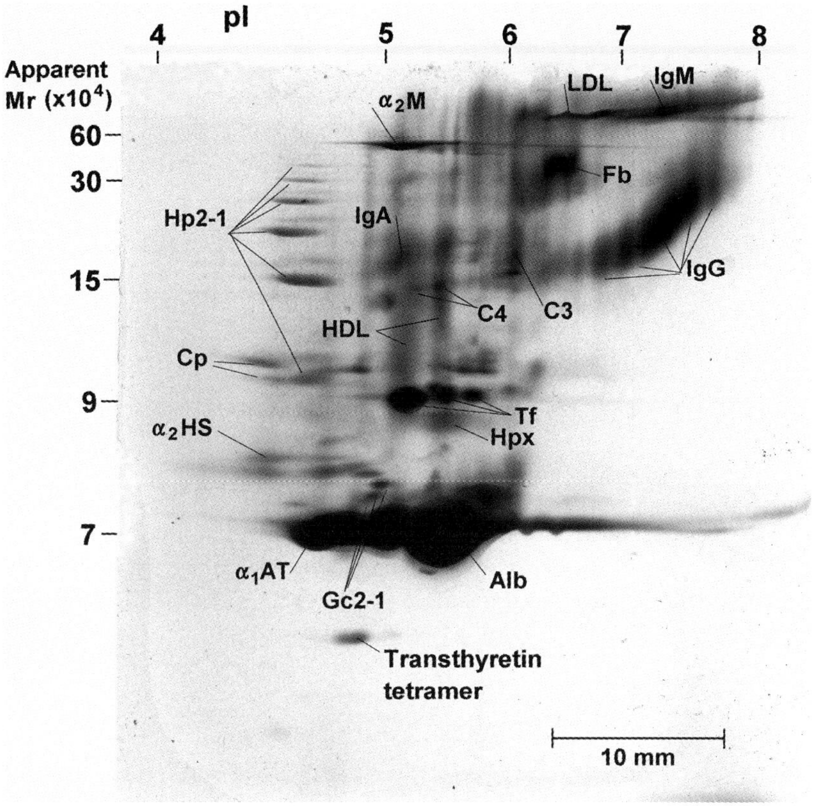

Fig. 14.42. Non-denaturing (type-I) 2-D map of human plasma proteins. A sample of normal human plasma was subjected to IEF in the absence of denaturants and then set on a micro-slab gel of polyacrylamide gradient (4.2%–17.85% *T* linear gradient, at 5% *C*) and the 2nd-D run also in the absence of denaturants. Abbreviations: α_2M, α_2-macroglobulin; LDL, low-density lipoprotein; IgM, immunoglobulin M; HP 2-1, haptoglobin phenotype 2-1, polymer series; IgA, immunoglobulin A; Fb, fibrinogen; IgG, immunoglobulin G; HDL, high-density lipoprotein; C4, complement factor C4; C3, complement factor C3; Cp, ceruloplasmin; Tf, transferrin; Hpx, hemopexin; α_2HS, α_2-HS glycoprotein; α_1AT, α_1-antitrypsin; Gc2-1, Gc-globulin phenotype 2-1; Alb, albumin. (From [274], with permission.)

(Dr. Weber GmbH, Kirchheim, Germany), a fractionation machine based also on two flows: a liquid curtain flowing downstream in a narrow gap in between two flat glass plates, coupled to a orthogonal electric field separating the macromolecules as they are passively transported by the liquid stream. At the bottom of this chamber, 80-collection tubings are to be found, able to individually collect just as many fractions. Burggraf et al. [264] have used this approach to pre-fractionate complex sample mixtures. Another potential approach could be the use of the Gradiflow (Gradipore, Sydney, Australia), a multifunctional electrokinetic membrane apparatus that can process and purify protein solutions based on differences of mobility, p*I* and size [265,266]. Although Corthals et al. [267] have reported its use for pre-fractionation prior to 2-D analysis, it is doubtful that it could achieve any of the unique fractionations reported for the isoelectric mem-

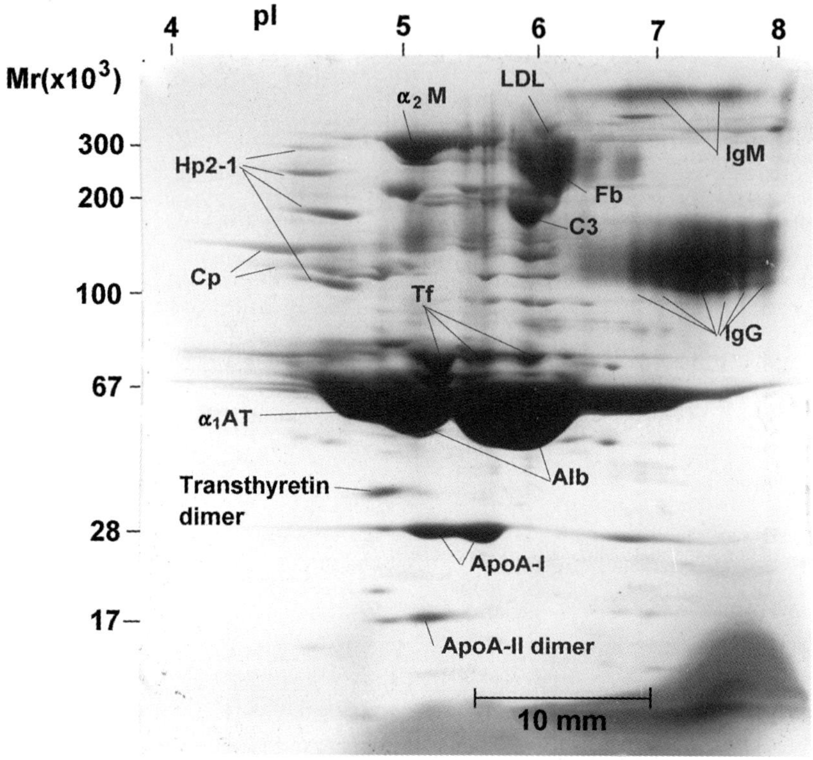

Fig. 14.43. Type-II (non-denaturing/SDS) 2-D map of human plasma proteins. The IEF conditions were the same as in Fig. 14.42. After the 1st-D step, the strip was equilibrated in plain 2% SDS (in the absence of disulphide bridges reducing agents) and then the 2nd-D was run in 0.1% SDS. All abbreviations as in Fig. 14.42. (From [274], with permission.)

brane apparatus, since this instrument can only use, typically, membranes based on size fractionation, which provide only crude resolution.

14.6. NON-DENATURING PROTEIN MAPS

Before closing this chapter on 2-D maps, a digression is due on native (non-denaturing) protein maps, as opposed to polypeptide maps, i.e. the standard maps obtained under fully denaturing conditions in both dimensions. Manabe's group has extensively developed this approach, which might have interesting practical implications, as outlined below. Starting already in 1979, Manabe et al. [268,269] proposed 2-D maps for analysis of proteins in various body fluids, including human plasma, under non-denaturing conditions, so as to acquire information on proteins under physiological conditions. Major plasma proteins could thus be identified by electrophoretic blotting followed by immunochemical staining, enabling them to construct a 'protein map' of human sera [270]. Enzyme activities could also be detected on such native 2-D maps

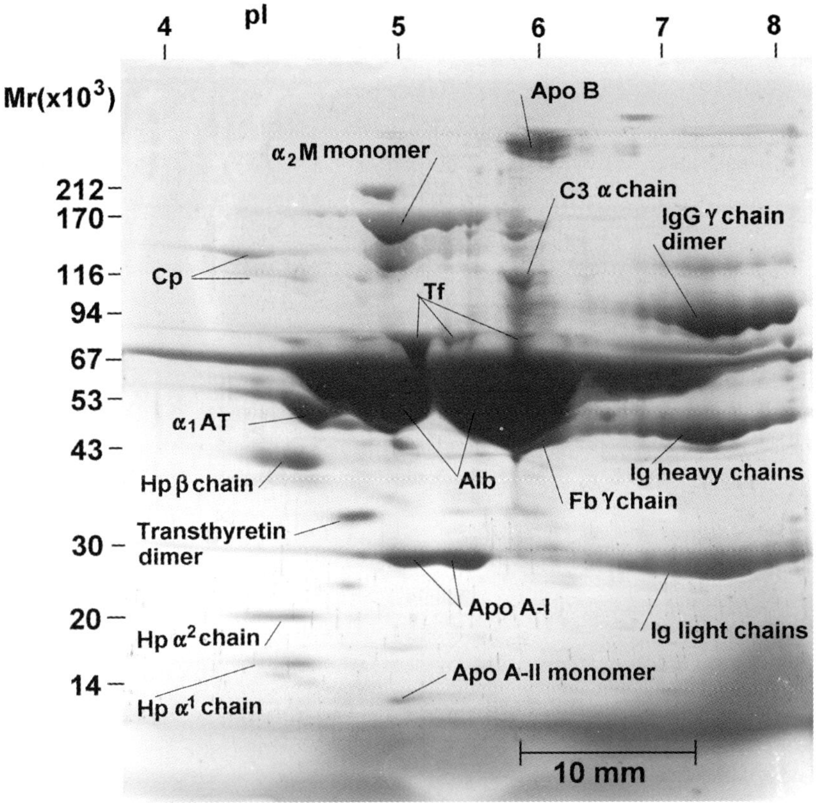

Fig. 14.44. Type-III (non-denaturing/reducing SDS) 2-D map of human plasma proteins. The IEF conditions were the same as in Fig. 14.42. After the 1st-D step, the strip was equilibrated in 2% SDS, 8 M urea and 5% β-mercaptoethanol and then the 2nd-D was run in 0.1% SDS. The size and the acrylamide gradient of the slab gel were the same as in Fig. 14.42. All abbreviations as in Fig. 14.42 (except for Apo B, apolipoprotein B-100). (From [274], with permission.)

[271]; additionally, the dissociation process of lipoproteins into apolipoproteins could be analysed by modifying the process of IEF gel equilibration [272,273]. More recently, Manabe et al. [274,275] have proposed four types of 2-D maps, for obtaining systematic information on proteins and their constituent polypeptides. In order to facilitate the analyses, a micro gel system was adopted for creating such a set of 2-D maps. In type-I 2-D map analysis, a plasma sample was first analysed under non-denaturing conditions in both dimensions, so as to characterise the properties of constituent proteins under physiological conditions (Fig. 14.42). In type-II 2-D maps, the sample was then analysed by employing non-denaturing IEF in the first dimension, followed by SDS-PAGE in the second dimension. In this second approach, the dissociation of non-covalently bound protein subunits could be revealed, as shown in Fig. 14.43. In type-III 2-D gel slabs, protein were separated again by non-denaturing IEF in the first dimension, just as performed in type-II gels; however, in the second-dimension SDS-PAGE, the denaturing solution of urea/SDS was added with mercaptoethanol, so as to induce the dissociation

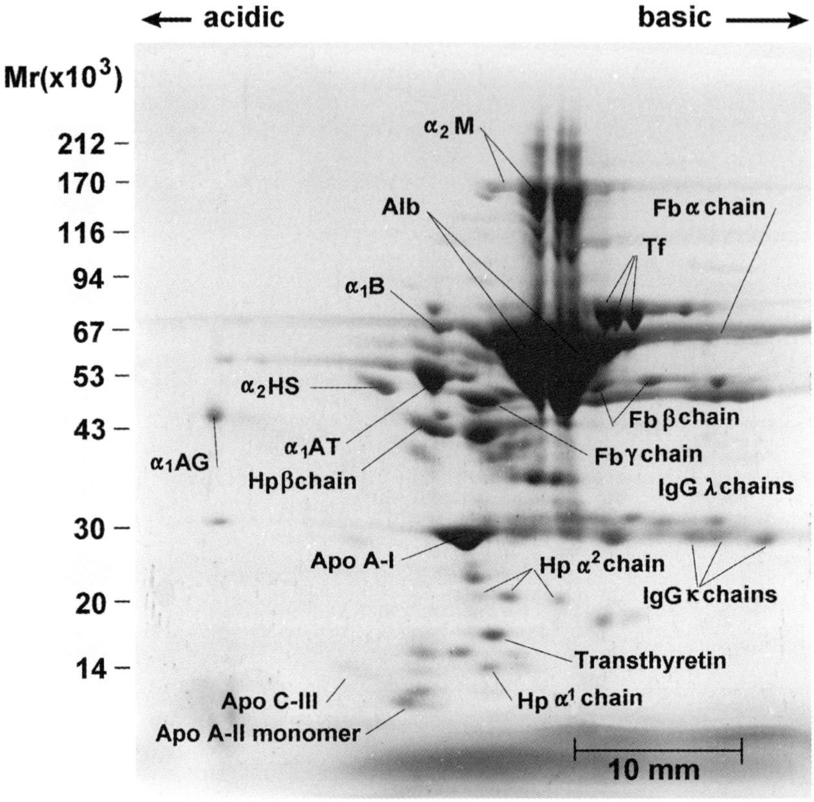

Fig. 14.45. Type-IV (denaturing in both dimensions) 2-D map of human plasma proteins. The plasma sample was treated with 8 M urea, 1% Nonidet P-40 (NP-40) and 10% β-mercaptoethanol. After the IEF step in urea/NP-40, the strip was equilibrated 2% SDS, 8 M urea and 5% β-mercaptoethanol and then the 2nd-D was run in 0.1% SDS. The size and the acrylamide gradient of the slab gel were the same as in Fig. 14.42. All abbreviations as in Fig. 14.42 (except for α_1B, α_1-glycoprotein B; α_1AG, α_1-acidic glycoprotein; Apo C-III, apolipoprotein C-III). (From [274], with permission.)

of disulphide bonded polypeptides (see Fig. 14.44). In type-IV 2-D mapping, full denaturing conditions were adopted in both dimensions, as routinely done in conventional proteome analysis (Fig. 14.45). These combined, four types of 2-D maps were found useful for starting the construction of a comprehensive database of plasma proteins combining the 'non-denaturing protein map' with the 'denaturing polypeptide map'. Such an approach should be applicable to the comprehensive analysis of other complex systems of soluble proteins, such as cytosol proteins extracted from various cells.

14.7. REFERENCES

1. M.R. Wilkins, K.L. Williams, R.D. Appel and D.F. Hochstrasser (Eds.), Proteome Research: New Frontiers in Functional Genomics, Springer, Berlin, 1997, 243 pp.

2. J.E. Celis, M. Ostergaard, N.A. Jensen, I. Gromova, H.H. Rasmussen and P. Gromov, FEBS Lett., 430 (1998) 64–72.
3. R.D. Appel, M.J. Dunn and D.F. Hochstrasser (Guest Eds.), Paper Symposium: Biomedicine and Bioinformatics, Electrophoresis, 18 (1997) 2703–2842.
4. R.D. Appel, M.J. Dunn and D.F. Hochstrasser (Guest Eds.), Paper Symposium: Biomedicine and Bioinformatics, Electrophoresis, 20 (1999) 3481–3686.
5. J.E. Celis (Guest Ed.), Paper Symposium: Protein Databases in Two Dimensional Electrophoresis, Electrophoresis, 10 (1989) 71–164.
6. J.E. Celis (Guest Ed.), Paper Symposium: Cell Biology, Electrophoresis, 11 (1990a) 189–280.
7. J.E. Celis (Guest Ed.), Paper Symposium: Two Dimensional Gel Protein Databases, Electrophoresis, 11 (1990b) 987–1168.
8. J.E. Celis (Guest Ed.), Paper Symposium: Two Dimensional Gel Protein Databases, Electrophoresis, 12 (1991) 763–996.
9. J.E. Celis (Guest Ed.), Paper Symposium: Two Dimensional Gel Protein Databases, Electrophoresis, 13 (1992) 891–1062.
10. J.E. Celis (Guest Ed.), Paper Symposium: Electrophoresis in Cancer Research, Electrophoresis, 15 (1994a) 307–556.
11. J.E. Celis (Guest Ed.), Paper Symposium: Two Dimensional Gel Protein Databases, Electrophoresis, 15 (1994b) 1347–1492.
12. J.E. Celis (Guest Ed.), Paper Symposium: Two Dimensional Gel Protein Databases, Electrophoresis, 16 (1995) 2175–2264.
13. J.E. Celis (Guest Ed.), Paper Symposium: Two Dimensional Gel Protein Databases, Electrophoresis, 17 (1996) 1653–1798.
14. J.E. Celis (Guest Ed.), Genomics and Proteomics of Cancer, Electrophoresis, 20 (1999) 223–428.
15. C. Damerval and D. de Vienne (Guest Eds.), Paper Symposium: Two Dimensional Electrophoresis of Plant Proteins, Electrophoresis, 9 (1988) 679–796.
16. M.J. Dunn (Guest Ed.), Paper Symposium: Biomedical Applications of Two-Dimensional Gel Electrophoresis, Electrophoresis, 12 (1991) 459–606.
17. M.J. Dunn (Guest Ed.), 2D Electrophoresis: from Protein Maps to Genomes, Electrophoresis, 16 (1995) 1077–1326.
18. M.J. Dunn (Guest Ed.), From Protein Maps to Genomes, Proceedings of the Second Siena Two-Dimensional Electrophoresis Meeting, Electrophoresis, 18 (1997) 305–662.
19. M.J. Dunn (Guest Ed.), From Genome to Proteome: Proceedings of the Third Siena Two-Dimensional Electrophoresis Meeting, Electrophoresis, 20 (1999) 643–1122.
20. M.J. Dunn (Guest Ed.), Proteomic Reviews, Electrophoresis, 21 (2000) 1037–1234.
21. L.A. Huber (Guest Ed.), Paper Symposium: Proteomics of Cell Organelles, Electrophoresis, 21 (2000) 3329–3528.
22. I. Humphery-Smith (Guest Ed.), Paper Symposium: Microbial Proteomes, Electrophoresis, 18 (1997) 1207–1497.
23. F. Lottspeich (Guest Ed.), Paper Symposium: Electrophoresis and Amino Acid Sequencing, Electrophoresis, 17 (1996) 811–966.
24. B. Tümmler (Guest Ed.), Microbial Genomes: Biology and Technology, Electrophoresis, 19 (1998) 467–624.
25. K.L. Williams (Guest Ed.), Strategies in Proteome Research, Electrophoresis, 19 (1998) 1853–2050.
26. P.G. Righetti, Immobilized pH Gradients: Theory and Methodology, Elsevier, Amsterdam, 1990, 400 pp.
27. M. Chevallet, V. Santoni, A. Poinas, D. Rouquie, A. Fuchs, S. Kieffer, M. Rossignol, J. Lunardi, J. Garin and T. Rabilloud, Electrophoresis, 19 (1998) 1901–1909.
28. B.R. Herbert, M.P. Molloy, B.A.A. Gooley, J. Walsh, W.G. Bryson and K.L. Williams, Electrophoresis, 19 (1998) 845–851.
29. T. Rabilloud, Anal. Chem., 72 (2000) 48A–55A.
30. P.H. O'Farrell, J. Biol. Chem., 250 (1975) 4007–4021.
31. J. Klose, Humangenetik, 26 (1975) 231–243.
32. G.A. Scheele, J. Biol. Chem., 250 (1975) 5375–5385.

33. U.K. Laemmli, Nature, 227 (1970) 680–685.
34. R.P. Tracy, R.M. Currie and D.S. Young, Clin. Chem., 28 (1982) 908–913.
35. A.H.M. Burghes, M.J. Dunn and V. Dubowitz, Electrophoresis, 3 (1982) 354–360.
36. R. Duncan and J.W.B. Hershey, Anal. Biochem., 11 (1984) 342–349.
37. G.H. Perdew, H.W. Schaup and D.P. Selivonchick, Anal. Biochem., 135 (1983) 453–459.
38. P.Z. O'Farrell, H.M. Goodman and P.H. O'Farrell, Cell, 12 (1977, 1133–1140.
39. R. Ruechel, J. Chromatogr., 132 (1977) 451–459.
40. H.M. Poehling, U. Wyss and V. Neuhoff, Electrophoresis, 1 (1980) 198–204.
41. N.L. Anderson and N.G. Anderson, Proc. Natl. Acad. Sci. USA, 74 (1977) 5421–5426.
42. N.L. Anderson and N.G. Anderson, Anal. Biochem., 85 (1978) 331–340.
43. N.L. Anderson and N.G. Anderson, Anal. Biochem., 85 (1978) 341–354.
44. K.E. Willard, C. Giometti, N.L. Anderson, T.E. O'Connor and N.G. Anderson, Anal. Biochem., 100 (1979) 289–298.
45. N.L. Anderson and N.G. Anderson, Clin. Chem., 28 (1982) 739–748.
46. N.L. Anderson and N.G. Anderson, Clin. Chem., 30 (1984) 1898–1905.
47. D.S. Young and N.G. Anderson (Guest Eds.), Special Issue: Two-Dimensional Gel Electrophoresis, Clin. Chem., 28 (1982) 737–1092.
48. J.S. King (Guest Ed.), Special Issue: Two-Dimensional Gel Electrophoresis, Protein Mapping, Clin. Chem., 30 (1984) 1897–2108.
49. B. Bjellqvist, K. Ek, P.G. Righetti, E. Gianazza, A. Görg, W. Postel and R. Westermeier, J. Biochem. Biophys. Methods, 6 (1982) 317–339.
50. R. Aebersold and J. Leavitt, Electrophoresis, 11 (1990) 517–527.
51. S.D. Patterson and R. Aebersold, Electrophoresis, 16 (1995) 1791–1814.
52. H.W. Lahm and H. Langen, Electrophoresis, 21 (2000) 2105–2114.
53. F. Lottspeich, T., Houthaeve and R. Kellner, in R. Kellner, F. Lottspeich and H.E. Meyer (Eds.), Microcharacterization of Proteins, Wiley–VCH, Weinheim, 1999, pp. 141–158.
54. M.R. Wilkins, C. Pasquali, R.D. Appel, K. Ou, O. Golaz, J.C. Sanchez, J.X. Yan, A.A. Gooley, G. Hughes, I. Humphrey-Smith, K.L. Williams and D.F. Hochstrasser, BioTechnology, 14 (1996) 61–65.
55. S.R. Pennington, M.R. Wilkins, D.F. Hochstrasser and M.J. Dunn, Trends Cell Biol., 7 (1997) 168–173.
56. R. Kellner, F. Lottspeich and H.E. Meyer, Microcharacterization of Proteins, Wiley–VCH, 1999, Weinheim.
57. T. Rabilloud, Proteome Research: Two-Dimensional Gel Electrophoresis and Identification Methods, Springer, Heidelberg, 2000, 248 pp.
58. R. Westermeier, Electrophoresis in Practice, VCH, Weinheim, 1997, pp. 213–228.
59. S. Hanash, in B.D. Hames (Ed.), Gel Electrophoresis of Proteins, Oxford Univ. Press, Oxford, 1998, pp. 189–212.
60. E. Gianazza, P. Giacon, B. Sahlin and P.G. Righetti, Electrophoresis, 6 (1985) 53–56.
61. E. Gianazza, S. Astrua-Testori and P.G. Righetti, Electrophoresis, 6 (1985) 113–117.
62. E. Gianazza, P. Giacon, S. Astrua-Testori and P.G. Righetti, Electrophoresis, 6 (1985) 326–331.
63. E. Gianazza and P.G. Righetti, J. Chromatogr., 193 (1980) 1–8.
64. P.G. Righetti, M. Chiari, P.K. Sinha and E. Santaniello, J. Biochem. Biophys. Methods, 16 (1988) 185–192.
65. A. Bossi, P.G. Righetti, G. Vecchio and S. Severinsen, Electrophoresis, 15 (1994) 1535–1540.
66. A. Bossi, C. Gelfi, A. Orsi and P.G. Righetti, J. Chromatogr. A, 686 (1994) 121–128.
67. P.K. Sinha, M. Praus, E. Köttgen, E. Gianazza and P.G. Righetti, J. Biochem. Biophys. Methods, 21 (1990) 173–179.
68. C. Tonani and P.G. Righetti, Electrophoresis, 12 (1991) 1011–1021.
69. P.G. Righetti and C. Tonani, Electrophoresis, 12 (1991) 1021–1027.
70. J.A. Lenstra and H. Bloemendal, Eur. J. Biochem., 135 (1983) 413–423.
71. M.P. Molloy, B.R. Herbert, B.J. Walsh, M.I. Tyler, M. Traini, J.C. Sanchez, D.F. Hochstrasser, K.L. Williams and A.A. Gooley, Electrophoresis, 19 (1998) 837–844.
72. M.P. Deutscher (Ed.), Guide to Protein Purification, Methods Enzymol., 182 (1990) 1–894.
73. N.L. Anderson and B.J. Hickman, Anal. Biochem., 93 (1979) 312–320.

74. B.J. Hickman, N.L. Anderson, K.E. Willard and N.G. Anderson, in B.J. Radola (Ed.), Electrophoresis '79, de Gruyter, Berlin, 1980, pp. 341–360.
75. S.L. Tollaksen, J.J. Edwards and N.G. Anderson, Electrophoresis, 2 (1981) 155–160.
76. J.D. Dignam, Methods Enzymol., 182 (1990) 194–203.
77. D.M. Bollag and S.J. Edelstein, Protein Methods, Wiley–Liss, New York, 1991.
78. T. Toda, Y. Ishijima, H. Matsushita, M. Yoshida and N. Kimura, Electrophoresis, 15 (1994) 984–987.
79. S.I. Kawaguchi and S. Karamitsu, Electrophoresis, 16 (1995) 1060–1066.
80. A.P. Teixera-Gomes, A. Cloeckaert, G. Bezard, G. Dubray and M.S. Zygmunt, Electrophoresis, 18 (1997) 156–162.
81. P. Gengenheimer, Methods Enzymol., 182 (1990) 174–193.
82. C. Theillet, F. Delpeyroux, M. Fiszman, P. Reigner and R. Esnault, Planta, 155 (1982) 478–485.
83. T.J. Wolpert and L.D. Dunkle, Proc. Natl. Acad. Sci. USA, 80 (1983) 6576–6580.
84. S.M. Jawinski, Methods Enzymol., 182 (1990) 154–174.
85. M. Cull and C.S. McHenry, Methods Enzymol., 182 (1990) 147–153.
86. D.W. Cleveland, S.G. Fischer, M.W. Kirscner and U.K. Laemmli, J. Biol. Chem., 252 (1977) 1102–1106.
87. C. Damerval, D. de Vienne, M. Zivy and H. Thiellement, Electrophoresis, 7 (1986) 52–54.
88. F.S. Wu and M.Y. Wang, Anal. Biochem., 139 (1984) 100–103.
89. P.A. Harrison and C.C. Black, Plant Physiol., 70 (1982) 1359–1366.
90. H.L.M. Granzier and K. Wang, Electrophoresis, 14 (1993) 56–64.
91. C. Colas de Francs, H. Thiellement and D. de Vienne, Plant Physiol., 78 (1985) 178–182.
92. E. Olivieri, B. Herbert and P.G. Righetti, Electrophoresis, 22 (2001) 560–565.
93. N.H.H. Heegaard and R. Poglod, Appl. Theor. Electrophoresis, 2 (1991) 109–127.
94. T. Rabilloud, T. Blisnick, M. Heller, S. Luche, R. Aebersold, J. Lunardi and C. Braun-Breton, Electrophoresis, 20 (1999) 3603–3610.
95. P. Holloway and P. Arundel, Anal. Biochem., 172 (1988) 8–15.
96. R. Flengsrud and G.A. Kobro, Anal. Biochem., 177 (1989) 33–36.
97. N.M. Matsui, D.M. Smith, K.R. Clauser, J. Fichmann, L.E. Andrews, C.M. Sullivan, A.L. Burlingame and L.B. Epstein, Electrophoresis, 18 (1997) 409–417.
98. A. Tsugita, M. Kamo, T. Kawakami and Y. Ohki, Electrophoresis, 17 (1996) 855–865.
99. F. Granier, Electrophoresis, 9 (1988) 712–806.
100. J.B. Clegg, M.A. Naughton and D. Weatherall, J. Mol. Biol., 19 (1966) 91–101.
101. H. Usada and K. Shimogawara, Plant Cell Physiol., 36 (1995) 1149–1155.
102. T. Rabilloud, Electrophoresis, 17 (1996) 813–829.
103. T. Manabe, E. Hayama and T. Okuyama, Clin. Chem., 28 (1982) 824–827.
104. T. Marshall and K.M. Williams, Electrophoresis, 13 (1992) 887–888.
105. T. Marshall, N.J. Abbott, P. Fox and K.M. Williams, Electrophoresis, 16 (1995) 28–31.
106. C.W. Heizmann, E.M. Arnold and C.C. Kuenzle, J. Biol. Chem., 255 (1980) 11504–11511.
107. E. Galante, T. Caravaggio and P.G. Righetti, Biochim. Biophys. Acta, 442 (1976) 309–315.
108. W.F. Glass, R.C. Briggs and L.S. Hnilica, Science, 211 (1981) 70–72.
109. M.M. Sanders, V.E. Groppi and E.T. Browning, Anal. Biochem., 103 (1980) 157–165.
110. J. Mohberg and H.P. Rusch, Arch. Biochem. Biophys., 134 (1969) 577–589.
111. M. Yoshida and K. Shimura, Biochim. Biophys. Acta, 263 (1972) 690–695.
112. E. Gianazza and P.G. Righetti, Biochim. Biophys. Acta, 540 (1978) 357–364.
113. P.G. Righetti, R. Brown and A.L. Stone, Biochim. Biophys. Acta, 542 (1978) 222–231.
114. J. Van Renswoude and C. Kemps, Methods Enzymol., 104 (1984) 329–339.
115. N.S. Radin, Methods Enzymol., 72 (1981) 5–7.
116. D. Wessel and U.I. Flugge, Anal. Biochem., 138 (1984) 141–143.
117. W. Menke and F. Koenig, Methods Enzymol., 69 (1980) 446–452.
118. H.S. Penefsky and A. Tzagoloff, Methods Enzymol., 22 (1971) 204–219.
119. G.F.L. Ames and K. Nikaido, Biochemistry, 15 (1976) 616–623.
120. F. Cremer and C. Van de Walle, Anal. Biochem., 147 (1985) 22–26.
121. P.G. Righetti, M. Chiari and C. Gelfi, Electrophoresis, 9 (1988) 65–73.

122. M.R. Wilkins, E. Gasteiger, J.C. Sanchez, A. Bairoch and D.F. Hochstrasser, Electrophoresis, 19 (1998) 1501–1505.
123. M.P. Molloy, Anal. Biochem., 280 (2000) 1–10.
124. T. Rabilloud and M. Chevallet, in T. Rabilloud (Ed.), Proteome Research: Two-Dimensional Gel Electrophoresis and Identification Methods, Springer, Heidelberg, 2000, pp. 9–29.
125. T. Rabilloud, C. Adessi, A. Giraudel and J. Lunardi, Electrophoresis, 18 (1997) 307–316.
126. L. Musante, G. Candiano and G.M. Ghiggeri, J. Chromatogr. A, 705 (1997) 351–356.
127. C. Pasquali, I. Fialka and L.A. Huber, Electrophoresis, 18 (1997) 2574–2581.
128. T. Rabilloud, Electrophoresis, 19 (1998) 758–760.
129. J.A. Gordon and W.P. Jencks, Biochemistry, 2 (1963) 47–57.
130. P. Hagel, J.J.T. Gerding, W. Fieggen and H. Bloemendal, Biochim. Biophys. Acta, 243 (1971) 366–373.
131. M.N. Jones, Int. J. Pharm., 177 (1999) 137–139.
132. D.F. Hochstrasser, M.G. Harrington, A.C. Hochstrasser, M.J. Miller and C.R. Merill, Anal. Biochem., 173 (1988) 424–435.
133. M.A. Rimpilainen and P.G. Righetti, Electrophoresis, 6 (1985) 419–422.
134. D. Satta, G. Schapira, P. Chafey, P.G. Righetti and J.P. Wahrmann, J. Chromatogr., 299 (1984) 57–72.
135. M.J. Dunn and A.H.M. Burghes, Electrophoresis, 4 (1983) 97–116.
136. E. Gianazza, T. Rabilloud, L. Quaglia, P. Caccia, S. Astrua-Testori, L. Osio, G. Grazioli and P.G. Righetti, Anal. Biochem., 165 (1987) 247–257.
137. T. Rabilloud, E. Gianazza, N. Cattò and P.G. Righetti, Anal. Biochem., 185 (1990) 94–102.
138. M.P. Molloy, B.R. Herbert, M.B. Slade, T. Rabilloud, A.S. Nouwens, K.L. Williams and A.A. Gooley, Eur. J. Biochem., 267 (2000) 1–12.
139. V. Santoni, T. Rabilloud, P. Doumas, D. Rouquie, A. Fuchs, S. Kieffer, J. Garin and M. Rossignol, Electrophoresis, 20 (1999) 705–711.
140. V. Santoni, S. Kieffer, D. Desclaux, F. Masson and T. Rabilloud, Electrophoresis, 21 (2000) 3329–3344.
141. J.P. Malone and M.R. Kramer, Abstract book of the 4th Siena Meeting: From Genome to Proteome: Knowledge, Acquisition and Representation, Siena, Sept. 4–7, 2000, p. 270.
142. P.G. Righetti, B. Barzaghi, E. Sarubbi, A. Soffientini and G. Cassani, J. Chromatogr., 470 (1989) 337–350.
143. E. Bordini, M. Hamdan and P.G. Righetti, Rapid Commun. Mass Spectrom., 13 (1999) 1818–1827.
144. P.G. Righetti, G. Tudor and E. Gianazza, J. Biochem. Biophys. Methods, 6 (1982) 219–227.
145. A. Goerg, W. Postel, J. Weser, S. Gunther, J.R. Strahler, S.M. Hanash and L. Somerlot, Electrophoresis, 8 (1987) 122–124.
146. A. Goerg, W. Postel and S. Gunther, Electrophoresis, 9 (1988) 531–546.
147. J.X. Yan, J.C. Sanchez, V. Rouge, K. Williams and D.F. Hochstrasser, Electrophoresis, 20 (1999) 723–726.
148. M.J. Dunn, Gel Electrophoresis: Proteins, Bio Sci. Publ., Oxford, 1993.
149. J.X. Yan, W.C. Kett, B.R. Herbert, A.A. Gooley, N.H. Packer and K.L. Williams, J. Chromatogr. A, 813 (1998) 187–200.
150. S. Caglio, M. Chiari and P.G. Righetti, Electrophoresis, 15 (1994) 209–214.
151. B. Herbert, M. Galvani, M. Hamdan, E. Olivieri, J. McCarthy, S. Pedersen, P.G. Righetti, Electrophoresis, 22 (2001) 2046–2057.
152. M. Galvani, M. Hamdan, B. Herbert and P.G. Righetti, Electrophoresis, 22 (2001) 2058–2065.
153. U.T. Ruegg, J. Rüdinger, Methods Enzymol., 47 (1977) 111–116.
154. B. Bjellqvist, J.C. Sanchez, C. Pasquali, F. Ravier, N. Paquet, S. Frutiger, C.J. Highes and D. Hochstrasser, Electrophoresis, 14 (1993) 1375–1378.
155. T. Rabilloud, C. Valette and J.J. Lawrence, Electrophoresis, 15 (1994) 1552–1558.
156. J. Schupbach, R.W. Ammann and A.U. Freiburghaus, Anal. Biochem., 196 (1991) 337–343.
157. J.C. Sanchez, V. Rouge, M. Pisteur, F. Ravier, L. Tonella, M. Moosmayer, M.R. Wilkins and D. Hochstrasser, Electrophoresis, 18 (1997) 324–327.
158. X. Zuo and D.W. Speicher, Electrophoresis, 21 (2000) 3035–3047.

159. M.R. Kramer and J.P. Malone, Abstract book of the 4th Siena Meeting: From Genome to Proteome: Knowledge, Acquisition and Representation, Siena, Sept. 4–7, 2000, p. 269.
160. T. Rabilloud, J.J. Pernelle, P. Wahrmann, C. Gelfi and P.G. Righetti, J. Chromatogr., 402 (1987) 105–113.
161. T. Rabilloud, C. Gelfi, M.L. Bossi and P.G. Righetti, Electrophoresis, 8 (1987) 305–312.
162. B.M. Gåveby, P. Pettersson, J. Andrasko, L. Ineva-Flygare, U. Johannesson, A. Görg, W. Postel, A. Domscheit, P.L. Mauri, P. Pietta, E. Gianazza and P.G. Righetti, J. Biochem. Biophys. Methods, 16 (1988) 141–164.
163. C. Gelfi and P.G. Righetti, J. Biochem. Biophys. Methods, 8 (1983) 156–171.
164. P.G. Righetti and C. Gelfi, J. Biochem. Biophys. Methods, 9 (1984) 103–119.
165. P.G. Righetti, E. Wenisch and M. Faupel, J. Chromatogr., 475 (1989) 293–309.
166. P.G. Righetti, E. Wenisch, A. Jungbauer, H. Katinger and M. Faupel, J. Chromatogr., 500 (1990) 681–696.
167. C. Chan, R.S. Warlow, P.H. Chapuis, R.C. Newland and E.L. Bokey, Electrophoresis, 20 (1999) 3467–3471.
168. P.G. Righetti, M. Chiari, E. Casale and C. Chiesa, Appl. Theor. Electrophor., 1 (1989) 115–121.
169. L.H. Choe and K.H. Lee, Electrophoresis, 21 (2000) 993–1000.
170. H. Nikaido, in F.C. Neidhardt (Ed.), *Escherichia coli* and *Salmonella* Cellular and Molecular Biology, ASM Press, Washington, DC, 1996, pp. 29–47.
171. M.P. Molloy, B.R. Herbert, K.L. Williams and A.A. Gooley, Electrophoresis, 20 (1999) 701–704.
172. B.R. Herbert, J.C. Sanchez and L. Bini, in M.R. Wilkins, K.L. Williams, R.D. Appel and D.F. Hochstrasser (Eds.), Proteome Research: New Frontiers in Functional Genomics, Springer, Berlin, 1997, pp. 13–33.
173. J.E. Celis, H.H. Rasmussen, P. Madsen, H. Leffers, B. Honorè, K. Dejgaard, B. Gesser, E. Olsen, P. Gromov, H.J. Hoffman, M. Nielsen, A. Celis, B. Basse, J.B. Lauridsen, G.P. Ratz, H. Nielsen, A.H. Andersen, E. Walbrum, I. Kjaegaard, M. Puype, J. Van Damme and J. Vandekerckhove, Electrophoresis, 13 (1992) 893–959.
174. M.R. Wilkins and A.A. Gooley, in M.R. Wilkins, K.L. Williams, R.D. Appel and D.F. Hochstrasser (Eds.), Proteome Research: New Frontiers in Functional Genomics, Springer, Berlin, 1997, pp. 35–64.
175. A.L. Burlingame, R.K. Boyd and S.J. Gaskell, Anal. Chem., 70 (1998) 647R–716R.
176. W. Blackstock, in W. Blackstock and M. Mann (Eds.), Proteomics: a Trend Guide, Elsevier, London, 2000, pp. 12–17.
177. R.C. Beavis and D. Fenyö, in W. Blackstock and M. Mann (Eds.), Proteomics: a Trend Guide, Elsevier, London, 2000, pp. 22–27.
178. S.P. Gygi and R. Aebersold, in W. Blackstock and M. Mann (Eds.), Proteomics: a Trend Guide, Elsevier, London, 2000, pp. 31–36.
179. B.L. Karger and W.S. Hancock (Eds.), High Resolution Separation and Analysis of Biological Macromolecules, Methods in Enzymology, Vol. 270, Part. A, Academic Press, San Diego, CA, 1996, 622 pp.
180. W.E. Seifert, Jr. and R.M. Caprioli, in B.L. Karger and W.S. Hancock (Eds.), High Resolution Separation and Analysis of Biological Macromolecules, Methods in Enzymology, Vol. 270, Part. A, Academic Press, San Diego, CA, 1996, pp. 453–485.
181. J.F. Banks, Jr. and C.M. Whitehouse, in B.L. Karger and W.S. Hancock (Eds.), High Resolution Separation and Analysis of Biological Macromolecules, Methods in Enzymology, Vol. 270, Part. A, Academic Press, San Diego, CA, 1996, pp. 486–518.
182. R.C. Beavis and B.T. Chait, in B.L. Karger and W.S. Hancock (Eds.), High Resolution Separation and Analysis of Biological Macromolecules, Methods in Enzymology, Vol. 270, Part. A, Academic Press, San Diego, CA, 1996, pp. 519–551.
183. J.E. Schwartz and I. Jardine, in B.L. Karger and W.S. Hancock (Eds.), High Resolution Separation and Analysis of Biological Macromolecules, Methods in Enzymology, Vol. 270, Part. A, Academic Press, San Diego, CA, 1996, pp. 552–586.
184. U. Bahr, M. Karas and F. Hillenkamp, in R. Kellner, F. Lottspeich and H.E. Meyer (Eds.), Microcharacterization of Proteins, Wiley–VCH, Weinheim, 1999, pp. 177–196.

185. B. Spengler, P. Chaurand and P. Bold, in R. Kellner, F. Lottspeich and H.E. Meyer (Eds.), Microcharacterization of Proteins, Wiley–VCH, Weinheim, 1999, pp. 197–212.
186. J.W. Metzger, C. Eckerskorn, C. Kempter and B. Behnke, in R. Kellner, F. Lottspeich and H.E. Meyer (Eds.), Microcharacterization of Proteins, Wiley–VCH, Weinheim, 1999, pp. 213–234.
187. J.W. Metzger, in R. Kellner, F. Lottspeich and H.E. Meyer (Eds.), Microcharacterization of Proteins, Wiley–VCH, Weinheim, 1999, pp. 235–244.
188. C. Siethoff, C., Lohaus and H.E. Meyer, in R. Kellner, F. Lottspeich and H.E. Meyer (Eds.), Microcharacterization of Proteins, Wiley–VCH, Weinheim, 1999, pp. 245–276.
189. W. Zhang, A.J. Czernik, T. Yungwirth, R. Aebersold and B.T. Chait, Protein Sci., 3 (1994) 677–685.
190. P.A. Binz, M. Muller, D. Walther, W.V. Bienvenut, R. Gras, C. Hoogland, G. Bouchet, E. Gasteiger, R. Fabbretti, S. Gay, P. Palagi, M.R. Wilkins, V. Rouge, L. Tonella, S. Paesano, G. Rossellat, A. Karmime, A. Bairoch, J.C. Sanchez, R.D. Appel and D.H. Hochstrasser, Anal. Chem., 71 (1999) 4981–4988.
191. B. Spengler, D. Kirsch, R. Kaufman and E. Jaeger, Rapid Commun. Mass Spectrom., 6 (1992) 105–108.
192. E. Bordini, M. Hamdan and P.G. Righetti, Rapid Commun. Mass Spectrom., 13 (1999) 2209–2215.
193. P. Roepstorff and J. Fohlman, J. Biomed. Mass Spectrom., 11 (1984) 601–610.
194. K.F. Medzihradszky, J.M. Campbell, M.A. Baldwin, A.M. Falick, P. Juhasz, M.L. Vestal and A.L. Burlingame, Anal. Chem., 72 (2000) 552–558.
195. A. Shevchenko, A. Loboda, A. Shevchenko, W. Ens and K.G. Standing, Anal. Chem., 72 (2000) 2132–2141.
196. A.N. Krutchinsky, W. Zhang and B.T. Chait, J. Am. Soc. Mass Spectrom., 11 (2000) 493–504.
197. T. Nohmi and J.B. Fenn, J. Am. Chem. Soc., 14 (1992) 3241–3246.
198. J.F. Banks, S. Shen, C.M. Whitehouse and J.B. Fenn, Anal. Chem., 66 (1994) 406–411.
199. J.A. Loo, C.G. Edmonds and R.D. Snith, Anal. Chem., 65 (1993) 425–430.
200. C.K. Meng, C.N. McEwen and B.S. Larsen, Rapid Commun. Mass Spectrom., 4 (1990) 151–155.
201. S.C. Beu, M.W. Senko, J.P. Quinn, F.M. Wampler III and F.W. McLafferty, J. Am. Soc. Mass Spectrom., 4 (1993) 557–561.
202. J.A. Loo, J.P. Quinn, S.I. Ryu, K.D. Henry, M.W. Senko and F.W. McLafferty, Proc. Natl. Acad. Sci. USA, 89 (1992) 286–291.
203. G.J. Van Berkel, G.L. Glish and S.A. McLucky, Anal. Chem., 62 (1990) 1284–1289.
204. J.C. Boyle and C.M. Whitehouse, Anal. Chem., 64 (1992) 2084–2088.
205. M. Labowski, C. Whitehouse and J. Fenn, Rapid Commun. Mass Spectrom., 7 (1993) 71–76.
206. M. Wilm and M. Mann, Anal. Chem., 68 (1996) 1–8.
207. G. Neubauer and M. Mann, Anal. Chem., 71 (1999) 235–242.
208. S. Geromanos, G. Freckleton and P. Tempst, Anal. Chem., 72 (2000) 777–790.
209. S.P. Gygi, Y. Rochon, B.R. Franza and R. Aebersold, Mol. Cell Biol., 19 (1999) 1720–1730.
210. B. Futcher, G.I. Latter, P. Monsardo, C.S. McLaughlin and J.I. Garrels, Mol. Cell Biol., 19 (1999) 7357–7368.
211. Y. Oda, K. Huang, F.R. Cross, D. Cowburn and B.T. Chait, Proc. Natl. Acad. Sci. USA, 96 (1999) 6591–6596.
212. L. Pasa-Tolic, P.K. Jensen, G.A. Anderson, M.S. Lipton, K.K. Peden, S. Martinovic, N. Tolic, J.E. Bruce and R.D. Smith, J. Am. Chem. Soc., 121 (1999) 7949–7950.
213. M. Kohlman, M. Macht, S. Deininger, A. Marquardt and M. Przybylski, Science, (2001) in press.
214. R.D. Smith, L. Pasa-Tolic, M.S. Lipton, P.K. Jensen, G.A. Anderson, Y. Shen, T.P. Conrads, H.R. Udseth, R. Harkewicz, M.E. Belov, C. Masselon and T.D. Veenstra, Electrophoresis, 22 (2001) 1652–1668.
215. M.P. Washburn and J.R. Yates, III, in W. Blackstock and M. Mann (Eds.), Proteomics: a Trend Guide, Elsevier, London, 2000, pp. 27–30.
216. J.H. Wahl, D.R. Goodlett, H.R. Udseth and R.D. Smith, Anal. Chem., 64 (1992) 3194–3196.
217. J.H. Wahl and R.D. Smith, J. Capillary Electrophor., 1 (1994) 62–71.
218. M.A. Moseley, L.J. Deterding, K.B. Tomer and J.W. Jorgenson, Anal. Chem., 63 (1991) 109–114.
219. L.J. Deterding, C.E. Parker, J.R. Perkins, M.A. Moseley, J.W. Jorgenson and K.B. Tomer, J. Chromatogr., 554 (1991) 329–338.

220. W. Weinmann, K. Baumeister, I. Kaufmann and M. Przybylski, J. Chromatogr., 628 (1993) 111–121.
221. W. Weinmann, C.E. Parker, K. Baumeister, C. Maier, K.B. Tomer and M. Przybylski, Electrophoresis, 15 (1994) 228–233.
222. T.J. Thompson, F. Foret, P. Vouros and B.L. Karger, Anal. Chem., 65 (1993) 900–906.
223. J.F. Banks Jr., J. Chromatogr. A, 712 (1995) 245–252.
224. A. Rosnack, J.G. Stroh, D.H. Singleton, B.C. Guarino and G.C. Andrews, J. Chromatogr., 675 (1994) 219–225.
225. F.Y.L. Hsieh, J. Cai and J. Henion, J. Chromatogr. A, A 679 (1994) 206–211.
226. D. Figeys and R. Aebersold, Electrophoresis, 19 (1998) 885–892.
227. W. Tang, A.K. Harrata and C.S. Lee, Anal. Chem., 69 (1997) 3177–3182.
228. L. Yang, C.S. Lee, S.A. Hofstadler, L. Pasa-Tolic and R.D. Smith, Anal. Chem., 70 (1998) 3235–3241.
229. P.K. Jensen, L. Pasa-Tolic, G.A. Anderson, J.A. Horner, M.S. Lipton, J.E. Bruce and R.D. Smith, Anal. Chem., 71 (1999) 2076–2084.
230. J.K. Eng, A.L. McCormack and J.R. Yates, J. Am. Soc. Mass Spectrom., 5 (1994) 976–989.
231. A.L. McCormack, D.M. Schieltz, B. Goode, S. Yang, G. Barnes, D. Drubin and J.R. Yates III, Anal. Chem., 69 (1997) 767–776.
232. P. Davidsson, A. Westman, M. Purchades, C.L. Nilsson and K. Blennow, Anal. Chem., 71 (1999) 642–647.
233. C.L. Nilsson, M. Puchades, A. Westman, K. Blennow and P. Davidsson, Electrophoresis, 20 (1999) 860–865.
234. M. Raida, P. Schulz-Knappe, G. Heine and W.G. Forssmann, J. Am. Soc. Mass Spectrom., 10 (1999) 45–54.
235. G.J. Opiteck, K.C. Lewis, J.W. Jorgenson and R.J. Anderegg, Anal. Chem., 69 (1997) 1518–1524.
236. G.J. Opiteck, S.M. Ramirez, J.W. Jorgenson and M.A. Moseley III, Anal. Biochem., 258 (1998) 349–361.
237. A.J. Link, E. Carmack and J.R. Yates III, Int. J. Mass Spectrom. Ion Proc., 160 (1997) 303–316.
238. A.J. Link, J. Eng, D.M. Schieltz, E. Carmack, G.J. Mize, D.R. Morris, B.M. Garvik and J.R. Yates III, Nature Biotec., 17 (1999) 676–682.
239. A. Bairoch, (1997) in M.R. Wilkins, K.L. Williams, R.D. Appel and D.F. Hochstrasser (Eds.), Proteome Research: New Frontiers in Functional Genomics, Springer, Berlin, 1997, pp. 93–148.
240. R.D. Appel, in M.R. Wilkins, K.L. Williams, R.D. Appel and D.F. Hochstrasser (Eds.), Proteome Research: New Frontiers in Functional Genomics, Springer, Berlin, 1997, pp. 149–175.
241. M.C. Peitsch and N. Guex, in M.R. Wilkins, K.L. Williams, R.D. Appel and D.F. Hochstrasser (Eds.), Proteome Research: New Frontiers in Functional Genomics, Springer, Berlin, 1997, pp. 177–186.
242. J.C. Sanchez, R.D. Appel, O. Golaz, C. Pasquali, F. Ravier, A. Bairoch and D.F. Hochstrasser, Electrophoresis, 16 (1995) 1131–1151.
243. C. Hoogland, J.C. Sanchez, L. Tonella, P.A. Binz, A. Bairoch, D.F. Hochstrasser and R.D. Appel, Nucleic Acids Res., 28 (2000) 286–288.
244. A. Bairoch and R. Apweiler, Nucleic Acids Res., 28 (1999) 45–48.
245. M.E. Crawford, M.E. Cusick and J.I. Garrels, in W. Blackstock and M. Mann (Eds.), Proteomics: a Trend Guide, Elsevier, London, 2000, pp. 17–21.
246. W. Fleischmann, S. Moeller, A. Gateau and R. Apweiler, Bioinformatics, 15 (1999) 228–233.
247. R.D. Appel, A. Bairoch and D.F. Hochstrasser, Trends Biochem. Sci., 19 (1994) 258–260.
248. R.D. Appel, J.C. Sanchez, A. Bairoch, O. Golaz, M. Miu, R. Vargas and D.F. Hochstrasser, Electrophoresis, 14 (1993) 1232–1238.
249. H. Sariouglu, F. Lottspeich, T. Walk, G. Jung and C. Eckerskorn, Electrophoresis, 21 (2000) 2209–2218.
250. G.L. Corthals, C.V. Wasinger, D.F. Hochstrasser and J.C. Sanchez, Electrophoresis, 21 (2000) 1104–1115.
251. S.P. Gygi, G.L. Corthals, Y. Zhang, Y. Rochon and R. Aebersold, Proc. Natl. Acad. Sci. USA, 97 (2000) 9390–9395.
252. M. Fountoulakis, H. Langen, S. Evers, C. Gray and B. Takacs, Electrophoresis, 18 (1997) 1193–1202.
253. M. Fountoulakis and B. Takàcs, Protein Expr. Purif., 14 (1998) 113–119.

254. M. Fountoulakis, H. Langen, C. Gray and B. Takacs, J. Chromatogr. A, 806 (1998) 279–291.
255. Fountoulakis, M.F. Takacs and B. Takacs, J. Chromatogr. A, 833 (1999) 157–168.
256. Fountoulakis, M.F. Takacs, P. Berndt, H. Langen and B. Takacs, Electrophoresis, 20 (1999) 2181–2195.
257. K. Karlsson, N. Cairns, G. Lubec and M. Fountoulakis, Electrophoresis, 20 (1999) 2970–2976.
258. M.G. Harrington, J.A. Coffman, F.J. Calzone, L.E. Hood, R.J. Britten and E.H. Davidson, Proc. Natl. Acad. Sci. USA, 89 (1992) 6252–6256.
259. P.G. Righetti, E. Wenisch and M. Faupel, in A. Chrambach, M.J. Dunn and B.J. Radola (Eds.), Advances in Electrophoresis, Vol. 5, VCH, Weinheim, 1992, pp. 159–200.
260. P. Wenger, M. de Zuanni, P. Javet, C. Gelfi and P.G. Righetti, J. Biochem. Biophys. Methods, 14 (1987) 29–43.
261. B.R. Herbert and P.G. Righetti, Electrophoresis, 21 (2000) 3639–3648.
262. A.C. Hochstrasser, R.W. James, D. Pometta and D. Hochstrasser, Appl. Theor. Electrophor., 1 (1991) 333–337.
263. N.B. Egen, M. Bliss, M. Mayerson, S.M. Owens, L. Arnold and M. Bier, Anal. Biochem., 172 (1988) 488–494.
264. D. Burggraf, G. Weber and F. Lottspeich, Electrophoresis, 16 (1995) 1010–1015.
265. J. Margolis, G. Corthals and Z.S. Horvàth, Electrophoresis, 16 (1995) 98–100.
266. Z.S. Horvàth, A.A. Gooley, C.W. Wrigley, J. Margolis and K.L. Williams, Electrophoresis, 17 (1996) 224–226.
267. G.L. Corthals, M.P. Molloy, R.B. Herbert, K.L. Williams and A.A. Zooley, Electrophoresis, 18 (1997) 317–323.
268. T. Manabe, K. Tachi, K. Kojima and T. Okuyama, J. Biochem., 85 (1979) 649–659.
269. T. Manabe, K. Kojima, S. Jitzukawa, T. Hoshino and T. Okuyama, J. Biochem., 89 (1981) 1317–1323.
270. T. Manabe, Y. Takahashi, N. Higuchi and T. Okuyama, Electrophoresis, 6 (1985) 462–467.
271. T. Kadofuku, T. Sato, T. Manabe and T. Okuyama, Electrophoresis, 4 (1983) 427–431.
272. T. Manabe, S. Visvikis, J. Steinmetz, M.M. Galteau, T. Okuyama and G. Siest, Electrophoresis, 8 (1987) 325–330.
273. T. Manabe, S. Visvikis, M.F. Dumon, M. Clerc and G. Siest, Electrophoresis, 8 (1987) 468–472.
274. T. Manabe, H. Mizuma and K. Watanabe, Electrophoresis, 20 (1999) 830–835.
275. T. Manabe, Electrophoresis, 21 (2000) 1116–1112.
276. J. Taylor, N.L. Anderson, A.E. Scandora Jr., K.E. Willard and N.G. Anderson, Clin. Chem., 28 (1982) 861–866.

Acknowledgements

The striking developments in the methodological section of this book, reported as part of the research from my group, in the field of immobilised pH gradients, would have been impossible without the heroic efforts of a group of close collaborators, among whom I would like to mention Drs. E. Gianazza, C. Gelfi and M. Chiari. For our own developments in the field of capillary isoelectric focussing and the use of isoelectric buffers I would like to express my appreciation of the work of Drs. C. Gelfi, A. Bossi, E. Olivieri, L. Castelletti and B. Verzola. Our recent work on 2-D map analysis and mass spectrometry would have been impossible without the help and close collaboration of Drs. M. Hamdan, B. Herbert, M. Galvani and A. Castagna. The research from my own group reported here has been supported over the years by grants from MURST (Ministero Università e Ricerca Scientifica e Tecnologica), 1999 (Protein Folding) and 2000 (New Techniques in Proteome Analysis) and by ASI (Agenzia Spaziale Italiana). Finally, I would like to thank the colleagues who have supplied me with the original photographs of their work, as reproduced in this volume.

P.G. Righetti

Abbreviations in Part II

α-LA	α-lactalbumin
β-Ala	β-alanine
β-LG	β-lactoglobulin
2-DE	two-dimensional electrophoresis
Aγ	foetal haemoglobin chains with Ala as residue 136
AMPS	2-acrylamido-2-methylpropane sulphonic acid
ASB14	amidosulphobetaine with a 14-carbon tail
BGE	background electrolyte
Bis	N,N'-methylene bisacrylamide
Bu_3PO	oxidized tributylphosphine
CA	carrier ampholytes
CCD	charge-coupled device
CHAPS	3-[(3-cholamidopropyl)dimethylammonium]-1-propane sulphonate
cIEF	capillary isoelectric focusing
CMC	critical micellar concentration
CPC	N-cetyl pyridinium chloride
CSF	cerebrospinal fluid
CSP	cold-shock proteins
CTAB	cetyltrimethylammonium bromide
Cys-A	cysteic acid
CZE	capillary zone electrophoresis
DATD	N,N'-diallyltartar diamide
DCLHb	diaspirin-cross-linked haemoglobin
DHEBA	N,N'-(1,2-dihydroxyethylene)bisacrylamide
DTE	dithioerythritol
DTT	dithiothreitol
EDTA	ethylendiamine tetracetic acid
EOF	electroendoosmotic flow
ESI	electrospray ionization
EST	expressed sequence tags
FAB	fast atom bombardment
FABP	fatty acid binding protein
Gγ	foetal haemoglobin chains with Gly as residue 136
Hb A	adult haemoglobin

Hb A1c	glycated adult haemoglobin
Hb F	foetal haemoglobin
Hb Fac	acetylated foetal haemoglobin
HEC	hydroxyethyl cellulose
HFP	hexafluoropropanol
HIC	hydrophobic interaction chromatography
I.D.	inner diameter
ICAT	isotope-coded affinity tags
IDA	imino diacetic acid
IEF	isoelectric focusing
IPC	isopicnic centrifugation
IPG	immobilized pH gradients
ITP	isotachophoresis
K_R	coefficient of retardation
LASPO	L-aspartate oxidase
LED	light-emitting diode
MALDI-TOF	matrix-assisted laser induced ionization-time of flight
MEKC	micellar electrokinetic chromatography
mequiv.	milli equivalents
MS	mass spectrometry
NDSB	non-detergent sulphobetaines
NEPHGE	non-equilibrium pH gradient electrophoresis
nESI	nano-electrospray ionization
PAA	polyacrylamide
PAGE	polyacrylamide gel electrophoresis
p*I*	isoelectric point
PMF	peptide mass fingerprinting
PMSF	phenylmethyl sulphonyl fluoride
PSD	post-source decay
PVDF	polyvinylidene difluoride
QAE	quaternary amino ethyl
RBC	red blood cell
r-DNA	recombinant DNA
R_f	relative mobility
S/N	signal to noise ratio
SCX	strong cation exchanger
SDS	sodium dodecyl sulphate
SSA	sulphosalicylic acid
TBP	tributylphosphine
TEMED	*N*,*N*,*N*′,*N*′-tetramethylethylene diamine
TEPA	tetraethylene pentamine
TFE	2,2,2-trifluoroethanol
TLCK	tosyl lysine chloromethyl ketone
TPCK	tosyl phenyl chloromethyl ketone
WWW	world wide web

Subject Index

C

D

E

N

O

P

R

S

T

U

V

W

Y

Z

JOURNAL OF CHROMATOGRAPHY LIBRARY

A Series of Books Devoted to Chromatographic and Electrophoretic Techniques and their Applications

Although complementary to the Journal of Chromatography, each volume in the library series is an important and independent contribution in the field of chromatography and electrophoresis. The library contains no material reprinted from the journal itself.

Other volumes in this series

Volume 1	**Chromatography of Antibiotics** (*see also* Volume 26) by G.H. Wagman and M.J. Weinstein
Volume 2	**Extraction Chromatography** edited by T. Braun and G. Ghersini
Volume 3	**Liquid Column Chromatography**. A Survey of Modern Techniques and Applications edited by Z. Deyl, K. Macek and J. Janák
Volume 4	**Detectors of Gas Chromatography** by J. Ševčík
Volume 5	**Instrumental Liquid Chromatography**. A Practical Manual on High-Performance Liquid Chromatographic Methods (*see also* Volume 27) by N.A. Parris
Volume 6	**Isotachophoresis**. Theory, Instrumentation and Applications by F.M. Everaerts, J.L. Beckers and Th.P.E.M. Verheggen
Volume 7	**Chemical Derivatization in Liquid Chromatography** by J.F. Lawrence and R.W. Frei
Volume 8	**Chromatography of Steroids** by E. Heftmann
Volume 9	**HPTLC – High Performance Thin-Layer Chromatography** edited by A. Zlatkis and R.E. Kaiser
Volume 10	**Gas Chromatography of Polymers** by V.G. Berezkin, V.R. Alishoyev and I.B. Nemirovskaya
Volume 11	**Liquid Chromatography Detectors** by R.P.W. Scott
Volume 12	**Affinity Chromatography** (*see also* Volume 55) by J. Turková
Volume 13	**Instrumentation for High-Performance Liquid Chromatography** edited by J.F.K. Huber
Volume 14	**Radiochromatography**. The Chromatography and Electrophoresis of Radiolabelled Compounds by T.R. Roberts
Volume 15	**Antibiotics**. Isolation, Separation and Purification edited by M.J. Weinstein and G.H. Wagman

Volume 16 — **Porous Silica**. Its Properties and Use as Support in Column Liquid Chromatography
by K.K. Unger

Volume 17 — **75 Years of Chromatography – A Historical Dialogue**
edited by L.S. Ettre and A. Zlatkis

Volume 18 — **Electrophoresis**. A Survey of Techniques and Applications
Part A: Techniques
Part B: Applications
edited by Z. Deyl

Volume 19 — **Chemical Derivatization in Gas Chromatography**
by J. Drozd

Volume 20 — **Electron Capture**. Theory and Practice in Chromatography
edited by A. Zlatkis and C.F. Poole

Volume 21 — **Environmental Problems Solving using Gas and Liquid Chromatography**
by R.L. Grob and M.A. Kaiser

Volume 22 — **Chromatography**. Fundamentals and Applications of Chromatographic and Electrophoretic Methods (*see also* Volume 51)
Part A: Fundamentals
Part B: Applications
edited by E. Heftmann

Volume 23 — **Chromatography of Alkaloids**
Part A: Thin-Layer Chromatography
by A. Baerheim-Svendsen and R. Verpoorte
Part B: Gas-Liquid Chromatography and High-Performance Liquid Chromatography
by R. Verpoorte and A. Baerheim-Svendsen

Volume 24 — **Chemical Methods in Gas Chromatography**
by V.G. Berezkin

Volumc 25 — **Modern Liquid Chromatography of Macromolecules**
by B.G. Belenkii and L.Z. Vilenchik

Volume 26 — **Chromatography on Antibiotics. Second, completely revised edition**
by G.H. Wagman and M.J. Weinstein

Volume 27 — **Instrumental Liquid Chromatography**. A Practical Manual on High-Performance Liquid Chromatographic Methods. **Second, completely revised edition**
by N.A. Parris

Volume 28 — **Microcolumn High-Performance Liquid Chromatography**
by P. Kucera

Volume 29 — **Quantitative Column Liquid Chromatography**. A Survey of Chemometric Methods
by S.T. Balke

Volume 30 — **Microcolumn Separations**. Columns, Instrumentation and Ancillary Techniques
by M.V. Novotny and D. Ishii

Volume 31 **Gradient Elution in Column Liquid Chromatography**. Theory and Practice
by P. Jandera and J. Churáček

Volume 32 **The Science of Chromatography**. Lectures Presented at the A.J.P. Martin Honorary Symposium, Urbino, May 27–31, 1985
edited by F. Bruner

Volume 33 **Liquid Chromatography Detectors. Second, completely revised edition**
by R.P.W. Scott

Volume 34 **Polymer Characterization by Liquid Chromatography**
by G. Glöckner

Volume 35 **Optimization of Chromatographic Selectivity**. A Guide to Method Development
by P.J. Schoenmakers

Volume 36 **Selective Gas Chromatographic Detectors**
by M. Dressler

Volume 37 **Chromatography of Lipids in Biomedical Research and Clinical Diagnosis**
edited by A. Kuksis

Volume 38 **Preparative Chromatography**
edited by B.A. Bidlingmeyer

Volume 39A **Selective Sample Handling and Detection in High-Performance Liquid Chromatography. Part A**
by R.W. Frei and K. Zech

Volume 39B **Selective Sample Handling and Detection in High-Performance Liquid Chromatography. Part B**
by K. Zech and R.W. Frei

Volume 40 **Aqueous Size-Exclusion Chromatography**
by P. Dubin

Volume 41 **High-Performance Liquid Chromatography of Biopolymers and Biooligomers**
Part A: Principles, Materials and Techniques
Part B: Separation of Individual Compounds Classes
by O. Mikeš

Volume 42 **Quantitative Gas Chromatography for Laboratory Analyses and On-line Process Control**
by G. Guiochon and C.L. Guillemin

Volume 43 **Natural Products Isolation**. Separation Methods for Antimicrobials, Antivirals and Enzyme Inhibitors
edited by G.H. Wagman and R. Cooper

Volume 44 **Analytical Artifacts**. GC, MS, HPLC, TLC and PC
by B.S. Middleditch

Volume 45A **Chromatography and Modification of Nucleosides Analytical Methods for Major and Modified Nucleosides – HPLC, GC, MS, NMR, UV and FT-IR**
edited by C.W. Gehrke and K.C.T. Kuo

Volume 45B **Chromatography and Modification of Nucleosides Biological Roles and Function of Modification** edited by C.W. Gehrke and K.C.T. Kuo